Wave Motion, Intelligent Structures and Nonlinear Mechanics

A Herbert Überall Festschrift Volume

SERIES ON STABILITY, VIBRATION AND CONTROL OF STRUCTURES

Series Editors: Ardéshir Guran & Daniel J. Inman

About the Series

Rapid developments in system dynamics and control, areas related to many other topics in applied mathematics, call for comprehensive presentations of current topics. This series contains graduate level text books, monographs, and a collection of thematically organized research or pedagogical articles addressing key topics in applied dynamics.

The material is ideal for a general scientific and engineering readership, and is also mathematically precise enough to be a useful reference for research specialists in mechanics and control, nonlinear dynamics, and in applied mathematics and physics.

Forthcoming volumes

Vol. 2: Nonlinear Dynamics — The Richard Rand 50th Birthday Volume
Editor: A. Guran

Vol. 3: Nonlinear Elasticity of Solids, Static and Dynamic Stability of Structures
Editor: A. Guran

Vol. 4: Structronics Systems: Active Structures, Devices, and Systems
Editors: H.-S. Tzou, A. Guran, G. Anderson, D. J. Inman, and M. C. Natori

Vol. 5: Acoustic Intractions with Submerged Elastic Structures
Editors: A. Guran and F. Ziegler

Vol. 6: Modelling and Control of Structural Systems
Editors: L. Faravelli, A. Guran, D. J. Inman and J.-N. Juang

Vol. 7: Dynamics of Friction and Damping: Modeling, Analysis, and Experiment
Editors: A. Guran, F. Pfeiffer, and K. Popp

Vol. 8: Nonlinear Dynamics of Gyroscopic Systems: Gyroelasticity, Oscillation of Rotors, Flow-Induced Vibration of Structures, and Mechanics of Axially Moving Continua
Editors: A. Guran and D. J. Inman

Vol. 9: Uncertainty Modelling in Stability, Vibration and Control of Structural Systems
Editors: A. Guran, A. Haldar, and B. Ayyub

SERIES ON STABILITY, VIBRATION AND CONTROL OF STRUCTURES

Vol. 1

Series Editors: **Ardéshir Guran & Daniel J Inman**

Wave Motion, Intelligent Structures and Nonlinear Mechanics

A Herbert Überall Festschrift Volume

Editors

Ardéshir Guran
The Catholic University of America
Washington, D.C.

Daniel J Inman
Virginia Polytechnic Institute &
State University, Blacksburg, Virginia

World Scientific
Singapore • New Jersey • London • Hong Kong

Published by

World Scientific Publishing Co. Pte. Ltd.
P O Box 128, Farrer Road, Singapore 9128
USA office: Suite 1B, 1060 Main Street, River Edge, NJ 07661
UK office: 57 Shelton Street, Covent Garden, London WC2H 9HE

WAVE MOTION, INTELLIGENT STRUCTURES AND NONLINEAR MECHANICS

ISBN 981-02-2981-X

Printed in Singapore by Uto-Print

This volume is dedicated to
Professor Herbert Überall
in honor of his many contributions
to Theoretical Physics and Applied Mathematics

Left to right: Professor Jean Ripoche, Mrs. Reyna T. Überall, Professor Gérard Maze, Mrs. Madeleine Ripoche, Professor Herbert Überall.

Dedication

A collection of articles, which are dedicated in honor of Professor Herbert M. Überall of the Catholic University of America on the occasion of his retirement, are being published in this as well as some future volumes of *Stability, Vibration, and Control of Structures.*

Professor Herbert Michael Überall has been an active researcher in the field of Theoretical Physics and Applied Mathematics for more than thirty years, making many fundamental and significant contributions to the literature in Theoretical Nuclear Physics, Electromagnetic Theory, Acoustics and Elastic Waves.

He was born in Neunkirchen, Austria. He was the only child of Michael and Stefanie Überall and he was raised in a Roman Catholic family. He attended the "Gymnasium" of Mathematics and Sciences and graduated with Matura, in June 1949. In the fall of 1949, he began his studies in Physics at the University of Vienna, Austria. Four years later he was awarded a Doctor of Philosophy "Sub auspiciis praesidentis" in Theoretical Nuclear Physics at the early age of twenty-two. Economically, times were very difficult in Europe after the second world war and young Herbert decided upon coming to the United States. Thus, he chose to write to H.A. Bethe (Professor of Physics at Cornell, later Nobel Laureate) seeking to work as one of his assistants. Professor Bethe wrote back to young Überall that he had no openings with his research group at Cornell at that time. In spite of this unfortunate situation, young Herbert accepted a position in Fort Monmouth, NJ, after an interview that he had with American recruiters in Vienna. In 1953, he sailed for the United States. A year after working in the United States, Dr. Überall visited Ithaca and asked Professor Bethe in person to accept him as his assistant. Bethe accepted him this time with the condition that young Herbert start another Ph.D. under his direct supervision. Two years later Überall was awarded a Ph.D. degree from Cornell (1956). His Ph.D. thesis of 1956 was groundbreaking, as it was there that he pointed out the effect of coherent bremsstrahlung, using crystal targets bombarded by an electron beam. After one year as Research Fellow at the University of Liverpool, England, he joined CERN, Geneva, Switzerland, as a Ford Foundation Fellow. From 1958 to 1960 he was a Research Physicist at Carnegie Institute of Technology, Pittsburg, PA, then he became an Assistant Professor of Physics at the University of Michigan, Ann Arbor, MI (1960-1964). He joined the Catholic University of America and was promoted to Associate Professor in 1964, and to Full Professor in 1965.

Professor Überall made extensive consulting contributions to government and industry, including: General Atomics, La Jolla, CA; Hydrospace Corp., Rockville, MD; Naval Research Laboratory, Washington, DC; Naval Ship Research and Development Center, Carderock, MD; Bolt Beranek and Newman, Inc., Arlington, VA; Naval Ocean Research and Development Activity, NSTL Station, MS. He was a Visiting Professor at the University of Paris in 1984-85, as well as the University of Le Havre, Le Havre, France, and the University of Bordeaux, Bordeaux, France.

Professor Überall has two children: daughter Bernadette is a financial analyst and lives in Paris, and son Bertrand is an applied mathematician working in Washington, DC. Professor Überall is married to the former Reyna Tosta who comes from a family of distinguished public servants in her native Honduras.

Professor Herbert Überall was the recipient of the outstanding Educators of America Award in 1973 and of the Invention Award, Naval Research Laboratory, in 1977. He received a Research Publication Award from the Naval Reseach Laboratory 1978, and from the David Taylor Research Center in 1984. He was the recipient of the Achievement Award of the Washington Academy of Sciences in 1984. He is a Fellow of the American Physical Society, a Fellow of the Acoustical Society of America, and a Fellow of the Institute of Electrical and Electronic Engineers (IEEE).

It is of interest to note that while Professor Überall is generally viewed as being one of the more mathematically inclined physicists in his field, he was also involved in several experimental endeavors. His work on Experiments and Analysis of Sound-Induced Structural Vibration (see for details the article by Herbert Überall, Jean Ripoche, and Gérard Maze in this volume) is an impressive use of very subtle and powerful mathematical techniques which he developed in 1977-78 in collaboration with L. Flax and L.R. Dragonette of the Naval Research Laboratory in Washington, DC. ("Resonance Scattering Theory") describing a large body of the experimentally observed phenomena. Indeed for this work he received an honorary doctorate degree (Doctor Honoris Causa) from the University of Le Havre, France in March 1987.

Professor Überall has made a most noteworthy and phenomenal contribution to the literature in theoretical physics and Applied Mathematics. He has been the author or editor of nine books, the most impressive of which is his comprehensive two-volume work entitled "Electron Scattering from Complex Nuclei". He has published close to four hundred journal papers, and has supervised more than forty Ph.D. students. His technical work addresses a broad spectrum of significant contributions in Nuclear Physics, Electromagnetic Theory, Acoustics, and Elastic Waves. In the area of Nuclear and Radiation Physics he pioneered the theory of coherent bremsstrahlung, and much of the early work in nuclear muon capture and neutrino reactions. He developed a multipolar collective nuclear model including spin flip, with application to electron scattering. He also developed a collective model of ion-ion scattering. In the area of Electromagnetic Theory he established the connection between the resonance scattering theory and the singularity expansion method (SEM); he also contributed to the theory of radio wave propagation in the earth ionosphere channel. In Acoustics he developed the theory of creeping waves in acoustic scattering. He pioneered the acoustic resonance scattering theory (RST), and contributed to the theory of sound channeling in the ocean. In Elasticity, he pioneered the resonance scattering theory of elastic waves from cavities and inclusions in solids.

Professor Überall is held in the highest esteem by his colleagues at CUA. He has always been kind and compassionate, and has given his honest and sincere advice to those seeking it. Beyond and above his technical accomplishments his abiding legacy to us all is the recollection of his gentle firmness and good humored determination, the guidance he gave to students and colleagues and his steadiness in the face of personal adversity and public acclaim alike.

Herbert Überall is a remarkable man. A man of wisdom, of a sympathetic and gentle nature, of integrity and of courage. I will always remember him as the most prominent colleague during my CUA days, a true scholar, a thoughtful person, and a gentleman — an abiding proof that these archaic-sounding virtues endure and continue to be valued.

It is with deep gratitude and warm feeling that the first volume in this series is dedicated to Professor Herbert Überall in honor of his many contributions to research in Theoretical Physics and Applied Mathematics and to mark his retirement from the Catholic University of America.

I gratefully acknowledge the generous help of Dr. Carl Werntz (Chairperson, Department of Physics, the Catholic University of America) in providing materials for this foreword. Drs. Hall Crannell and Lawrence Fagg (CUA, Physics department) were also very helpful in polishing this homage. Special thanks go to Drs. Jan Achenbach, Laszlo Adler, Juan Arvelo, Ezekiel Bahar, X. L. Bao, Terence Bazov, Wolfgang Boerner, Giorgio Borgiotti, Anders Boström, Sam Brown, Homer Bucker, Courtney Burroughs, Rafael Carbó-Fite, J.M. Conoir, Greg Cowart, Michel de Billy, Harry de Ferrari, D. Decultot, Pier Paolo Delsanto, J. DeSanto, Marc Deschamps, Joseph Dickey, Richard Doolittle, Louis Dragonette, Donald Dudley, Nader Engheta, Paul Etter, Richard Farrell, Ralph Fiorito, Frederick Fisher, Louis Fishman, George Frisk, Alain Gérard, Kenneth Gilbert, Stewart Glegg, Yanping Guo, Yueping Guo, Henning Harmuth, Barbara Howell, William Howell, Claude Horton, Anton Haug, Sabih Hayek, George Herrmann, Akira Ishimaru, J.L. Izbicki, Arthur Jordan, Miguel Junger, Joseph Keller, Aleksander Klauson, Ralph Kleinman, Robert Kouyoumjian, William Kuperman, Francine Luppé, Patricio Laura, F. Lecroq, F. Léon, Oswald Leroy, Leonid Lyamshev, L. Medgyesi-Mitschang, Fouad Major, Philip Marston, Walter Mayer, Gérard Maze, Suzanne McDaniel, John McNicholas, Nicholas Nicholas, Barbara Merchant, Jaan Metsaveer, Edmund Miller, Russel Miller, Frédéric Molinet, Philip Moser, Anton Nagl, Werner Neubauer, Susan Numrich, Marshall Orr, Antares Parvulescu, Allan Pierce, Douglas Photiadis, Bernard Poirée, Gérard Quentin, Ronald Radlinski, P.K. Raju, Carlos Ranz, Edward Reiss, P. Rembert, Jean Ripoche, Martine Rousseau, J.-L. Rousselot, Wolfgang Sachse, William Sáenz, Tapan Sarkar, Angie Sarkissian, J. P. Sessarego, Morris Schulkin, Aleksandr, Basil and Yuri Stoyanov, Boris Belinskiy, Marylin Talmant, Douglas Taylor, Bernard Tocquet, Alexandra Tolstoy, Victor Twersky, Piere-Georgio Uslenghi, Peter Uginčius, , Vijay and Vesundara Varadan, Dimitri Vassiliev, James Wait, Eric Walton, Ferris Webster, Michael Werby, Jianrem Yuan, Kwang-Bock Yoo, Manell Zakharia for having provided additional information. Finally, I wish to express my thanks to Drs. Franz Ziegler (Head, Institut für Allgemeine Mechanik, Technical University of Vienna, Austria) and Hans Irschik (Inst. Tech. Mech.,Johannes Kepler University, Linz, Austria) for their support in organizing the Herbert Überall festschrift volumes.

Ardéshir Guran
Series Editor-in-Chief
Washington, D.C.

Preface

In this volume on *Stability, Vibration, and Control of Structures*, research articles on certain selected topics have been compiled to form an integrated set. The first three articles are in the area of Wave Motion and Acoustics. First of all Alexander Vakakis presents an overview of recent analytical, numerical and experimental results on nonlinear mode localization and nonlinear motion confinement in discrete and continuous oscillators. The paper by Herbert Überall, Gerard Maze and Jean Ripoche addresses theoretical and experimental work on the scattering of acoustic waves from submerged elastice objects. A. Safaai-Jazi considers the propagation of elastic waves in fiber optic type acoustic waveguides.

Several chapters in this volume treat the topic of Modeling and Control of Dynamic Systems. Kon-Well Wang introduces a new structural vibration control technique, using piezoelectro materials with real-time adaptable electrical networks. The paper by Bingen Yang considers a transfer function method to study eigenvalue inclusion principle and vibration control of gyroscopic dynamic systems. Nonlinear control and boundary transition of distributed systems is presented by Horn-Sen Tzou, David Johnson, and Jack Liu. Minh Phan, Jer-Nan Juang and David Hyland consider identification and control of dynamic systems by using Neural Networks.

The last three chapters in this volume deal with the topic of Nonlinear Mechanics. Tondl and Nabergoj investigate the stability of the semi-trivial solution of an autoparametric system consisting of several masses and springs arranged in a chain, their results enables us to find the optimum tuning of pendulum subsystems attached to the masses of the excited subsystem. The chapter by Inna Sharf gives a critical review of several publications dealing with the issue of geometric stiffening as it arises in the context of multibody dynamics. This volume also contains papers which addresses some of the nonlinear aspects of fluid-solid oscillations; thus, the chapter by Volker Mehl and Jörg Wauer examines the flow instability between coaxial rotating cylinders with a flexible support. They elegantly solve the more general form of Taylor problem in hydrodynamics by including the effects of flexibility of the supports of the rotating inner cylinder in formulation of the problem. It is hoped that *Stability, Vibration, and Control of Structures* will stimulate exchange of ideas among those working in mechanics and control and moreover will be useful for those entering the field.

We warmly thank all contributors for their cooperation. This project was completed while one of us (Arde Guran) was a guest of the Department of Electrical Engineering, University of Maryland at College Park, and an invited Professor at Technical University of Hamburg, Harburg. He would like to thank Professor Nariman Farvardin (Chaiperson, EE department at UMCP), and Professor Edwin Kreuzer (Vice President, TUH) for their warm hospitality and their ever-present patience and good humor to ensure the success of the project from the moment of its inception.

We also want to thank all those staff at World Scientific Publishing who so kindly contributed to preparing and producing this volume.

Ardéshir Guran
Hamburg, Germany

Daniel J. Inman
Blacksburg, VA

Contributors

David Hyland
Government Aerospace System Division
Harris Corporation
Melbourne, FL 32902
USA

David Johnson
Department of Mechanical Engineering
Center for Robotics and Manufacturing Systems
University of Kentucky
Lexington, KY 40506
USA

Jer-Nan Juang
Structural Dynamics Branch
NASA Langley Research Center
Hampton, VA 23681-0001
USA

Jack Liu
Active Noise and Vibration Tech.
Phoenix, AZ 85040
USA

Gerard Maze
LAUE, URA CNRS 1373
University of Le Havre
76610 Le Havre
France

Volker Mehl
Institut für Technische Mechanik
Universität Karlsruhe
Kaiserstraße 12
Karlsruhe D-76128
Germany

Minh Phan
Department of Mechanical and Aerospace Engineering
Princeton University
Princeton, NJ 08544
USA

Radeslav Nabergoj
Department of Naval Architecture, Ocean and Environmental Engineering
Via Valesio 10
34127 Trieste
Italy

Jean Ripoche
LAUE, URA CNRS 1373
University of Le Havre
76610 Le Havre
France

Ahmad Safaai Jazi
Bradly Department of Electrical Engineering
Virginia Polytechnic Institute and State University
Blacksburg, VA 24061
USA

Inna Sharf
Department of Mechanical Engineering
University of Victoria
Victoria, British Columbia
Canada V8W 3P6

Aleš Tondl
Zborovská 41
1500 Praha
Czech Republic

Horn-Sen Tzou
Department of Mechanical Engineering
Center for Robotics and Manufacturing Systems
University of Kentucky
Lexington, KY 40506
USA

Herbert Überall
Department of Physics
Catholic University of America
Washington, DC 20064
USA

Alexander Vakakis
Department of Mechanical and Industrial Engineering
University of Illinois at Urbana-Champaign
1206 W. Green Street
Urbana, IL 61821
USA

Kon-Well Wang
Department of Mechanical Engineering
The Pennsylvania State University
University Park, PA 16802-1400
USA

Jörg Wauer
Institut für Technische Mechanik
Universität Karlsruhe
Kaiserstraße 12
Karlsruhe, D-76128
Germany

Bingen Yang
Department of Mechanical Engineering
University of Southern California
Los Angeles, CA 90089
USA

Contents

Chapter 9: A Survey of Geometric Stiffening in Multibody Dynamics Formulations
Inna Sharf

Chapter 10: Flow Instability Between Coaxial Rotating Cylinders with a Flexible Support
Volker Mehl and Jörg Wauer

Wave Motion, Intelligent Structures and Nonlinear Mechanics

A Herbert Überall Festschrift Volume

Wave Motion, Intelligent Structures and Nonlinear Mechanics, pp. 1–34
edited by A. Guran and D. J. Inman
Series on Stability, Vibration and Control of Structures: Vol. 1

NONLINEAR MODE LOCALIZATION AND PASSIVE MOTION CONFINEMENT IN DISCRETE AND CONTINUOUS SYSTEMS

ALEXANDER F. VAKAKIS
Department of Mechanical & Industrial Engineering
University of Illinois at Urbana - Champaign
1206 W.Green Street, Urbana, IL 61821

ABSTRACT

Asymptotic analysis of localized nonlinear normal modes (NNMs) in symmetric discrete and continuous oscillators is carried out. These are periodic motions where all positional variables of a system vibrate in a periodic, synchronous oscillation. It is shown, that in contrast to linear theory where such localized modes exist only in the presence of weak structural disorders, nonlinear mode localization can occur in perfectly symmetric, weakly coupled nonlinear systems. It is analytically and numerically shown that systems possessing localized NNMs also possess motion confinement properties: disturbances generated by external impulses remain spatially confined instead of "spreading" through the entire structure. A first experimental verification of the nonlinear motion confinement phenomenon is presented. Finally, it is shown how the concept of NNM can be extended to analytically study (standing or travelling) waves with spatially localized envelopes in certain nonlinear partial differential equations. In the context of NNM theory, such standing nonlinear waves can be regarded as NNMs of continuous systems defined over unbounded spatial domains.

1. Introduction

In a number of recent works, the phenomenon of mode localization in disordered linear periodic systems was investigated[1-5]. In these works it was shown that the (spatially extended) normal modes of weakly coupled, symmetric linear systems become localized when weak perturbations of the periodicity are introduced. Linear mode localization was detected when the coupling between subsystems was of the order or smaller than the "spread" in natural frequencies of the component systems[1-5]. The phenomenon of mode localization in discrete periodic oscillators with nonlinear stiffnesses was analytically and numerically studied in works by Vakakis and co-workers[6-9]. This was accomplished using the notion of nonlinear normal mode[10-14] (NNM), i.e., of a free motion during which all positional variables of a discrete system oscillate equiperiodically, reaching their extreme values at the same instant of time.

Recently, the notion of NNM was extended to one-dimensional nonlinear continuous systems[15-18]. A series of additional works investigated linear and nonlinear localized normal modes in periodic continuous systems, such as truss structures or cyclic antennas[3,18-22]. It was shown that, continuous systems possessing localized normal modes also possess *motion confinement properties*[23]; that is, motions generated by external impulsive excitations remain passively confined in a limited number of substructures instead of "spreading" through the entire periodic structure. The beneficial effects of the motion confinement phenomenon should be evident: structures whose inherent dynamics lead to spatial motion confinement of externally generated disturbances are more amenable to control than structures with no such dynamic characteristics. In other words, the controllability of structures with motion confinement properties is greatly improved, a feature which can find application on improved and novel passive or active vibration and shock isolation designs of coupled periodic structures.

This work provides an overview of recent analytical, numerical and experimental results on nonlinear mode localization and nonlinear motion confinement in discrete and continuous oscillators. Asymptotic techniques for studying the localized NNMs of periodic discrete and continuous oscillators are presented, and the theoretical results are verified by numerical integrations. A review of the nonlinear motion confinement properties of a continuous flexible assembly is given, and the analytical findings are verified experimentally. Finally, it is shown how the concept of localized NNM can be employed to analyze waves with spatially localized envelopes in a certain class of nonlinear partial differential equations.

2. Discrete Oscillators

2.1. *Existence of Localized NNMs*

The asymptotic methodology for computing localized NNMs in discrete oscillators will be demonstrated by means of an example[7]. Consider the n degree-of-freedom (DOF) cyclic nonlinear structure depicted in figure 1. It consists of n identical, nonlinear single-degree-of-freedom substructures coupled by means of weak nonlinear stiffnesses. The equations of motion are given by:

$$\ddot{x}_i = f_i(x_1,...,x_n) = f_i^{(u)}(x_i) + \varepsilon\, f_i^{(c)}(x_{i-1}, x_i, x_{i+1}) \,, \quad i=1,...,n \tag{1}$$

where $f_i^{(u)}$ is the restoring force of the i-th uncoupled substructure, $\varepsilon f_i^{(c)}$ is the weak restoring force due to coupling between substructures i-1, i and i+1, $|\varepsilon|<<1$ is a small parameter of perturbation order, and $f_i^{(u)}=f_j^{(u)}$, $f_i^{(c)}=f_j^{(c)}$, $\forall\, i,j\in[1,...,n]$. Furthermore, system (1) is assumed to be conservative. Functions $f_i^{(u)}$ and $f_i^{(c)}$ are assumed to have the following properties:

(i) $f_i^{(u)}(0)=0,\ f_i^{(u)}(-x_i)=-f_i^{(u)}(x_i)$
(ii) $f_i^{(c)}(x_{i-1}, x_i, x_{i+1}) = \hat{f}_i^{(c)}(x_i-x_{i-1}) + \hat{f}_i^{(c)}(x_i-x_{i+1})$
(iii) $\hat{f}_i^{(c)}(0)=0,\ \hat{f}_i^{(c)}(-u)=-\hat{f}_i^{(c)}(u)$

With (i), the stiffnesses of the uncoupled substructures are assumed to be symmetric, i.e., to respond to an equal amount to tension and compression. With assumption (ii), the total stiffness interaction between three substructures i-1, i and i+1 can be decomposed into mutual individual interactions between neighboring substructures. The individual restoring forces are then symmetric by (iii). Finally, the functions $f_i^{(u)}$ and $f_i^{(c)}$ are assumed to be sufficiently smooth and to possess Taylor series expansions about the zero value of their arguments. Under these conditions, it can be shown[7] that the dynamical system represented by Eq. (1) possesses NNMs, i.e., periodic motions passing through the origin of the configuration space with all positional variables reaching their extreme values simultaneously. These modes are represented in the configuration space of the system by modal lines, and can be classified as *similar NNMs* (with straight modal lines), or *nonsimilar NNMs* (with modal curves). Similar modes can only occur in systems with special symmetries. Nonsimilar normal modes are expressed by relations of the form:

$$x_i = \hat{x}_i(x_1), \quad i=1,...,n, \quad x_1 \equiv \hat{x}_i(x_1) \tag{2}$$

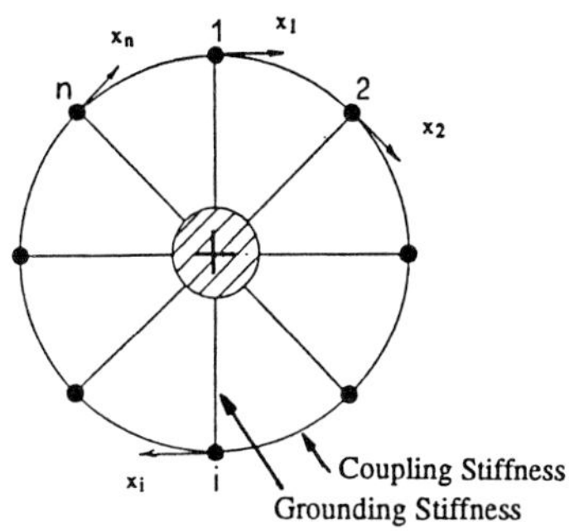

Figure 1. The nonlinear discrete cyclic system.

where coordinate x_1 is used to parametrize the nonsimilar motion. The functions $\hat{x}_i(\bullet)$ (which generally have a nonlinear dependence on x_1), define a curve in the n - dimensional configuration space of the oscillator.

Expressing the time derivatives in the equations of motion in terms of the functions $\hat{x}_i(\bullet)$ and the coordinate x_1, one finds that the following (n-1) functional equations must be satisfied in order for a nonsimilar normal mode to exist[7]:

$$2\left[h - V(\hat{x}_1(x_1),...,\hat{x}_n(x_1))\right]\left[1 + \sum_{k=1,k\neq 1}^{n}\left(\frac{d\hat{x}_k}{dx_1}\right)^2\right]^{-1}\frac{d^2\hat{x}_i}{dx_1^2} +$$

$$+ f_1(\hat{x}_1(x_1),...,\hat{x}_n(x_1))\frac{d\hat{x}_i}{dx_1} = f_i(\hat{x}_1(x_1),...,\hat{x}_n(x_1)) \tag{3}$$

where, i = 2,...,n, h is the total (fixed) energy, and V is the potential energy of the system. Since the functional equations are singular at the maximum equipotential surface V=h, it is necessary to develop asymptotic approximations for the modal lines; these approximations will be valid in open neighborhoods of the origin of the configuration space, and will be analytically continued up to the maximum equipotential surface[12,24]. Therefore, it is necessary to impose the following additional set of (n-1) functional boundary orthogonality conditions which hold at the points of intersection of the modal line with the maximum equipotential surface (X_1 is the amplitude of oscillation of the parametrizing coordinate x_1):

$$f_1(\hat{x}_1(X_1),...,\hat{x}_n(X_1))\left\{\frac{d\hat{x}_i}{dx_1}\right\}_{x_1=X_1} = f_i(\hat{x}_1(X_1),...,\hat{x}_n(X_1)), \quad i=2,...,n \tag{4}$$

Equations (3) and (4) can asymptotically solved by employing the methodology first introduced by Manevitch and Mikhlin[24]. The asymptotic solutions are developed for the specific set of initial conditions $x_i(0) = X_i$, $\dot{x}_i(0) = 0$, $i=1,2,...,n$, but the same methodology can be applied for systems with different initial conditions. The modal line (2) is expressed as:

$$\hat{x}_i(x_1) = \sum_{k=0}^{\infty} \hat{x}_i^{(k)}(x_1) \quad , \quad i=2,...,n \tag{5}$$

where $\hat{x}_i^{(k)}(x_1)=O(\varepsilon^k)$. Series (5) is substituted into Eq.(3,4), the coefficients of respective powers of ε are set equal to zero, and successive approximations to the modal line (2) are computed. It is assumed that the zero-th order approximation in (5) corresponds to a similar normal mode; as a result, the nonsimilar normal modes computed by Eq.(3,4) result as perturbations of similar normal modes. Details of the calculations can be found in the paper by Vakakis and Cetinkaya[7]. The nonsimilar normal modes can be asymptotically approximated as:

$$x_i = \hat{x}_i(x_1) = \alpha_i\, x_1 + \varepsilon \sum_{m=0}^{M} a_{im}^{(1)}\, x_1^{2m+1} + O(\varepsilon x_1^{2M+3}, \varepsilon^2)\,, \quad i=2,...,n \tag{6}$$

where $\alpha_i = 0$ or ± 1 when the restoring force of the i-th uncoupled substructure $f_i(u)$ is nonlinear, and α_i assumes arbitrary values when $f_i(u)$ is linear. Coefficients $a_{im}^{(1)}$ depend on X_1, and their explicit expressions are listed in section 2.4 (Appendix).

It is concluded that close to each similar mode $\hat{x}_i^{(0)}=\alpha_i x_1$, a nonsimilar mode exists, with its modal curve computed by Eq. (6). Using the derived asymptotic approximations for the modal curves, one can compute the time response of the parametrizing coordinate, $x_1=x_1(t)$, and the frequency of free oscillation of the nonsimilar mode, by substituting Eq. (6) into the equations of motion (1) and integrating by quadratures. Suppose now that $\alpha_i = 0$, $i=2,...,n$. Then the asymptotic theory predicts the following motions for the corresponding nonsimilar mode:

$$x_1 = O(1),\ x_2 = O(\varepsilon),\ x_n = O(\varepsilon),\ x_i = O(\varepsilon^2)\,, \quad i=3,...,n-2 \tag{7}$$

Hence, it is proven that the perfectly periodic, weakly nonlinear, cyclic structure possesses a localized nonsimilar NNM during which the motion is mainly confined to only one of its coordinates, the remaining coordinates oscillating with amplitudes of at least $O(\varepsilon)$. It can be shown that this localized mode exists only in the nonlinear cyclic structure; when no nonlinearity exists, mode localization can be achieved only when the cyclic periodicity is perturbed by weak substructure "mistunings." A stability analysis indicates that the localized normal mode is orbitally stable and thus physically realizable[7]. In addition to the strongly localized nonsimilar mode (7), there exist additional weakly localized modes during which two or more coordinates oscillate with amplitudes of O(1), whereas the remaining coordinates are of at least $O(\varepsilon)$. These modes are not considered in this work.

2.2. Transition from Mode Localization to Mode Non-localization

The strongly localized mode (7) is realized only for sufficiently small coupling between substructures. In the work by Vakakis et.al.[8] the transition from mode localization to mode non-localization for increasing substructure coupling is studied using the method of multiple-scales. In that work it is proven that when the ratio of coupling over nonlinear forces exceeds certain threshold values the nonlinear mode localization phenomenon is eliminated. Assuming only cubic stiffness nonlinearities in Eq. (1), one sets $f_i^{(u)}(x_i)=-(x_i+\varepsilon\mu x_i^3)$, $\varepsilon f_i^{(c)}(x_{i-1}, x_i, x_{i+1}) = -\varepsilon[k\,(x_i-x_{i+1})+k(x_i-x_{i-1})]$, to obtain the following equations of motion for the n-DOF cyclic structure of Fig.1:

$$\ddot{x}_i + x_i + \varepsilon\mu\, x_i^3 + \varepsilon k\,(x_i-x_{i+1}) + \varepsilon k\,(x_i-x_{i-1}) = 0\,,\quad i=1,...,n \tag{8}$$

with the understanding that $x_{n+1}\equiv x_1$, $x_0\equiv x_n$, and $|\varepsilon|<<1$. Employing the method of multiple-scales[25], the approximate solutions of Eq. (8) are expressed as[8]:

$$x_i = (1/2)\, a_i(\varepsilon t)\, \exp\{j(\beta_i(\varepsilon t)+t)\} + cc + O(\varepsilon)\,,\quad i=1,...,n \tag{9}$$

where $j=(-1)^{1/2}$, (cc) denotes the complex conjugate, and the slowly varying complex amplitudes $a_i(\varepsilon t)$ and $\beta_i(\varepsilon t)$ are governed by the following set of modulation equations:

$$\begin{aligned} a_1' - (k/2)\, a_{i+1}\sin\phi_{i+1} - (k/2)\, a_{i-1}\sin\phi_i \quad &, i=1,...,n \\ \phi_i' = (3\mu/8)\,(a_i^2-a_{i-1}^2) - (k/2)\left[\,(a_{i+1}/a_i)\cos\phi_{i+1} - (a_i/a_{i-1})\cos\phi_i\,\right] - \\ - (k/2)\left[\,(a_{i-1}/a_i)\cos\phi_i - (a_{i-2}/a_{i-1})\cos\phi_{i-1}\,\right] \quad &, i=2,...,n \end{aligned} \tag{10}$$

where $\phi_j=\beta_j-\beta_{j-1}$, $\phi_1=-\sum_{k=2}^{n}\phi_k$, and primes denote differentiation with respect to the slow time scale (εt). Equations (10) describe (correct to $O(\varepsilon)$) the free oscillations of the cyclic system on the isoenergetic manifold. By imposing the stationarity conditions $a_i' = 0$, $\phi_i' = 0$, $i=1,...,n$, one determines the values of the amplitudes a_i and ϕ_i corresponding to synchronous periodic free motions, i.e., NNMs of the cyclic system. It can be shown[8], that there exist branches of localized NNM solutions satisfying the following symmetry relations:

<u>For n=(2p+1)</u>

$a_2=a_n$, $a_3=a_{n-1}$, $a_4=a_{n-2}$, ... , $a_{p+1}=a_{p+2}$

$\phi_i = \pi$, $i=1,...,n$, $i\neq p+2$, $\phi_{p+2} = -2p\pi$

<u>For n=2(p-1)</u>

$a_2=a_n$, $a_3=a_{n-1}$, $a_4=a_{n-2}$, ... , $a_{p-1}=a_{p+1}$

$\phi_i = \pi$, $i=1,...,n$, $i\neq p$, $\phi_p = -(2p-3)\pi$

Taking into account these symmetry relations and imposing the stationarity conditions in (10), one can numerically compute the localized branches of solutions. It turns out that at a

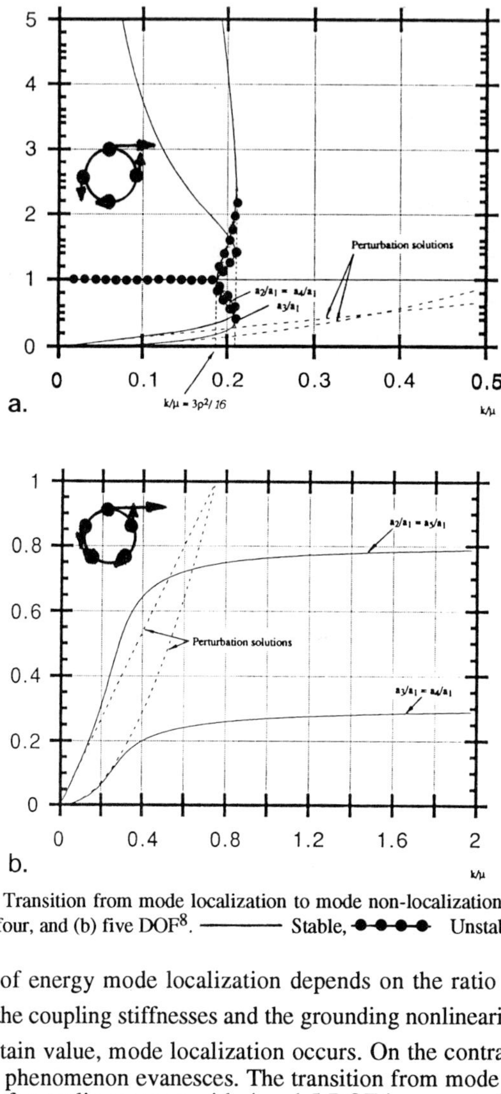

Figure 2. Transition from mode localization to mode non-localization in discrete cyclic systems with, (a) four, and (b) five DOF[8]. ———— Stable, ●●●● Unstable NNMs.

fixed level of energy mode localization depends on the ratio (k/μ), i.e., on the relative strength of the coupling stiffnesses and the grounding nonlinearities. When (k/μ) decreases below a certain value, mode localization occurs. On the contrary, as (k/μ) increases, the localization phenomenon evanesces. The transition from mode localization to mode non-localization for cyclic systems with 4 and 5 DOF is presented in figure 2. Note that, as (k/μ) increases the branches of localized modes become non-localized and, either coalesce with an antisymmetric NNM of the system in a degenerate double inverse pitchfork bifurcation (system with 4 DOF), or reach non-localized asymptotic limits (system with 5 DOF).

2.3. Passive Nonlinear Motion Confinement

In order to demonstrate the effect of the localized modes on the forced transient dynamics of the nonlinear cyclic structure, the response of a cyclic system with fifty DOF to external impulsive loading was determined by finite-element computations[8]; an impulse of magnitude 1 was applied for a duration of $\Delta t=0.2$ seconds to subsystem 1. The structural parameters were chosen equal to, $\varepsilon\mu=0.3$, $\varepsilon k=0.05$, and $k/\mu=0.166$, to ensure that the structure possesses localized NNMs. The dynamic response is depicted in Fig. 3 at time instants t=0.2, 500 and 1000sec.

For comparison purposes, the respective responses of the linear system (corresponding to $\varepsilon\mu=0$) are also shown. In the nonlinear system the energy of the impulse remains confined to the element of the structure where it is originally applied, in contrast to the linear system where this energy "spreads" to the remaining part of the structure. This passive motion confinement in the nonlinear system is attributed to its stable nonlinear localized modes which have no counterparts in linear theory. *Thus, the numerical simulations show that localized modes in a structure can have profound effects on its forced dynamic response; in particular, confinement of motions generated by external impulses in the weakly coupled nonlinear cyclic systems is detected, a phenomenon with profound implications regarding their passive or active vibration suppression.*

2.4. Appendix: Coefficients $a_{is}^{(1)}$ in equation (6).

The coefficients $a_{is}^{(1)}$ are computed using the following recursive formula[7]:

$$4\,h\,(2s+1)\,s\,a_{is}^{(1)} =$$

$$= 4\sum_{p=1}^{s-1}\{A_{p-1}\,a_{i(s-p)}^{(1)}\,[2(s-p)+1](s-p)\} - \sum_{p=1}^{s}\{B_{p-1}\,[2(s-p+1)-1] +$$

$$+ C_{i(p-1)}]\,a_{i(s-p)}^{(1)}\} - D_{i(s-1)}\,,\quad i=2,\ldots,n,\quad s=1,2,\ldots \tag{A-1}$$

where

$$A_k = f_{1,(2k+1)}^{(u)}(0)\,(1+\sum_{p=2}^{n}\alpha_p^{2k+2})\,/\,(2k+2)!$$

$$B_k = f_{1,(2k+1)}^{(u)}(0)\,(1+\sum_{p=2}^{n}\alpha_p^{2})\,/\,(2k+1)!$$

$$C_{ik} = -\,(1+\sum_{p=2}^{n}\alpha_p^{2})\,f_{1,(2k+1)}^{(u)}(0)\,\alpha_i^{2k}\,/\,(2k)!$$

$$D_{ik} = \varepsilon\,(1+\sum_{p=2}^{n}\alpha_p^{2})\,[\,\alpha_i F_{1,(2k+1)}^{(c)}(0;\,\alpha_n,\alpha_1,\alpha_2)\, -$$

$$-\,F_{1,(2k+1)}^{(c)}(0;\,\alpha_{i-1},\alpha_i,\alpha_{i+1})\,]\,/\,(2k+1)! \tag{A-2}$$

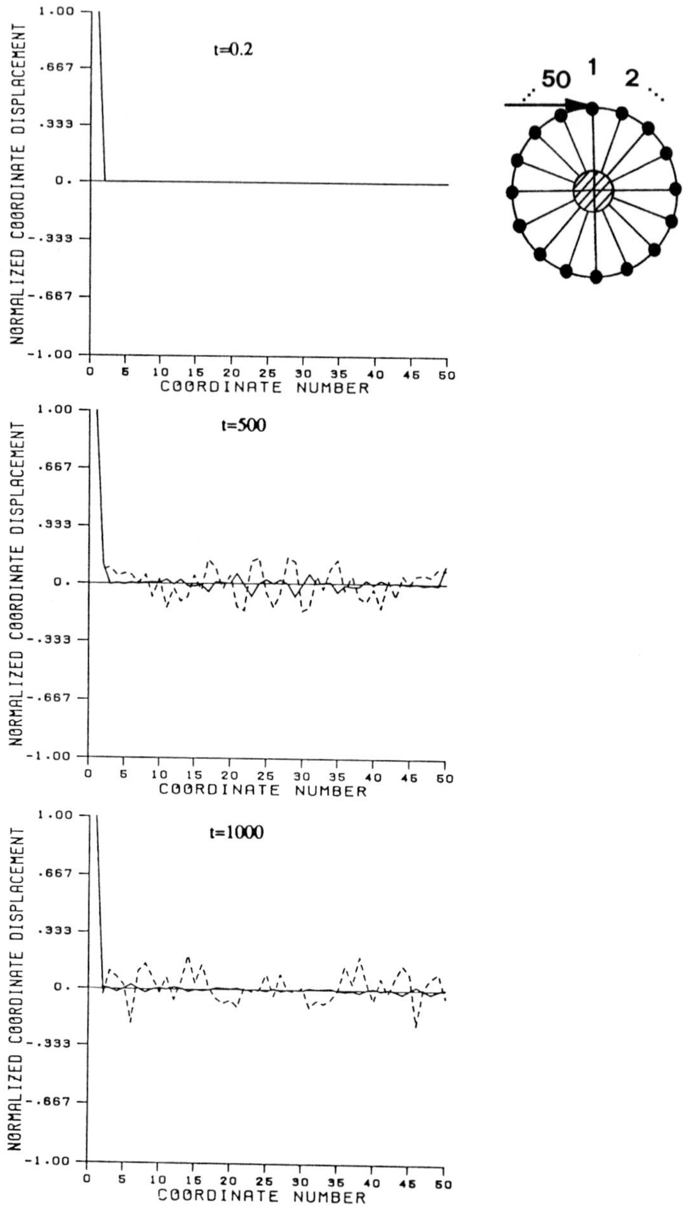

Figure 3. Nonlinear motion confinement of impulsive responses due to localized NNMs.

and $f_{1,(2k+1)}^{(u)}$, $F_{1,(2k+1)}^{(c)}$ are coefficients of the Taylor series expansions of stiffnesses $f_i^{(u)}$ and $f_i^{(c)}$, defined by the relations:

$$f_i^{(u)}(x_i) = \sum_{k=0}^{\infty} f_{i,(2k+1)}^{(u)}(0)\, x_i^{2k+1} / (2k+1)!$$

$$f_i^{(c)}(\alpha_{i-1}x_1, \alpha_i x_1, \alpha_{i+1}x_1) \equiv F_i^{(c)}(x_1; \alpha_{i-1}, \alpha_i, \alpha_{i+1}) =$$
$$= \sum_{k=0}^{\infty} F_{i,(2k+1)}^{(c)}(0; \alpha_{i-1}, \alpha_i, \alpha_{i+1})\, x_i^{2k+1} / (2k+1)! \qquad \text{(A-3)}$$

Using (A-1), coefficients $a_{is}^{(1)}$, s=1,2,..., are expressed in terms of the leading (linear) coefficient $a_{i0}^{(1)}$. This coefficient is evaluated as follows[7]:

$$a_{i0}^{(1)} =$$
$$= -\sum_{k=0}^{\infty} X_1^{2k}\{D_{ik}+B_k X_1^2[3K_{i1}^{(1)}+5K_{i2}^{(1)}X_1^2] +$$
$$+ C_{ik}X_1^2[K_{i1}^{(1)}+K_{i2}^{(1)}X_1^2]\}[\sum_{k=0}^{\infty} X_1^{2k}\{B_k[1+3L_{i1}^{(1)}X_1^2+5L_{i2}^{(1)}X_1^4]+$$
$$+ C_{ik}[1+ L_{i1}^{(1)}X_1^2 +L_{i2}^{(1)} X_1^4]\}]^{-1} + O(\varepsilon x_1^7) \qquad \text{(A-4)}$$

Using (A-1) and (A-4), coefficients $a_{i1}^{(1)}$ and $a_{i2}^{(1)}$ are evaluated as follows:

$$a_{i1}^{(1)} = L_{i1}^{(1)}\, a_{i0}^{(1)} + K_{i1}^{(1)}$$

$$a_{i2}^{(1)} = L_{i2}^{(1)}\, a_{i0}^{(1)} + K_{i2}^{(1)} \qquad \text{(A-5)}$$

where,

$$L_{i1}^{(1)} = -(B_0+C_{i0}) / 12h\,,\quad K_{i1}^{(1)} = -D_{i0} / 12h\,,$$

$$L_{i2}^{(1)} = -[\,(12A_0-3B_0-C_{i0})(B_0+C_{i0}) + 12h(B_1+C_{i1})\,] / 480h^2\,,$$

$$K_{i2}^{(1)} = -[\,(12A_0-3B_0-C_{i0})D_{i0} + 12hD_{i1}\,] / 480h^2 \qquad \text{(A-6)}$$

3. Continuous Oscillators

3.1. Study of Localized NNMs

To extend the analysis in coupled flexible systems, consider the weakly coupled assembly depicted in Fig. 4, consisting of two isotropic, linearly elastic beams of identical material properties and dimensions, coupled by means of a linear stiffness. Systems of more complicated geometry can be analyzed similarly. Assuming no out-of-plane components of motion and sufficiently large beam flexibility, the nonlinear relation between curvature and transverse displacement and the longitudinal inertia of the beams give rise to geometric nonlinearities which can greatly influence the dynamic response[26]. Assuming that the beams are rigidly fixed to a non-moving rigid base, and that weak coupling stiffness $K=\varepsilon k$ exists, where $|\varepsilon|<<1$, a rescaling of the transverse displacements, $v_p \rightarrow \varepsilon^{1/2} v_p$, leads to the following governing equations of motion[26]:

$$
\begin{aligned}
v_{ptt} + v_{pxxxx} &= \\
&= -\varepsilon\{ v_{px} [v_{px} v_{pxx}]_x + (1/2)\, v_{px} \int_1^x \int_0^s v_{px}^2\, du\,]_{tt}\, ds \}_x - \\
&\quad - \varepsilon\, (kL^4/EI)\, \{ v_p(l/L,t) - v_m(l/L,t) \}\, \delta(x-l/L) + F_p(x,t) \\
u_p(x,t) &= O(\varepsilon)\ ,\ p,\ m=1,2\ ,\ p\neq m
\end{aligned}
\qquad (11)
$$

where v_p and u_p denote the transverse and longitudinal displacements of beam p, and x the arclength per unit length of the two beams. In Eq.(11), t is the scaled time, defined by $t=\tau\{EI/\rho L^4\}^{1/2}$, where τ represents physical time, E the modulus of elasticity of the material of the beams, I the moment of inertia of the cross section of the beams about axes orthogonal to the plane of their motion, and ρ the material density per unit length. No shear deformations of the cross-section, nor any rotary inertia effects are taken into account. The first nonlinear term in the right-hand-side of Eq. (11) is due to the nonlinear relation between the curvature and the transverse displacement, and the second nonlinear term represents the nonlinear coupling effects due to the nonlinear longitudinal inertia of the beam.

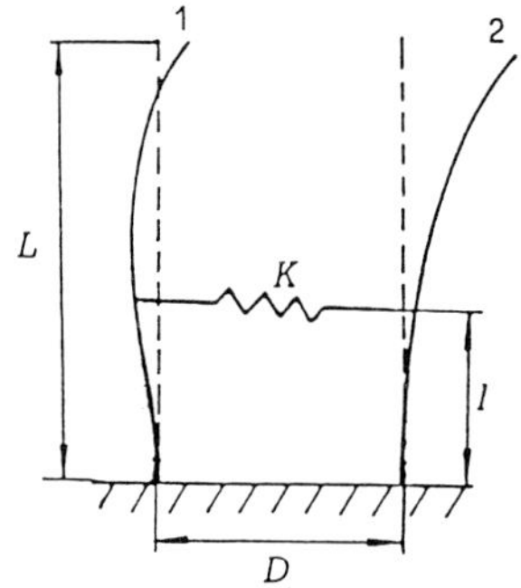

Figure 4. The flexible nonlinear assembly.

For small values of ε, equations (11) form a set of weakly nonlinear, and weakly coupled partial differential equations. This set can be discretized by expressing the transverse displacements $v_p(x,t)$ in the following series form:

$$v_p(x,t) = \sum_{m=1}^{n} \phi_m(x)\, q_{pm}(t)\,,\ p=1,2 \tag{12}$$

where the functions $\phi_m(x)$ are the normalized cantilever eigenfunctions of the linear parts of (11) (corresponding to $\varepsilon=0$), and $q_{pm}(t)$ new generalized coordinates. Substituting (12) into (11), premultiplying by the i-th normalized eigenfunction $\phi_i(x)$, integrating from $x = 0$ to $x = 1$ with respect to the spatial variable, and using the orthogonality properties of the linearized eigenfunctions, the following set of ordinary differential equations for $q_{pi}(t)$ is obtained:

$$\ddot{q}_{pi} + \omega_i^2 q_{pi} =$$

$$= -\varepsilon\{\sum_{b=1}^{n}\sum_{k=1}^{n}\sum_{l=1}^{n}[a_{iblk}\, q_{pb}\, q_{pk}\, q_{pl} + \frac{b_{ibkl}}{2} q_{pb}\, [q_{pk}\, q_{pl}]_{tt}\,] +$$

$$+ \sum_{k=1}^{n} \gamma_{ki}\,(q_{pk} - q_{(p+1)k})\} + F_{pi}(t) \tag{13}$$

where the first subscript of q_{pi} represents the beam number, $p = 1, 2$, $p = 3 \equiv 1$, and the second subscript denotes the order of the linearized mode-shape, $i = 1,2...n$. The various terms in (13) are defined as follows:

$$a_{ibkl} = \int_0^1 \phi_i\, [\phi_b'(\phi_k'\phi_l'')']'\, dx,\quad b_{ibkl} = \int_0^1 \phi_i\, [\phi_b' \int_1^x \int_0^\zeta \phi_k'\phi_l'\, d\lambda\, d\zeta]'\, dx,$$

$$\gamma_{ki} \equiv \frac{k\,L^4}{E\,I}\, \phi_k(l/L)\, \phi_i(l/L),\quad F_{pi}(t) = \int_0^1 F_p(x,t)\, \phi_i\, dx \tag{14}$$

where prime denotes differentiation with respect to the argument. A detailed dynamic analysis of the unforced system (13) was carried out in the work by King and Vakakis[22] by employing the method of multiple-scales. The responses q_{pi} in Eq.(13) are expressed in the form:

$$q_{pi}\,(t) = q_{pi}\,(T_0,T_1) = A_{pi}(T_1)e^{j\omega_i T_0} + \bar{A}_{pi}(T_1)\, e^{-j\omega_i T_0} + O(\varepsilon) \tag{15}$$

where $T_0 = t$ and $T_1 = \varepsilon t$ are fast and slow time-scales, $A_{ip}(T_1)$ are complex amplitudes, $\overline{(\bullet)}$ denotes the complex conjugate, and ω_i is the natural frequency of the i-th linearized cantilever mode. The complex amplitudes in (15) are computed by substituting (15) into

(13) and eliminating secular terms[25] of the O(ε) equations, i.e., terms which lead to unbounded and, thus, non-uniformly valid solutions in time. At this point, it is noted that a low-order *internal resonance* exists between the second and third flexural cantilever modes $\phi_2(x)$ and $\phi_3(x)$, since their corresponding linearized natural frequencies are nearly integrably related: $\omega_3 \sim 3\omega_2$. *Such a low-order internal resonance is well known[25] to lead to nonlinear transfer of energy between the associated linearized modes and hence, is expected to influence the nonlinear mode localization in the flexible system under consideration.* To study the effects of this low-order "internal resonance" one introduces a detuning parameter, σ, defined as:

$$\omega_3 = 3\omega_2 + \varepsilon\sigma. \tag{16}$$

Parameter σ quantifies the closeness of the multiples of the natural frequencies of the modes participating in the internal resonance. Truncating expression (13) to three modes per beam (n=3), setting $A_{pi}(T_1) = (1/2)\, a_{pi}(T_1)\, e^{j\theta_{pi}(T_1)}$, where a_{pi} and θ_{pi} are real amplitudes and phases, and eliminating secular terms, the following set of differential equations governing the amplitude- and phase-modulations is derived:

(Mode 1)

$$\omega_1 a_{11}' = (1/2)\,\gamma_{11}\, a_{21} \sin(\theta_{21}-\theta_{11})$$

$$\omega_1 a_{21}' = -\,(1/2)\,\gamma_{11}\, a_{11} \sin(\theta_{21}-\theta_{11})$$

$$\omega_1 a_{11}\theta_{11}' = (1/2)\gamma_{11}\,(a_{11}-a_{21}\cos(\theta_{21}-\theta_{11}))-(a_{11}/8)\sum_{k=1}^{3}\eta_{1k} a_{1k}^2$$

$$\omega_1 a_{21}\theta_{21}' = (1/2)\gamma_{11}(a_{21}-a_{11}\cos(\theta_{11}-\theta_{21}))-(a_{21}/8)\sum_{k=1}^{3}\eta_{1k} a_{2k}^2 \tag{17i}$$

(Mode 2)

$$\omega_2 a_{12}' = (1/2)\,\gamma_{22}\, a_{22} \sin(\theta_{22}-\theta_{12}) + $$
$$+ (\xi/8)\, a_{13}\, a_{12}^2 \sin(\theta_{13}-3\theta_{12}+\sigma_1 T_1)$$

$$\omega_2 a_{22}' = (1/2)\,\gamma_{22}\, a_{12} \sin(\theta_{12}-\theta_{22}) + $$
$$+ (\xi/8)\, a_{23}\, a_{22}^2 \sin(\theta_{23}-3\theta_{22}+\sigma_1 T_1)$$

$$\omega_2 a_{12}\theta_{12}' = (1/2)\gamma_{22}(a_{12}-a_{22}\cos(\theta_{22}-\theta_{12}))-(a_{12}/8)\sum_{k=1}^{3}\eta_{2k}\, a_{1k}^2 -$$
$$-(\xi/8)\, a_{13}\, a_{12}^2 \cos(\theta_{13}-3\theta_{12}+\sigma_1 T_1)$$

$$\omega_2 a_{22}\theta_{22}' = (1/2)\gamma_{22}(a_{22}-a_{12}\cos(\theta_{12}-\theta_{22}))-(a_{22}/8)\sum_{k=1}^{3}\eta_{2k} a_{2k}^2 -$$
$$-\,(\xi/8)\, a_{23}\, a_{22}^2 \cos(\theta_{23}-3\theta_{22}+\sigma_1 T_1) \tag{17ii}$$

(Mode 3)

$$\omega_3 a_{13}' = (1/2)\,\gamma_{33}\, a_{23} \sin(\theta_{23}-\theta_{13}) -$$

$$- (\delta/8)\, a_{12}^3 \sin(\theta_{13}-3\theta_{12}+\sigma_1 T_1)$$

$$\omega_3 a_{23}'=(1/2)\gamma_{33} a_{13}\sin(\theta_{13}-\theta_{23})-(\delta/8)a_{22}^3\sin(\theta_{23}-3\theta_{22}+\sigma_1 T_1)$$

$$\omega_3 a_{13}\theta_{13}'=(1/2)\gamma_{33}(a_{13}-a_{23}\cos(\theta_{23}-\theta_{13}))-(a_{13}/8)\sum_{k=1}^{3}\eta_{3k} a_{1k}^2-$$

$$- (\delta/8)a_{12}^3\cos(\theta_{13}-3\theta_{12}+\sigma_1 T_1)$$

$$\omega_3 a_{23}\theta_{23}'=(1/2)\gamma_{33}(a_{23}-a_{13}\cos(\theta_{13}-\theta_{23}))-(a_{23}/8)\sum_{k=1}^{3}\eta_{3k} a_{2k}^2-$$

$$- (\delta/8)\, a_{22}^3 \cos(\theta_{23}-3\theta_{22}+\sigma_1 T_1) \qquad (17iii)$$

In (17), primes denote differentiation with respect to the slow time T_1. The various parameters appearing in (17) are defined as:

$$\eta_{ii} \equiv 2\omega_i^2 b_{iiii}-3a_{iiii},\quad \eta_{pi} \equiv 2b_{pipi}\,(\omega_p^2+\omega_i^2)-2(a_{ppii}+a_{piip}+a_{pipi})$$

$$\zeta \equiv b_{2232}\,(\omega_3 - \omega_2)^2 + 2\omega_2^2\, b_{2322} - a_{2223} - a_{2232} - a_{2322},$$

$$\delta \equiv 2\omega_2^2\, b_{3222} - a_{3222} \qquad (18)$$

The free nonlinear periodic solutions of the system are investigated by replacing the angle-variables θ_{pi} with the new phase-variables $\Phi_1 = \theta_{21} - \theta_{11}$, $\Phi_2 = \theta_{22} - \theta_{12}$, $\Phi_3 = \theta_{23} - \theta_{13}$, and $\Psi_1 = \theta_{13} - 3\theta_{12} + \sigma T_1$, and reducing equations (17) to a set of ten *autonomous* ordinary differential equations of first order. The periodic solutions of the system are then obtained by imposing stationarity conditions on the amplitude- and phase-variables, i.e., by setting $a_{pi}' = 0$, $\Phi_i' = 0$, and $\Psi_1' = 0$ in the reduced set of equations, and solving the resulting set of stationary algebraic equations. This calculation was performed in the work by King and Vakakis[22], where it was found that, *certain of the periodic solutions of the system are localized.* During such motions (localized normal oscillations), the modal amplitudes of one beam are much larger in magnitude than the corresponding amplitudes of the other beam, and thus, the vibrational energy is spatially confined and nearly restricted to only one of the two subsystems. In a localized normal oscillation, the motions of the two beams are approximately given by:

$$v_p(x,t) \approx \sum_{i=1}^{3} [a_{pi}^*\cos(\omega_i t + \theta_{pi}^*(\varepsilon t))+O(\varepsilon)\,]\,\phi_i(x),\ p = 1, 2 \qquad (19)$$

where a_{pi}^* and $\theta_{pi}^*(\varepsilon t)$ denote the localized modal amplitudes and angles, obtained by solving the stationary equations. The stability of the periodic solutions (19) can be studied by Floquet analysis, i.e., by forming the appropriate system of linear variational equations in terms of the amplitude and phase modulations and computing the eigenvalues of the associated Floquet matrix.

It can be shown that there exist two basic classes of localized nonlinear modes in the system. The first category involves participation of only the first cantilever mode of the two blades, i.e., $a_{11}^* \neq 0$, $a_{21}^* \neq 0$, $a_{pi}^* = 0$, p=1,2, i=2,3, and the localized modes are depicted in Fig. 5a. Note that nonlinear localization depends on the ratio $r=\gamma_{11}/|\eta_{11}|$. From previous definitions it can be seen that γ_{11} is a parameter related to the strength and position of the coupling stiffness, whereas η_{11} relates to the geometric nonlinearities of the beam. As $r \to 0$, the bifurcating modes become localized, $\lim_{r\to 0}\{a_{11}^*/a_{21}^*\}=0$ or ∞, and the energy of the corresponding free motion is mainly confined to only one of the two beams (cf. Fig. 5a). When r increases, the localized branches become non-localized and eventually coalesce with the antisymmetric mode in a hamiltonian pitchfork bifurcation. *This bifurcation point can be regarded as the point of generation of the nonlinear mode localization phenomenon.* In physical terms, parameter r represents the ratio of coupling over nonlinear forces, and thus, the plot in Fig. 5a indicates that *when both beams oscillate in their first bending mode, nonlinear mode localization occurs only when the coupling forces are weak, and/or the beam nonlinearities are strong.*

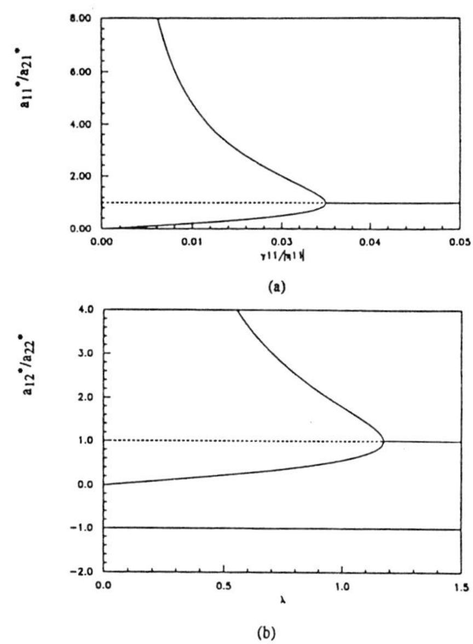

Figure 5. Nonlinear mode localization when beams oscillate in (a) their first cantilever mode, and (b) in their second and third cantilever modes. ——— Stable, - - - - - Unstable NNMs.

The second class of nonlinear localized modes involves participation of only the second and third cantilever modes, i.e., $a_{11}^* = a_{21}^* = 0$, $a_{pi}^* \neq 0$, p=1,2, i=2,3. In this case, the "internal resonance" between modes 2 and 3 greatly affects the topology of the localized branches. For c = l / L = 0.7650 (where l is the coupling position), the branches of localized modes are depicted in Fig. 5b. In each of the two diagrams the ratio of the modal amplitudes (a_{1i}^*/a_{2i}^*), i=2,3 is plotted versus the parameter $\lambda = (kL^4/EI)/a_{3223}$. Again, it is observed that $\lim_{\lambda\to 0}\{a_{1i}^*/a_{2i}^*\}=0$ or ∞, i=2,3, i.e., that as the coupling decreases and/or the nonlinearity increases nonlinear mode localization occurs. As λ increases, the localized modes become non-localized, until they coalesce with the symmetric mode in a pitchfork bifurcation. An interesting feature of mode localization in the presence of internal resonance is its essential dependence on the position of the coupling stiffness. Indeed, when c=l/L is close to 0.783, the value corresponding to the node of the second (lower) cantilever mode, a complicated sequence of bifurcations of certain solution branches takes place[22]. Additionally, it can be shown that *high modes are more susceptible to nonlinear mode localization than lower ones.*

3.2. Nonlinear Passive Motion Confinement Due to Mode Localization

It will now be shown that the stable nonlinear localized modes in the system of Fig. 4 give rise to passive confinement of externally induced impulses. To show this, it will be assumed that system (11) is forced by an impulse excitation of the form:

$$F_1(x,t) = (1/\varepsilon)\, f_1(t)\, \phi_1(x) \text{ for } 0 \le t < \varepsilon D,\quad F_1(x,t) = 0 \text{ for } t \ge \varepsilon D$$
$$F_2(x,t) = 0\,,\quad 0 \le t < \infty \tag{20}$$

Expressing the displacements $v_p(x,t)$, p=1,2, in terms of the linearized cantilever modes, the discretized set of forced ordinary differential equations (13) is obtained, with forcing terms of the form,

$$(1/\varepsilon)\, F_{pi}(t) = (1/\varepsilon) \int_0^1 f_1(t)\, \phi_i^2(x)\, dx\, \delta_{p1}\, \delta_{i1} \text{ for } 0 \le t < \varepsilon D\,,$$
$$F_{pi}(t) = 0 \text{ for } t \ge \varepsilon D \tag{21}$$

Hence, it is assumed that only the first cantilever mode of beam 1 is *directly* excited by the impulsive force (20). *It is of interest to study the transfer of the energy of the impulse from the directly excited mode to the other modes of the system; clearly, spatial motion confinement of the external impulse is achieved only when minimal amounts of vibrational energy eventually "leak" to the modes of the unforced beam 2.* Previously it was found that the first cantilever mode is not nonlinearly coupled with other higher modes. Hence, no energy transfer is expected to occur from the directly excited first mode of beam 1, to any other higher modes of the two beams, and the only possible energy exchange is anticipated to take place between the first cantilever modes of the two beams. This theoretical prediction will be verified in the following analysis. For the sake of simplicity, the analysis is limited to n=3 cantilever modes per beam, and two phases of the motion are considered.

<u>Phase 1, $0 \le t < \varepsilon D$</u>. During this phase, the applied force is non-zero, and the response is asymptotically approximated by introducing the new time T, defined by $t=\varepsilon T$. The range of

values of the new time variable during this phase of the motion is, $0 \leq T < D$. Expressing the time derivatives in (13) in terms of the new variable T, the governing equations for the modal amplitudes are expressed as follows:

(Mode 1)

$$\begin{aligned} &d^2q_{p1}/dT^2= \\ &=\varepsilon\{-\sum_{b=1}^{3}\sum_{k=1}^{3}\sum_{l=1}^{3}\frac{b_{1bkl}}{2}q_{pb}\,[q_{pk}\,q_{pl}]_{TT}+\hat{p}_1(T)\delta_{p1}\}- \\ &-\varepsilon^2\omega_1^2\,q_{p1}-\varepsilon^3\{\sum_{b=1}^{3}\sum_{k=1}^{3}\sum_{l=1}^{3}a_{1blk}\,q_{pb}\,q_{pk}\,q_{pl}+ \\ &+\sum_{k=1}^{3}\gamma_{ik}(q_{pk}-q_{(p+1)k}))\},\qquad p=1,2 \end{aligned} \tag{22}$$

with similar expressions holding for the modal displacements q_{p2}, and q_{p3}, p=1,2, of the higher modes[23]. Complementing (22), is the set of initial conditions $q_{pi}(0)=0$, $d(q_{pi}(0))/dT=0$, p=1,2, i=1,2,3, since at t=T=0 the system is assumed to be at rest. In equations (22), all depended variables are functions of T, and the new forcing function is defined as,

$$\hat{p}_1(T) \equiv F_{11}(\varepsilon T) \tag{23}$$

The responses of system (22) are analytically approximated using regular perturbation expansions of the form, $q_{pi}(T)=\sum_{m=1}^{\infty}\varepsilon^m q_{pi}^{(m)}(T)$. By matching the coefficients of respective powers of ε, one determines the various orders of approximation. Omitting the calculations[23], and employing the original time variable, the responses of the system during this phase of the motion are computed as:

$$\begin{aligned} &q_{11}(t)=\varepsilon\int_0^{(t/\varepsilon)}\int_0^{\eta}\hat{p}_1(s)ds\,d\eta+O(\varepsilon^3)\,,\quad \dot{q}_{11}(t)=\int_0^{(t/\varepsilon)}\hat{p}_1(s)ds+O(\varepsilon^2) \\ &q_{ij}(t)=O(\varepsilon^4),\quad \dot{q}_{ij}(t)=O(\varepsilon^3)\,,\quad \text{otherwise} \end{aligned} \tag{24}$$

Hence, the analysis predicts that for $0 \leq t < \varepsilon D$ (the duration of the impulse), the response of the system is mainly determined by the impulse itself and not by any structural parameters (the system "does not have time to oscillate"). The resulting physical motions of the two beams are given by:

$$v_1(x,t) \approx \phi_1(x)q_{11}(t)+O(\varepsilon^3),\ v_2(x,t) \approx O(\varepsilon^4),\quad 0 \leq t < \varepsilon D \tag{25}$$

<u>Phase 2, $t \geq \varepsilon D$.</u> During this phase of the motion, the impulse ceases to apply, and, *the system performs free oscillations,* with initial conditions determined from the solutions of

phase 1, Eq. (24). Introducing the time translation $\hat{t}=t-\varepsilon D$, the governing equations (13) are expressed as:

$$\ddot{q}_{pi} + \omega_i^2 q_{pi} = -\varepsilon\{\sum_{b=1}^{3}\sum_{k=1}^{3}\sum_{l=1}^{3}[a_{iblk}\, q_{pb}\, q_{pk}\, q_{pl} + \frac{b_{ibkl}}{2}\, q_{pb}\, d^2(q_{pk}\, q_{pl})/d\hat{t}^2\,]+$$
$$+\sum_{k=1}^{3}\gamma_{ik}\,(q_{pk} - q_{(p+1)k}))\}\quad, p=1,2,\ i=1,2,3 \qquad (26)$$

where differentiation is carried out with respect to $\hat{t}$, and $q_{pi} = q_{pi}(\hat{t})$. The set of initial conditions complementing (26) is, $q_{11}(0)=\varepsilon\int_0^D\int_0^\eta \hat{p}_1(s)ds\,d\eta+O(\varepsilon^3)$, $\dot{q}_{11}(0)=\int_0^D \hat{p}_1(s)ds +O(\varepsilon^2)$, $q_{pi}(0)=O(\varepsilon^4)$, and $\dot{q}_{pi}(0)=O(\varepsilon^3)$, otherwise. Since equations (26) represent *free* oscillations of the system, their solutions can be analytically approximated by a multiple-scales singular perturbation analysis. Omitting the details of this calculation, the resulting physical motions of the beams are computed as:

$$v_p(x,t) = \sum_{i=1}^{3}\,[a_{pi}(\varepsilon(t-\varepsilon D))\cos(\omega_i(t-\varepsilon D) + \theta_{pi}(\varepsilon(t-\varepsilon D))) + O(\varepsilon)]\,\phi_i(x),$$
$$t\geq \varepsilon D,\ p = 1, 2 \qquad (27)$$

where, $a_{12}(T_1) = a_{13}(T_1) = a_{22}(T_1) = a_{23}(T_1) = 0$, $\forall\, T_1$, and the amplitudes and phases of the first modes of the two beams are computed by the following set of equations:

$$a_{11}' = (\gamma/2\omega)\, a_{21}\sin(\theta_{21}-\theta_{11})$$
$$a_{21}' = -\,(\gamma/2\omega)\, a_{11}\sin(\theta_{21}-\theta_{11})$$
$$a_{11}\theta_{11}' = [(3\alpha/8\omega_1)-(\beta\omega_1/4)]a_{11}^3+(\gamma/2\omega)a_{11}-(\gamma/2\omega)a_{21}\cos(\theta_{21}-\theta_{11})$$
$$a_{21}\theta_{21}' = [(3\alpha/8\omega_1)-(\beta\omega_1/4)]a_{21}^3+(\gamma/2\omega)a_{21}\ (\gamma/2\omega)a_{11}\cos(\theta_{21}-\theta_{11}) \qquad (28i)$$

where primes denote differentiation with respect to the "slow time" T_1, and the simplified notation, $\alpha\equiv a_{1111}$, $\beta\equiv b_{1111}$, and $\gamma\equiv\gamma_{11}$, is adopted from now on. Define at this point the quantity $J=\gamma/([(3\alpha/2)-(\beta\omega_1^2)]\rho)$, where $\rho=(2\omega_1)^{-1}\int_0^D \hat{p}_1(s)\,ds$; *parameter J is recognized as the ratio of the coupling over the nonlinear forces.* For J<<1, the response of the system can be analytically approximated[23]. This is achieved by setting,

$$a_{p1}(T_1)= \sum_{m=0}^{\infty} J^m a_{p1}^{(m)}(T_1),\ \theta_{p1}(T_1)= \sum_{m=0}^{\infty} J^m\, \theta_{p1}^{(m)}(T_1)\,,\quad p=1, 2 \qquad (28ii)$$

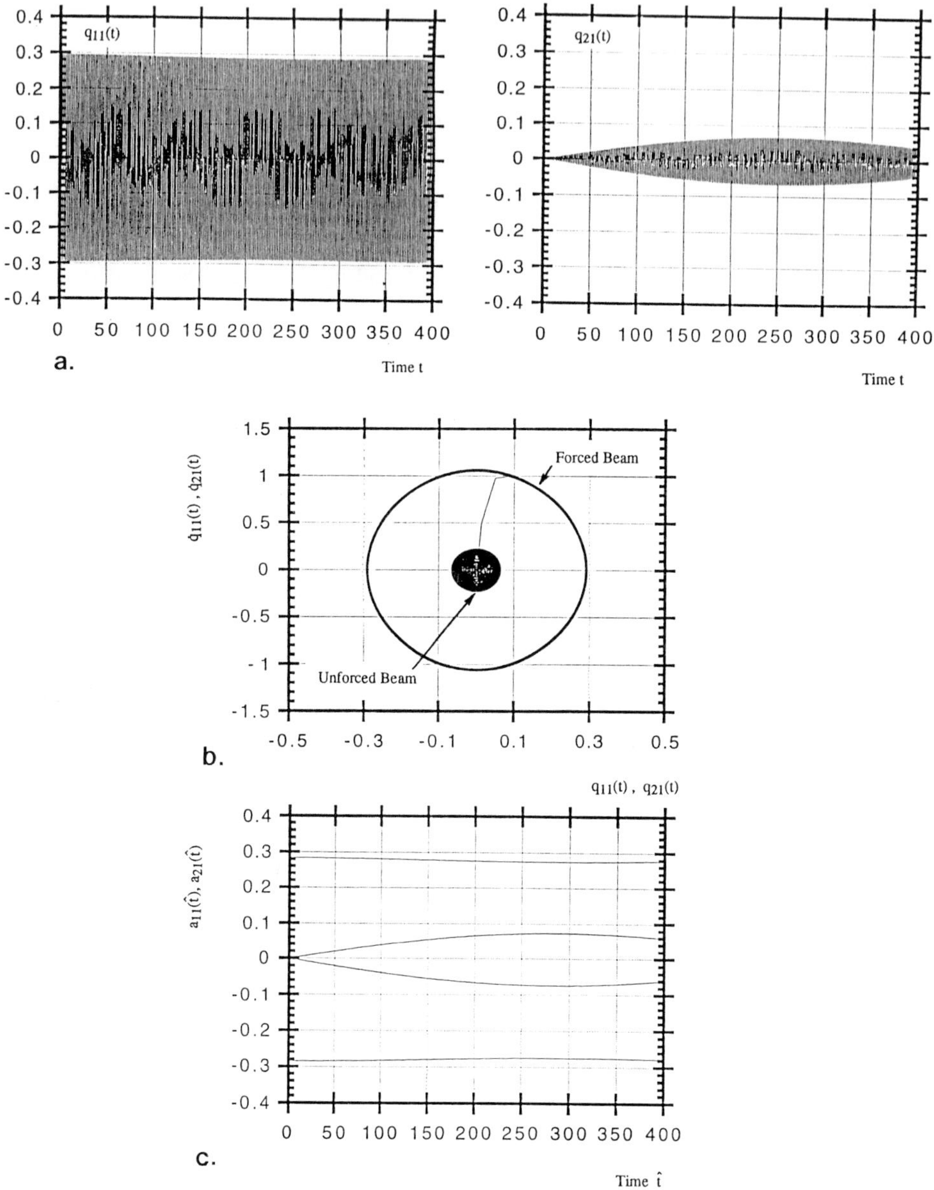

Figure 6. Single-mode oscillations, amplitude modulations computed by, (a) direct integrations of the equations of motion, time responses, (b) direct integrations of the equations of motion, responses in a projection of the phase space, (c) theoretical results, eq.(29).

and substituting into (28i). Matching terms proportional to the same power of J, leads to the following analytical expressions for the amplitude modulations:

$$a_{11}(T_1) = \pm\{2\rho - (J^2/\rho)\sin^2([(3\alpha/4\omega_1) - (\beta\omega_1/2)]\rho^2 T_1) + O(J^3)\}$$
$$a_{21}(T_1) = \pm\{-2J\sin([(3\alpha/4\omega_1) - (\beta\omega_1/2)]\rho^2 T_1) + O(J^2)\} \quad (J<<1) \qquad (29)$$

Similar analytical expressions hold for the angles $\theta_{p1}(T_1)$. Solutions (29) indicate that, *when J<<1, the transient impulsive response of the system is mainly confined to the directly excited blade, since $a_{11}(T_1) >> a_{21}(T_1)$.* Based on previous results, the condition J<<1 (weak coupling and/or strong nonlinearities) is also recognized as the condition for the existence of localized nonlinear modes in the system. It is therefore concluded that, *when the system possesses nonlinear localized modes, it also possesses passive motion confinement properties: when an external impulse acts at one beam of the system, the disturbance remains spatially confined to its point of application, and only a small portion of the injected energy "leaks" to the unforced beam.* Moreover, from Eq. (29), it is predicted that the responses of the forced and unforced beams contain $O(J^2)$ and O(J)-amplitude modulations respectively. This is consistent with energy conservation considerations. Indeed, from (29), a direct computation gives, $a_{11}^2(T_1) + a_{21}^2(T_1) = 4\rho^2 + O(J^3)$. Correct to $O(\varepsilon)$, this relation can be shown to represent conservation of energy of the free, undamped system under consideration.

To verify the results of the multiple-scales analysis, an impulsive distributed excitation of magnitude $F_1(x,t) = 5\,\phi_1(x) \Rightarrow (1/\varepsilon)\,F_{11}(t) = \int_0^1 10\,\phi_i\,dx$, and duration $\varepsilon D=0.1$sec was applied to beam 1, and the structural parameters were assigned the values, $\varepsilon=0.5$, $a_{1111}=40.44$, $b_{1111}=4.59$, $\gamma_{11}=0.02$, $\omega_1=3.51$ (rad/sec), J=0.0367<<1; the response was numerically computed by directly integrating the differential equations of motion (22) with $q_{12}=q_{13}=q_{22}=q_{23}=0$, and assuming zero initial conditions at t=0. In figure 6a the responses of the two beams are shown as functions of time, and in figure 6b, a projection of the phase space of the motion is depicted. For comparison purposes, the theoretically predicted amplitude modulations (29) are presented in figure 6c. Clearly, the asymptotic theory agrees well with the numerical computations for this low value of J. Note, that in the absence of nonlinearities, all injected energy is continuously transferred between the forced and unforced beams, in the well-known beat phenomenon. *Hence, the detected passive motion confinement phenomenon is solely attributed to the geometric nonlinearities of the system.*

4. Experimental Study of Nonlinear Motion Confinement

4.1. Experimental Fixture

As shown in previous sections, an interesting feature of systems with localized modes is that they possess motion confinement properties. Such systems can be designed so that *energy induced from an external impulsive excitation is mainly confined to the directly excited substructure instead of spreading through the entire system.* The principal aim of this section is to experimentally verify the existence of the nonlinear motion confinement

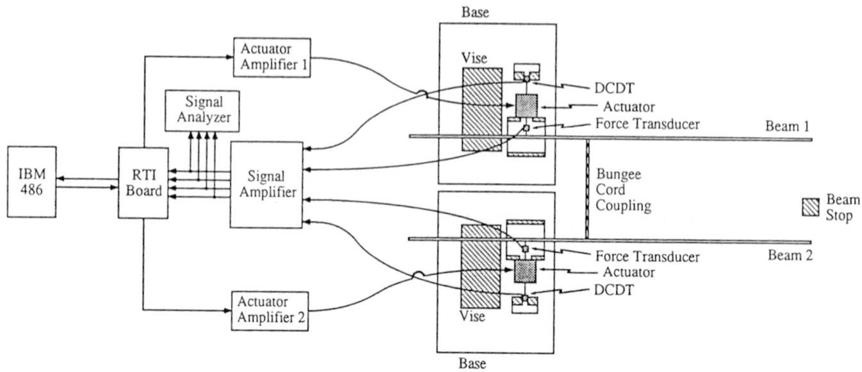

Figure 7. Schematic Diagram of The Experimental Fixture.

phenomenon. The work reported herein is a summary of the M.Sc. Thesis by Aubrecht[27], and that work should be consulted for more information about the experimental fixture and the procedure followed.

The experimental apparatus is similar to the theoretical model of Fig. 4. It consists of two coupled cantilever beams and is depicted in Fig. 7. Aluminum beams of identical material properties and dimensions were used; each beam was 1m in length, 0.0445m in width and 0.00635m in thickness. Vises were employed to simulate rigidly clamped boundary conditions, and each cantilever beam was rigidly attached to a steel plate which was firmly fixed to a rigid concrete foundation. Modal testing was employed to establish that the beams were linear and possessed nearly identical dynamic properties[27]. Since the experimental beams possessed relatively small flexibility no geometric nonlinearities of the type described in section 3 occurred. Instead, nonlinearities in the experimental fixture were *actively induced* by means of two control actuators simulating cubic grounding stiffnesses. Direct coupling between beams was achieved using an elastic cord.

Since the aim of the experiment was to verify the nonlinear motion confinement phenomenon, a linearity check was first performed by modal-testing the beams with actuators-off at different forcing levels and comparing the corresponding Frequency Response Functions (FRFs)[27]. It was verified that the measured FRFs remained unaltered for different forcing levels, and, hence, the linearity of the beam dynamics was established.

The equipment used for data acquisition and the control of the actuators consisted of a computer (data processor and collector), an Analog Devices RTI-815 input/output board, amplifiers, actuators, displacement transducers, and force transducers. The analog input/output board was used both to collect data, as well as to provide control signals to the actuator amplifiers. Motion Science Inc., DHB series, Mosfet pulse width modulated switching amplifiers were used to amplify the voltage commands from the RTI board, producing current command signals for the actuators. One amplifier was used for each beam, such that the beams could be controlled independent of each other. Two actuators, one for each beam, were used in conjunction with these amplifiers. These were electro-mechanical linear force motors supplied by Moog, Inc., and produced a force output proportional to a current command. Two displacement transducers were attached to the armature of each actuator, through the back ends of the actuators. This enabled the position of each actuator armature to be recorded and to be used for control purposes. Collocated with the DCDTs and the actuators were force transducers, which were placed between each beam and its corresponding control actuator. This permitted direct measurements of the forces applied by the actuators to the beams, enabling these forces to be used for control purposes. The force transducers (PCB model 208A02) were piezoelectric transducers designed to measure compressive and tensile forces over a dynamic range of from 10 to 5000 lbf. A computer code in C language was used to interface with the RTI board (the input/output board), enabling both data acquisition and control of the actuators.

In the experimental fixture of Fig. 7 the nonlinearities required to produce the transient motion confinement phenomenon were actively induced using the control and data acquisition apparatus described earlier. This was performed by using the actuators to simulate cubic nonlinear grounding stiffnesses. Hence, the overall objective of the employed control methodologies was to achieve a cubic relationship between the displacements of the actuators at their points of attachments to the beams and the force outputs of the actuators In the control methodology, the displacement of the actuator was sensed, cubed, multiplied by a gain, and fed as a control voltage to the actuator amplifiers. The position loop, therefore, was closed. However, in this approach no attempt was made to match the *desired* force output to the *actual* force between the beam and the actuator. That is, the force loop was not closed. For this reason, this technique was termed the *open loop control.* An alternative methodology, termed the closed loop control, also was explored, which implemented closed position *and* force loops[27]. The closed loop control methodology had the advantages of producing a very consistent active nonlinear stiffness, however, it had one very strong limitation. The maximum cubic gain attainable with closed loop control was significantly smaller that that attainable with open loop control. This problem stemmed from the fact that, with closed loop control the controller was attempting to drive the force error signal to zero, whereas the open loop controller was not subject to such an obligation[27].

Initial experimental trials with the open loop control methodology indicated that, on average, it approximated satisfactory the desired cubic force law. However, it also possessed certain drawbacks. First, this control law was not very robust, and the force output was not extremely consistent from oscillation to oscillation. Second, it was not possible to fix *a priori* the gain on the cubic force law. Prior to an experimental run, the open loop gain was set at what was thought to be an appropriate value. The experimentally obtained actuator force vs. actuator displacement curve was later used to determine the actual cubic gain. Note, however, that while it was not possible to specify *a priori* the gain of the cubic force for open loop control, the resulting cubic gain remained essentially constant throughout the experiment. For the reasons above, *the open loop control was chosen over the closed loop control for the localization experiments.* Even though the open loop control approximated the desired cubic grounding stiffness less accurately than the closed loop control, it was able to produce a higher cubic gain with adequate performance, a feature which was instrumental for the generation and detection of the nonlinear transient motion confinement phenomenon[27].

Finally, notch filters for eliminating unwanted unstable dynamics in the integrated controller-beam system had to be designed and implemented[27]. These filters suppressed unstable dynamic response due to the excitation by the controllers of the second and third cantilever modes of the two beams.

4.2. Experimental Results

The existence of nonlinear motion confinement in the experimental fixture was experimentally verified by performing a series of transient vibration tests. In each test an initial deflection was imparted on beam #2, termed "directly excited beam," (cf. Fig.7) by imposing a fixed initial tip displacement equal to 4 inches; no initial deflection was imposed on beam #1, termed "non-directly excited beam." The ensuing free transient vibrations of beams #1 and #2 were analyzed by recording the force and displacement signals from the force transducers and DCDTs, respectively, located at the positions of the actuators. This enabled the recording of the transient responses of the beams and, in addition, the experimental determination of the (linear or nonlinear) relationship between the force applied by the actuator and the corresponding beam deflection at the position of the actuator. These experimentally determined force-deflection relationships were used to quantify the active grounding stiffness nonlinearities induced by the actuators to the beams.

A typical set of linear and nonlinear results is presented in Figures 8 and 9, where the transient responses of the two beams with coupling stiffness k=1.01 lbs/in, and actuators-off and -on, respectively, are depicted[27]. Considering the envelopes of the beam responses of the linear assembly (actuators-off, fig. 8d), it is observed that at t ~ 6.5 sec near equipartition of instantaneous vibrational energy occurs between beams #1 and #2, indicating that a significant amount of vibrational energy "leaks" from the directly excited to the unexcited beam in the classical (linear) beat-phenomenon. In addition, a careful examination of the time traces of figure 8c reveals that *a continuously varying phase difference between the transient motions of the two beams exists.* This nonconstant phase difference is due to the participation of both in-phase and out-of-phase (linear) normal modes of the assembly in the transient responses[27]. The transient responses of the system with actuators-on (and active grounding stiffness nonlinearities) are depicted in figure 9. It is noted that *no equipartition on of instantaneous energy occurs in this case, since smaller amounts of vibrational energy "leak" to beam #1.* Hence, a much greater portion of the induced vibrational energy remains confined to beam #2 and *nonlinear transient motion confinement occurs.* The energy "leakage" from the directly excited to the unexcited beam can be further reduced by increasing the gain (strengthening) of the active nonlinear stiffnesses. Comparing the envelopes depicted in figures 8d and 9d one observes *a diminishing of the period of the beat of the nonlinear response of beam #1, compared to that of the linear one.* This feature is due to hardening-stiffness effects in the transient nonlinear response, and is in full qualitative agreement with the theoretical predictions. Moreover, consideration of the relative phase differences between the time traces of beams #1 and #2 (fig. 9c) indicates that, *in contrast to the linear case, the nonlinear beams vibrate in a nearly anti-phase transient oscillation, a strong indication that a strongly localized out-of-phase NNM participates in the nonlinear transient responses.*

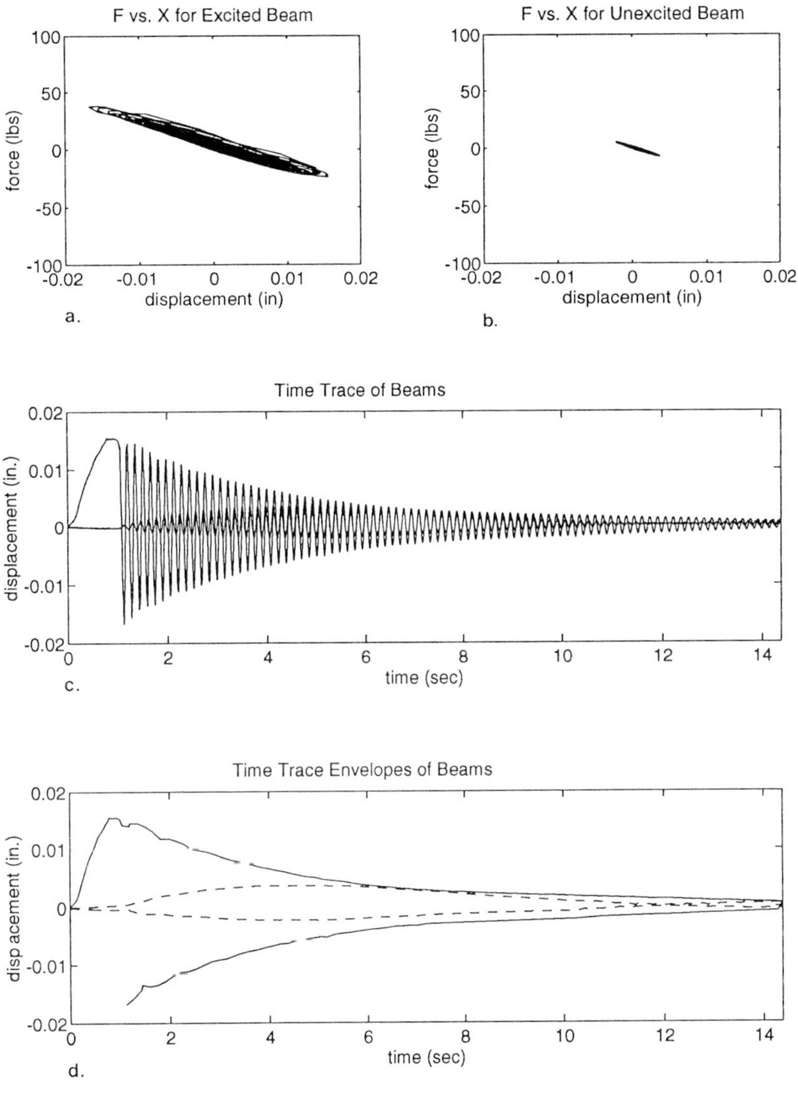

Figure 8. Transient Responses at the Positions of Actuators for the System with Actuators-off: (a), (b) Force versus Deflection Relationships, (c) Beam Deflections, (d) Envelopes of Beam Vibrations.

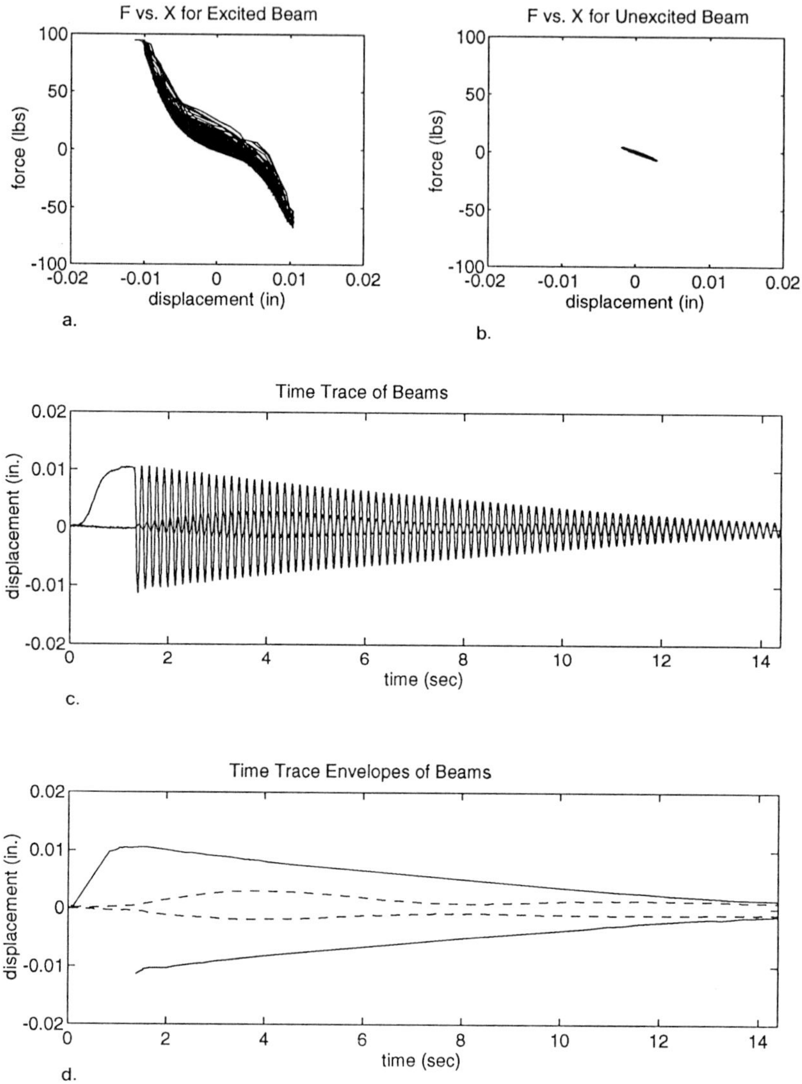

Figure 9. Transient Responses at the Positions of Actuators for the System with Actuators-on: (a), (b) Force versus Deflection Relationships, (c) Beam Deflections, (d) Envelopes of Beam Vibrations.

As mentioned earlier, the nonlinear motion confinement phenomenon is of considerable practical importance since the controllability of a structure possessing localized NNMs is greatly enhanced. The results reported in this section indicate that the nonlinear motion confinement phenomenon can be induced in practical flexible structures through active means. In such structures vibration and shock isolation can be achieved by employing the following new design concept:

- Initially, disturbances generated by external excitations become spatially localized by means of actively induced nonlinearities.
- Once spatial motion confinement is achieved, the confined unwanted disturbances are eliminated through active control.

Current research focuses on the implementation of the above design concept with flexible structures of more complicated geometry. Future applications are anticipated in the areas of vibration and shock isolation of mechanical components, rotordynamics (cyclic bladed assemblies), large space structures, and layered composite media.

5. Waves with Solitary Envelopes

5.1. Preliminary Notions

In this final section it will be shown how the concept of NNM can be extended to study waves with localized (solitary) envelopes in a class of nonlinear partial differential equations. The asymptotic methodology followed resembles the one employed for studying the localized modes of the cyclic discrete system in section 2. Travelling waves in nonlinear partial differential equations have been studied in numerous previous works. A *solitary wave* is a travelling wave possessing a constant, spatially localized form which decays or reaches a constant value for large values of the spatial independent variable. A remarkable property[28] of some solitary waves is their property to interact "elastically" with each other: after a collision of two or more of such waves, each wave retains its original form, and the only effect of the collision is an alteration of the relative phasing between participating waves. Such, "elastically" interacting solitary solutions are termed *solitons*. Certain nonlinear partial differential equations possess only solitary waves (but not solitons), whereas other (as the KdV and sine-Gordon equations) possess soliton solutions. The theory of solitons and solitary waves has applications in various engineering fields, including plasma physics[29], superconducting Josephson junctions[30], and solid state physics[31]. In addition to solitons and solitary waves nonlinear partial differential equations also admit envelope solitons. These are stationary or travelling wave motions whose envelopes decay (*breather solitons* [32]) or reach nonzero limits (*dark* or *hole solitons* [33]) as $x\rightarrow\pm\infty$, where x is the spatial coordinate. An envelope soliton is a stationary or travelling periodically oscillating wavepacket, which possesses a spatially localized envelope or a spatially localized slope envelope. Envelope solitons can interact "elastically" with other envelope solitons or with other types of solitons.

In what follows, a new approach for studying travelling or stationary waves with spatially localized envelopes is described. As an example of application of this methodology, a specific nonlinear partial differential equation is considered and the amplitude distributions of stationary (nonpropagating) localized waves are asymptotically computed. Propagating, weakly modulated waves are then computed by imposing Lorentz coordinate transformations. The technique presented herein can be regarded as a direct extension of previously described NNM methodologies employed for analyzing NNMs of discrete and (bounded, one-dimensional) continuous nonlinear oscillators. In the context of these previous methods, the stationary wave solutions computed herein can be regarded as *localized NNMs of unbounded, continuous, one-dimensional systems.* The analysis presented is based on the work by King and Vakakis[34], and that work should be consulted to obtain a more detailed exposition of the analysis.

5.2. General Analytical Formulation

The partial differential equations considered herein of the general form:

$$u_{tt} = \mathsf{L}^{(0)}[u(x,t)] + \varepsilon\mathsf{L}^{(1)}[u(x,t)]\,, \quad -\infty < x < +\infty \tag{30}$$

where $\mathsf{L}^{(0)}[\bullet]$ and $\varepsilon\mathsf{L}^{(1)}[\bullet]$ are integro-differential operators acting on the dependent variable u(s,t), x and t are the spatial and temporal variables, respectively, and ε is a perturbation parameter ($|\varepsilon|<<1$). It is assumed, (a) that equation (1) possesses a time-independent first integral of motion, (b) that for $\varepsilon=0$ the system is separable in space and time, and, (c) that u(x,t) is a sufficiently smooth function of its variables, so that all derivatives appearing in the following analysis can be defined. Note that operator $\mathsf{L}^{(0)}[u(x,t)]$ may be linear or nonlinear; the only requirement posed is that it is separable in space and time.

First, *stationary (non-travelling)*, time-periodic solutions of Eq. (30) are sought, satisfying conditions of the form:

$$\lim_{x\to\pm\infty} u(x,t) = u^{\pm}, \quad u(x,t+T) = u(x,t) \tag{31}$$

where T is the period of the oscillation. It is noted that only a very restricted class of equations (30) admit solutions that satisfy conditions (31)[35,36]. During such stationary solutions, the value of the dependent variable at an *arbitrary* point, u(x,t), can be related to its value at a prescribed *reference* point $x=x_o$, $u_o(t)\equiv u(x_o,t)$, by a functional relation of the form:

$$u(x,t) = U(x,u_o(t)) \tag{32}$$

In writing Eq. (32) it is assumed that $u_o(t)$ is not identically equal to zero. The function $U(x,u_o(t))$ will be termed the *amplitude modulation function*, since it represents the amplitude modulation of the stationary solution. Introducing a new set of independent variables $(x,t)\to(x,u_o(t))$, and assuming that a first integral of motion for (30) exists, one can eliminate the time dependence from (1) and construct a functional equation that must be satisfied by the amplitude modulation function $U(x,u_o(t))$. Using the functional relation (32), the partial derivatives of the displacement with respect to time are expressed as:

$$\partial u/\partial t = (\partial U/\partial u_o)u_{o,t}\,, \quad \partial^2 u/\partial t^2 = (\partial^2 U/\partial u_o{}^2)(u_{o,t})^2 + (\partial U/\partial u_o)u_{o,tt}\,, \ldots \tag{33}$$

Throughout the following exposition, subscripts t and x denote differentiation with respect to the temporal and spatial variables, respectively. The first integral of Eq. (30) is assumed to be of the form:

$$E = (1/2)\int_{-\infty}^{+\infty} u_t^2\, ds + \hat{\mathcal{L}}^{(0)}[u(x,t)] + \varepsilon\, \hat{\mathcal{L}}^{(1)}[u(x,t)] \tag{34}$$

where $\hat{\mathcal{L}}^{(0)}[u(x,t)]$ and $\varepsilon\hat{\mathcal{L}}^{(1)}[u(x,t)]$ denote the energy terms due to operators $\mathsf{L}^{(0)}[\bullet]$ and $\mathsf{L}^{(1)}[\bullet]$ respectively. Typically, these terms involve integrations with respect to the spatial variable, with integrands containing partial derivatives of the displacement. Taking into account Eq.(34), one obtains the following relations:

$$E = (1/2)(u_{o,t})^2 \int_{-\infty}^{+\infty} (\partial U/\partial u_o)^2 ds + \hat{\mathsf{L}}^{(0)}[U,u_o(t)] + \varepsilon\, \hat{\mathsf{L}}^{(1)}[U,u_o(t)] \Rightarrow$$

$$\Rightarrow u_{o,t}(t) = f\,(\,U(x,u_o(t))\,;\,\varepsilon\,) \tag{35}$$

where $\hat{\mathsf{L}}^{(0,1)} \equiv \{\hat{\mathcal{L}}^{(0,1)}\}_{[u(s,t)\,=\,U(s,uo(t))]}$, and function $f(\bullet)$ denotes the symbolic solution for $u_{o,t}(t)$. Note, that when system (30) reaches its maximum potential energy, $V=V_{max}=E$, the time derivative $u_{o,t}(t)$ vanishes. Assuming that at the point of maximum potential energy the displacement of the reference point attains the value $u_o(t)=u_o^*$, at that point, the following relations must be satisfied:

$$f\,(\,U(x,u_o^*)\,;\,\varepsilon\,) = 0$$

$$E = V_{max} = \hat{\mathsf{L}}^{(0)}[U(x,u_o^*),u_o^*] + \varepsilon\, \hat{\mathsf{L}}^{(1)}[U(x,u_o^*),u_o^*] \tag{36}$$

Substituting (32) and (33) into the equation of motion (30), taking into account (35), and noting that the second time derivative of u(x,t) at the reference point $x=x_o$ can be expressed as $u_{o,tt} = \left\{ \mathsf{L}^{(0)}[u(x,t)] + \varepsilon \mathsf{L}^{(1)}[u(x,t)] \right\}_{x=xo}$, one obtains the following equation governing the amplitude modulation function $U(x,u_o(t))$:

$$f^2(U(x,u_o(t));\,\varepsilon)\,\frac{\partial^2 U}{\partial u_o^2} + \left\{ \underline{\mathsf{L}}^{(0)}[U(x,u_o)] + \varepsilon \underline{\mathsf{L}}^{(1)}[U(x,u_o)] \right\}_{x=xo} \frac{\partial U}{\partial u_o} =$$
$$= \underline{\mathsf{L}}^{(0)}[U(x,u_o)] + \varepsilon \underline{\mathsf{L}}^{(1)}[U(x,u_o)] \tag{37}$$

with

$$\lim_{x\to\pm\infty} U(x,u_o(t)) = u^{\pm} \tag{38}$$

In (37), operators $\underline{\mathsf{L}}^{(0)}[\bullet]$, and $\underline{\mathsf{L}}^{(1)}[\bullet]$ are derived from the original operators $\mathsf{L}^{(0)}[\bullet]$, and $\mathsf{L}^{(1)}[\bullet]$ by transforming the set of independent variables $(x,t)\to(x,u_o(t))$. *If* equation (30) admits a stationary time-periodic wave as a solution, its corresponding amplitude modulation is computed by solving (37) and (38). *Therefore, the problem of existence of stationary time-periodic waves in Eq.(30), is reduced to the equivalent problem of determining the solutions of Eq. (37) and (38).* The solution of this system of equations is complicated by the fact that, since $f^2(U(x,u_o^*);\varepsilon)=0$, equation (37) becomes singular when $u_o(t)=u_o^*$, i.e., when system (30) reaches its maximum potential energy value. Therefore, the solutions for $U(x,u_o(t))$ must first be asymptotically approximated in open intervals $|u_o(t)|<|u_o^*|$, and then analytically continued up to the maximum potential energy value

$u_o(t)=u_o^*$. This can be accomplished by requiring that the following additional condition is satisfied when the system reaches its maximum potential energy:

$$\left[\left\{\underline{L}^{(0)}[U(x,u_o)] + \varepsilon\underline{L}^{(1)}[U(x,u_o)]\right\}_{x=xo}\frac{\partial U}{\partial u_o} - \underline{L}^{(0)}[U(x,u_o)] - \varepsilon\underline{L}^{(1)}[U(x,u_o)]\right]_{uo=uo^*} = 0 \tag{39}$$

The asymptotic methodology for solving the set of equations (37-39) was described in the work by King and Vakakis[34], and it will be demonstrated by computing the localized waves of a specific nonlinear partial differential equation. Once the amplitude modulation function, $U(x,u_o(t))$, of the stationary wave is asymptotically approximated, the reference displacement, $u_o(t)$, is computed by substituting the relation $u(x,t)=U(x,u_o(t))$ into the equation of motion, and evaluating the resulting expression at $x=x_o$. Determination of $u_o(t)$ leads to the complete description of the stationary time-periodic wave. *Nonstationary (travelling) waves* can then be computed from the derived expressions of stationary solitons if appropriate Lorentz coordinate transformations can be applied to the system.

5.3. Application of the Methodology

The general analytical methodology will be demonstrated by studying the following equation:

$$u_{tt} = - u - \varepsilon\lambda u_{xx} - \varepsilon\alpha u^3 = 0 \;, \quad -\infty < x < +\infty \tag{40}$$

This equation arises in the "continuum-limit approximation" of weakly-modulated anti-phase motions of an infinite chain of coupled nonlinear oscillators with cubic stiffness nonlinearities[9,37]. In Eq. (40), λ and α are assumed to be positive scalars and $|\varepsilon|<<1$. A first integral of the motion can be easily computed as:

$$E = (1/2)\int_{-\infty}^{+\infty} (\, u_t^2 + u^2 - \varepsilon\lambda u_x^2 + (\varepsilon\alpha/2)u^4 \,)\, dx \tag{41}$$

Stationary, time-periodic solutions of Eq. (40) are sought in the form of the NNM (32), subject to the limiting and periodicity conditions (31). Following the previously described methodology, one obtains the following equation satisfied by the amplitude modulation function:

$$\frac{\int_{-\infty}^{+\infty} (\, [U^2(x,u_o^*)-U^2] - \varepsilon\lambda[U_x^2(x,u_o^*)-U_x^2] + (\varepsilon\alpha/2)[U^4(x,u_o^*)-U^4] \,)\, dx}{\int_{-\infty}^{+\infty} (\partial U/\partial u_o)^2 dx}\frac{\partial^2 U}{\partial u_o^2} +$$

$$+ [\, - U - \varepsilon\lambda U_{xx} - \varepsilon\alpha U^3 \,]_{x=xo}\frac{\partial U}{\partial u_o} = - U - \varepsilon\lambda U_{xx} - \varepsilon\alpha U^3 \tag{42}$$

which must be solved with the following two additional conditions:

$$\lim_{x\to\pm\infty} U(x,u_o(t)) = u^{\pm} \tag{43}$$

$$\left[\, [\, - U - \varepsilon\lambda U_{xx} - \varepsilon\alpha U^3 \,]_{x=xo}\frac{\partial U}{\partial u_o} + U + \varepsilon\lambda U_{xx} + \varepsilon\alpha U^3 \right]_{uo=uo^*} = 0 \tag{44}$$

An asymptotic solution of Eq. (42-44) is now derived in the form of power series. Away from the maximum potential energy state, the solution is expressed in the following series form:

$$U(x,u_o(t)) = \sum_{k=0}^{\infty} \varepsilon^k U^{(k)}(x,u_o) \tag{45}$$

where,

$$U^{(0)}(x,u_o) = a_1^{(0)}(x)u_o(t) \tag{46}$$

$$U^{(k)}(x,u_o) = \sum_{m=0}^{\infty} a_m^{(k)}(x)\, u_o^m(t)\,, \quad k \geq 1 \tag{47}$$

The form of $U^{(0)}(x,u_o)$ arises from the fact that, for $\varepsilon=0$ equation (40) is separable in space and time. Moreover, due to the compatibility relation $U(x,u_o(t)) \equiv u_o(t)$, the following conditions must be satisfied:

$$a_1^{(0)}(x_o) = 1, \text{ and } a_m^{(k)}(x_o) = 0,\ m=0,1,2,\ldots,\ k \geq 1 \tag{48}$$

The various orders of approximation are computed by substituting Eq. (45) into Eq. (42-44) and matching terms of various orders in ε.

<u>O(1) Approximation</u>

The O(1) term of the amplitude modulation $U^{(0)}(x,u_o)=a_1^{(0)}(x)u_o(t)$ satisfies the relation:

$$-U^{(0)}(x_o,u_o)\frac{\partial U^{(0)}}{\partial u_o}(x,u_o) = -U^{(0)}(x,u_o) \Rightarrow [\,1-a_1^{(0)}(x_o)\,]\, a_1^{(0)}(x)\, u_o(t) = 0 \tag{49}$$

Assuming that the product $(a_1^{(0)}(x)u_o(t))$ is not identically zero, relation (49) leads to the requirement that, $a_1^{(0)}(x_o)=1$. This result is identical to the first of the compatibility relations (48), and the O(1) analysis does not provide any further information on the form of $a_1^{(0)}(x)$. This quantity is computed by considering $O(\varepsilon)$ terms in (42-44).

<u>O(ε) Approximation</u>

Setting $U(x,u_o(t))=a_1^{(0)}(x)u_o(t)+\varepsilon U^{(1)}(x,u_o(t))$ and retaining only terms of $O(\varepsilon)$, one obtains the following equation for $U^{(1)}(x,u_o(t))$:

$$\begin{aligned}&(u_o^{*2}-u_o^2(t))\frac{\partial^2 U^{(1)}}{\partial u_o^2} - a_1^{(0)}(x_o)u_o(t)\frac{\partial U^{(1)}}{\partial u_o} + \\ &+ [\,-U^{(1)}(x_o,u_o(t)) - \lambda a_1^{(0)''}(x_o)u_o(t) - \alpha a_1^{(0)3}(x_o)u_o^3(t)\,]\frac{\partial U^{(0)}}{\partial u_o} = \\ &= -U^{(1)}(x,u_o(t)) - \lambda a_1^{(0)''}(x)u_o(t) - \alpha a_1^{(0)3}(x)u_o^3(t)\end{aligned} \tag{50}$$

and the following condition when the system reaches its maximum potential energy:

$$\Big\{ - a_1^{(0)}(x_o)\, u_o(t) \frac{\partial U^{(1)}}{\partial u_o} + [\, - U^{(1)}(x_o,u_o(t)) - \lambda a_1^{(0)''}(x_o)u_o(t) - \alpha a_1^{(0)3}(x_o)u_o^3(t)\,] \frac{\partial U^{(0)}}{\partial u_o} + $$
$$+ U^{(1)}(x,u_o(t)) + \lambda a_1^{(0)''}(x)u_o(t) + \alpha a_1^{(0)3}(x)u_o^3(t) \Big\}_{uo=uo^*} = 0 \qquad (51)$$

A solution for $U^{(1)}(x,u_o(t))$ is sought in the form (47) with k=1, which upon substitution into Eq. (50-51) and matching of respective powers of $u_o(t)$, leads to the following equations for the spatial distributions $a_m^{(1)}(x)$:

$$a_1^{(0)''}(x) - [\, a_1^{(0)''}(x_o) + (3\alpha/4\lambda)\, u_o^{*2}\,]\, a_1^{(0)}(x) + (3\alpha/4\lambda)\, u_o^{*2}\, a_1^{(0)3}(x) = 0$$

$$a_2^{(1)}(x) = 0$$

$$a_3^{(1)}(x) = (\lambda/6u_o^{*2})\, [\, a_1^{(0)''}(x_o)a_1^{(0)}(x) - a_1^{(0)''}(x)\,]$$

$$a_4^{(1)}(x) = 0$$

$$\bullet \quad \bullet \quad \bullet \qquad (52)$$

In Eq. (52) only terms up to $O(u_o^3(t))$ were included. Complementing the above set of equations is the following set of compatibility and limiting conditions:

$$a_1^{(0)}(x_o) = 1, \text{ and } a_m^{(1)}(x_o) = 0,\ m=0,1,2,... \qquad (53)$$

$$\lim_{x\to\pm\infty} a_1^{(0)}(x) = c^{(0)\pm},\ \lim_{x\to\pm\infty} a_m^{(1)}(x) = c^{(1m)\pm},\ m=0,1,2,.... \qquad (54)$$

Conditions (53) guarantee that, to this order of approximation, $U(x_o,u_o(t)) \equiv u_o(t)$, whereas relations (54) guarantee the attainment by $U(x,u_o(t))$ of constant limits as $x\to\pm\infty$. Note that the linear $O(\varepsilon)$ approximation $a_1^{(1)}(x)$ is not computed at this order of approximation, since it cancels from the $O(\varepsilon)$ equations (50-51). To compute $a_1^{(1)}(x)$ one must consider $O(\varepsilon^2)$ terms in the functional equations[34].

Solving Eq. (52-54), one obtains the following expressions for the spatial distributions $a_1^{(0)}(x)$ and $a_3^{(1)}(x)$:

$$a_1^{(0)}(x) = \text{sech}\left((3\alpha/8\lambda)^{1/2}\, u_o^*\, (x-x_o)\right)$$

$$a_3^{(1)}(x) = -\,(\alpha/8)\, \text{sech}\left((3\alpha/8\lambda)^{1/2}\, u_o^*\, (x-x_o)\right) \left[1 - \text{sech}^2\left((3\alpha/8\lambda)^{1/2}\, u_o^*\, (x-x_o)\right)\right] \qquad (55)$$

which satisfy the limiting conditions $\lim_{x\to\pm\infty} a_1^{(0)}(x) = \lim_{x\to\pm\infty} a_3^{(1)}(x) = 0$. Similar expressions can be derived for higher approximations $a_m^{(1)}(x)$, m=5,... if terms of $O(u_o^5(t))$ or higher, are included in the $O(\varepsilon)$ analysis.

<u>$O(\varepsilon^2)$ Approximation</u>

The quantity $a_1^{(1)}(x)$ is computed by expressing the amplitude modulation in the form, $U(x,u_o(t)) = a_1^{(0)}(x)u_o(t)+\varepsilon(a_1^{(1)}(x)u_o(t)+a_3^{(1)}(x)u_o^3(t)+O(u_o^5(t)))+\varepsilon^2 U^{(2)}(x,u_o(t))$, and retaining only terms of $O(\varepsilon^2)$ in (42-44). Expressing the $O(\varepsilon^2)$ correction to the amplitude modulation according to (47) with k=2, one obtains the following differential equation for $a_1^{(1)}(x)$:

$$\begin{aligned}(4\lambda/3)\, a_1^{(1)''}(x) + [\, 3\alpha u_o^{*2} a_1^{(0)2}(x) - (4\lambda/3)a_1^{(0)''}(x) - \alpha u_o^{*2}\,]\, a_1^{(1)}(x) = \\ = -\,[2(K_1 u_o^{*2}+K_2 u_o^{*4}) + 3u_o^{*2}(\lambda a_1^{(0)''}(x_o) + \alpha u_o^{*2}) + 3\alpha u_o^{*4} a_1^{(0)2}(x)]\, a_3^{(1)}(x) - \\ -\,[\,-(4\lambda/3)\, a_1^{(1)''}(x_o) - \lambda u_o^{*2}\, a_3^{(1)''}(x_o)\,]\, a_1^{(0)}(x)\end{aligned} \tag{56}$$

where the scalars K_1 and K_2 are independent of x, and are defined in terms of the previously computed approximations $a_1^{(0)}(x)$ and $a_3^{(1)}(x)$ as follows:

$$K_1 = \frac{\int_{-\infty}^{+\infty} -\lambda a_1^{(0)'2}(x)\,dx}{\int_{-\infty}^{+\infty} a_1^{(0)'2}(x)\,dx}\,, \quad K_2 = \frac{\int_{-\infty}^{+\infty} (2a_1^{(0)}(x)a_3^{(1)}(x)+(\alpha/2)a_1^{(0)4}(x))\,dx}{\int_{-\infty}^{+\infty} a_1^{(0)'2}(x)\,dx} \tag{57}$$

Equation (56) is solved subject to the additional conditions:

$$a_1^{(1)}(x_o) = 0, \text{ and } \lim_{x\to\pm\infty} a_1^{(1)}(x) = c^{(11)\pm} \tag{58}$$

Introducing the following coordinate transformations,

$$\begin{aligned} z = (3\alpha/8\lambda)^{1/2}\, u_o^*\, (x-x_o),\ \phi(z) = a_1^{(1)}(z\, (3\alpha/8\lambda)^{-1/2}\, u_o^{*-1}+x_o) \\ \theta(z) = a_1^{(0)}(z\, (3\alpha/8\lambda)^{-1/2}\, u_o^{*-1}+x_o) \equiv \operatorname{sech} z, \\ v(z) = a_3^{(1)}(z\, (3\alpha/8\lambda)^{-1/2}\, u_o^{*-1}+x_o) \equiv -\,(\alpha/8)\operatorname{sech} z\, (\, 1-\operatorname{sech}^2 z\,) \end{aligned} \tag{59}$$

equations (56) and (58) are transformed as follows:

$$\phi''(z) - (\, 1 - 6\operatorname{sech}^2 z\,)\, \phi(z) = F_1\, v''(z) + F_2\, v(z) + F_3\, v(z)\theta^2(z) + F_4\, \theta(z) \equiv G(z)$$

$$\phi(0) = 0, \text{ and } \lim_{z\to\pm\infty} \phi(z) = c^{(11)\pm} \tag{60}$$

where the quantities F_i are defined as, $F_1 = -3u_o^{*2}/4$, $F_2 = -(1/\alpha u_o^{*2})\,[\,4\,(K_1 u_o^{*2} + K_2 u_o^{*4}) + (9u_o^{*2}/4)\,\theta''(0) + 6u_o^{*2}\,]$, $F_3 = -6u_o^{*2}$, $F_4 = \phi''(0)+ (3u_o^{*2}/4)v''(0)$. Note that coefficients F_1, F_2 and F_3 depend on the parameters u_o^{*2}, α, and on the previously computed approximations v(z) and $\theta(z)$, whereas, *coefficient F_4 depends on the solution itself, and thus is yet undetermined.* The term $\phi''(0)$ in F_4 is chosen so that the limiting requirements (60) for $\phi(z)$ are satisfied. The solution of the linear, parameter-dependent, nonhomogeneous ordinary differential equation (60) which decays at infinity was computed in explicit form by King and Vakakis[34] in the following form:

$$\phi(z)=\gamma\sinh z/\cosh^2 z + \Big[[(\alpha u_o^{*2}+K_1+u_o^{*2}K_2)/24]\, z\cosh z - (\alpha u_o^{*2}/48)\sinh z\Big] \tanh z\, \mathrm{sech}^2 z \tag{61}$$

from which, the coefficient $a_1^{(1)}(x)$ was computed as, $a_1^{(1)}(x) = \phi((3\alpha/8\lambda)^{1/2}\, u_o^*\, (x-x_o))$.

Employing all previous findings, one derives the following asymptotic approximation for the amplitude modulation $U(x,u_o(t))$ of the stationary spatially localized wave of Eq.(40):

$$U(x,u_o(t)) = \Big\{ \mathrm{sech} z + \\ + \varepsilon\Big[[(\alpha u_o^{*2}+K_1+u_o^{*2}K_2)/24]\, z\cosh z - (\alpha u_o^{*2}/48)\sinh z\Big] \tanh z\, \mathrm{sech}^2 z \Big\}\, u_o(t) + \\ - \varepsilon\,(\alpha/8)\,\mathrm{sech} z\,(1-\mathrm{sech}^2 z)\, u_o^3(t) + O(\varepsilon u_o^5(t),\varepsilon^2),\quad z = (3\alpha/8\lambda)^{1/2}\, u_o^*\,(x-x_o) \tag{62}$$

To compute the reference displacement $u_o(t)$, relation (62) is substituted into the equation of motion (40), and the resulting expression is evaluated at the reference point $x=x_o$. The resulting equation possesses the solution[34]:

$$u_o(t) = u_o^*\, \mathrm{cn}(pt,k) \tag{63}$$

where $\mathrm{cn}(\bullet,\bullet)$ is the elliptic cosine, its argument p is computed as $p = [1+\varepsilon\lambda a_1^{(0)\prime\prime}(x_o)+\varepsilon^2\lambda a_1^{(1)\prime\prime}(x_o) + [\varepsilon\alpha+\varepsilon^2\lambda a_3^{(1)\prime\prime}(x_o)]\, u_o^{*2}\,]^{1/2}$, and the elliptic modulus k is given by, $k^2 = [\varepsilon\alpha+\varepsilon^2\lambda a_3^{(1)\prime\prime}(x_o)]\, u_o^{*2} / 2p^2$. This result completes the solution for the stationary, spatially localized wave (stationary breather), with frequency of oscillation $\omega = \omega(u_o^*) = \pi p / 2K(k)$, where $K(\bullet)$ is the complete elliptic integral of the first kind. *Travelling breathers* of equation (40) are computed from the previous solution, by imposing the Lorentz coordinate transformation:

$$\hat{u}(x,t) = u\left(\frac{x+(\varepsilon\lambda)^{1/2}vt}{(1+v^2)^{1/2}}, \frac{t-v(\varepsilon\lambda)^{-1/2}x}{(1+v^2)^{1/2}}\right) \Rightarrow \\ \Rightarrow \hat{u}(x,t) = U\left(\frac{x+(\varepsilon\lambda)^{1/2}vt}{(1+v^2)^{1/2}}, u_o\Big(\frac{t-v(\varepsilon\lambda)^{-1/2}x}{(1+v^2)^{1/2}}\Big)\right) \tag{64}$$

where v is the travelling wave velocity (group velocity), and is related to the frequency, ω, and wavelength, k, of the propagating wavepacket by the relation:

$$\omega = \omega(v,u_o^*) = \pi p(1+v^2)^{1/2} / 2K(k) \tag{65}$$

where p and k were defined earlier. Note, that the stationary solution (62-63) can be regarded as a special case of the travelling solution (65) with v=0.

5.4. Conclusions

Summarizing, a methodology for analyzing waves with spatially localized envelopes in a class on nonlinear partial differential equations was presented. For stationary periodic motions, the response of the system at an arbitrary point was expressed as function of a reference response $u_0(t)$. This enabled the elimination of the time dependence from the problem. The singular equation governing the amplitude modulation function was solved asymptotically, by approximating the function in power series and by analytically continuing the obtained results up to the regular singular point of the problem. The resulting stationary breather corresponded to a periodic motion of the system, with all points vibrating in-unison, i.e., performing synchronous motions. In the terminology of the previous sections, this stationary breather can be regarded as a nonlinear normal mode (NNM). Using the analytic approximation of the stationary breather, travelling breathers were computed by employing a Lorentz coordinate transformation.

An interesting extension of the presented methodology would be the modification of the definition of the amplitude modulation function to account for possible *phase shifts* between responses of various points of the system. In the present formulation, $U(x,u_0(t))$ models only in-unison (synchronous) motions. Moreover, current work by the author and his co-workers, aims to extend the presented methodology to partial differential equations which possess *dissipation*, and are *non-separable* when the small parameter of the problem is set to zero.

6. Acknowledgements

This work was funded by NSF Grant No. MSS 92-07318, and by an NSF Young Investigator Award. These Grants are monitored by Dr. Devendra Garg. The author is grateful to Moog, Inc., for kindly providing on loan the control actuators used in the experimental fixture. Additionally, the author would like to thank his graduate research assistants Mr. Melvin King, Mr. Johannes Aubrecht, Mr. Tariq Nayfeh, and Mr. Cetin Cetinkaya, whose dedicated work was instrumental for completing the various sections of this work.

7. References

1. C.H. Hodges, *J. Sound Vib.* **82** (1982), 411-424.
2. C. Pierre and E.H. Dowell, *J. Sound Vib.* **114** (1987), 549-564.
3. P.J. Cornwell and O.O. Bendiksen, *AIAA J.* **27** (1989), 219-226.
4. C. Pierre and P. Cha, *AIAA J.*, **27** (1989), 227-241.
5. C. Pierre, *J. Sound Vib.* (1994) (in press).
6. A.F. Vakakis, *J. Sound Vib.* **158** (1992), 341-361.
7. A.F. Vakakis and C. Cetinkaya, *SIAM J. Appl. Math.* **53** (1993), 265-282.
8. A.F. Vakakis, T.A. Nayfeh and M.E. King, *J. Appl. Mech.* **60** (1993), 388-397.
9. A.F. Vakakis, M.E. King, and A. Pearlstein, *Int. J. Non-linear Mech.* **29** (1994), 429-447.
10. R.M. Rosenberg, *Adv. Appl. Mech.* **9** (1966), 155-242.
11. R.H. Rand, *J. Appl. Mech.* **38** (1971), 561.
12. R.H. Rand, *Int. J. Non-Linear Mech.* **6** (1971), 545-547.

13. A.F. Vakakis, *Analysis and Identification of Linear and Nonlinear Normal Modes in Vibrating Systems*, Ph.D. Thesis, California Institute of Technology, Pasadena, California (1990).
14. N. Boivin, C. Pierre and S.W. Shaw, in *Nonlinear Vibrations* (ASME DE-Vol. 54, 1993).
15. S.W. Shaw and C. Pierre, *J. Sound Vib.* (1993), (in press).
16. S.W. Shaw and C. Pierre, in *Nonlinear Vibrations* (ASME DE-Vol. 50 and AMD-Vol. 144, 1992).
17. M.E. King and A.F. Vakakis, *J. Vibr. Acoust.* **116** (1994), 332-340.
18. M.E. King and A.F. Vakakis, *Int. J. Solids Str.* (1994), (in press).
19. C. Pierre, D.M. Tang and E.H. Dowell, *AIAA J.* **25** (1987), 1249-1257.
20. O.O. Bendiksen, *AIAA J.* **25** (1987), 1241-1248.
21. D. Bouzit and C. Pierre, in *Structural Dynamics of Large Scale and Complex Systems* (ASME DE-Vol. 59, 1993).
22. M.E. King and A.F. Vakakis, *ZAMM* (1994) (in press).
23. A.F. Vakakis, *AIAA J.* (1994) (in press).
24. L. Manevitch and I. Mikhlin, *PMM* **36** (1972), 1051-1058.
25. A.H. Nayfeh and D. Mook, *Nonlinear Oscillations* (Wiley Interscience, New York, 1985).
26. M.R.M. Crespo Da Silva and C.C. Glynn, *J. Str. Mech.* **6** (1978), 437-461.
27. J. Aubrecht, *Nonlinear Mode Localization and Motion Confinement in Continuous Oscillators*, M.Sc. Thesis, University of Illinois at Urbana-Champaign, Urbana, Illinois (1994).
28. N.J. Zabusky and M.D. Kruskal, *Phys. Rev. Lett.* **15** (1965), 240-243.
29. A. Hasegava, *Plasma Instabilities and Nonlinear Effects* (Springer Verlag, New York, 1975).
30. P.S. Lomdahl, *J. Stat. Phys.* **39** (1985), 551-561.
31. H. Thomas (ed.), *Nonlinear Dynamics in Solids* (Spinger Verlag, New York, 1992).
32. G.L. Lamb, *Elements of Soliton Theory* (John Wiley & Sons, New York, 1980).
33. S. Pnevmatikos (ed.), *Singularities and Dynamical Systems* (Mathematical Studies **103**, North Holland Publishers, Amsterdam, 1985).
34. M.E. King and A.F. Vakakis, *Wave Motion* **19** (1994), 391-405.
35. A. Weinstein, *Commun. Math. Phys.* **99** (1985), 385-388.
36. M. Kruskal and H. Segur, *Phys. Rev. Lett.* **58** (1987), 747.
37. A.M. Kosevich and A.S. Kovalev, *Sov. Phys. - JETP* **40** (1975), 891-896.

Wave Motion, Intelligent Structures and Nonlinear Mechanics, pp. 35–91
edited by A. Guran and D. J. Inman
Series on Stability, Vibration and Control of Structures: Vol. 1

EXPERIMENTS AND ANALYSIS OF SOUND-INDUCED STRUCTURAL VIBRATIONS

HERBERT ÜBERALL
Department of Physics, Catholic University of America
Washington, DC 20064, USA and
LAUE, URA CNRS 1373, University of Le Havre
76610 Le Havre, France

and

GERARD MAZE and JEAN RIPOCHE
LAUE, URA CNRS 1373, University of Le Havre
76610 Le Havre, France

ABSTRACT

Vibrations of elastic structures take place in the form of normal modes, which appear at certain well-defined (complex) eigenfrequencies, or "natural frequencies." The excitation of such mechanical vibrations by external mechanical forces, or else by incident acoustic waves (the subject of the present article) will exhibit resonances at these eigenfrequencies which can be exceedingly prominent for highly elastic objects. Resonant acoustic excitation, as described by the "Resonance Scattering Theory" (RST) [L. Flax, L R. Dragonette and H. Überall, *J. Acoust. Soc. Am.* **63** (1978) 723-731] has been the subject of numerous experimental investigations since about 1980, often employing an experimental method termed MIIR [G. Maze and J. Ripoche, *Rev. Phys. Appl.* **18** (1983) 319-326] . These theoretical and experimental approaches, and their results, will be outlined in the present expository article.

1. Introduction

The earliest experimental work on the scattering of acoustic waves from submerged elastic objects seems to be that of Faran[1], carried out in the years after World War II and observing the acoustic echoes from solid elastic (metallic) cylinders and spheres

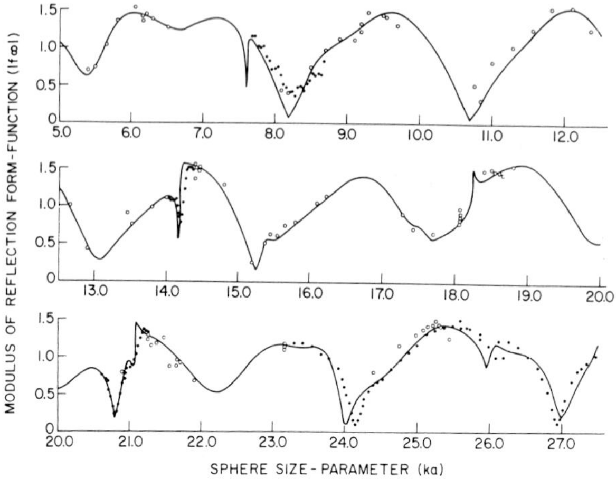

Fig. 1. Theoretical (solid curve) and measured (points) form function vs. frequency parameter ka, for a solid aluminum sphere in water. From Ref. 2.

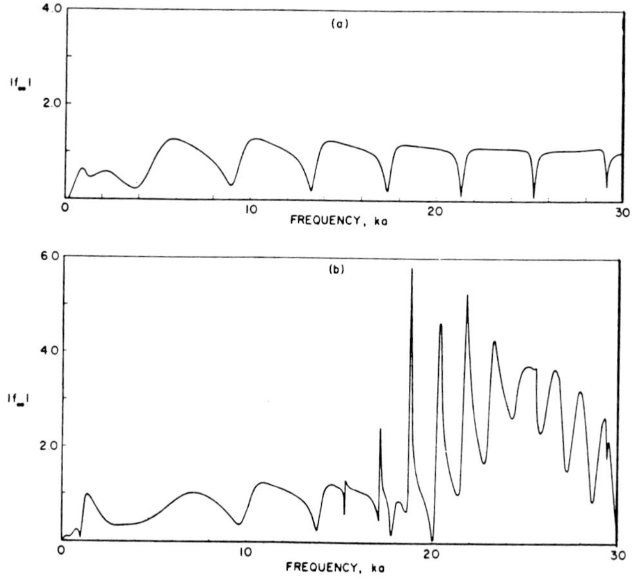

Fig. 2. Plot of the calculated form function vs. ka for a spherical aluminum shell in water: (a) very thin shell; (b) not so thin shell. From Ref. 3.

suspended in a water tank. In these studies, Faran already noted the excitation of resonances corresponding to structural vibrations of the elastic target objects, and their occurrence at well-defined eigenfrequencies of the normal modes of vibration. In later experimental[2] and theoretical work[3] that exhibited the normalized acoustic 180°-backscattering amplitude (termed the "form function") as a function of frequency, the resonant excitation of the eigenvibrations of spherical targets was seen to manifest itself through the appearance of regularly-spaced interference minima superimposed on a non-resonant background, together with other resonance features. This is shown in Fig. 1 for the form function of a solid aluminum sphere in water[2], both calculated (solid curve) and measured (points), or of a spherical air-filled aluminum shell in water[3], Fig. 2, calculated for (a) a very thin shell, and (b) a not so thin shell. These results are plotted versus the non-dimensional frequency parameter ka, where k is the wave number of the sound incident through the ambient fluid and a the radius of the sphere (or, outer radius of the shell).

However, the excitations of the target resonances could also be seen from the effects they had on the shape of incident harmonic wave trains. Figure 3(a) shows the incident wave train in Neubauer's experiments[2], and Figs. 3 (b) and (c) the backscattered wave trains for the aluminum sphere of Fig. 1 at (b) ka = 20.78, and (c) ka = 21.21, both being resonance frequencies as seen in Fig. 1. The change of shape of the reflected compared to the incident pulse shape is striking. The central constriction of the pulse in Fig. 3(b) is connected to the fact that the form function at ka = 20.78 has a deep interference minimum; but the principal feature of both return pulses is the tails they acquired upon reflection, representing the ringing of the corresponding resonance.

A calculation of the shape of wave trains after backscattering was carried out for metal spheres in water by Hickling[4]. Figure 4 shows a short (5-cycle) incident pulse (top), with its spectrum overlaid over the calculated form function of a target iron sphere at ka = 24.5 (top, left) and ka = 25.5 (top, right). The situation here is different from that of Fig. 3: in that case, the pulse was long and its narrow spectrum (not shown) covered either the ka = 20.78 or the ka = 21.21 resonance exclusively. Here (Fig. 4, top) the spectrum of the pulse is wide and overlaps several of the resonances visible in the form function. Their concerted action causes an appearance of the echo pulse that consists of a time sequence of individual pulses, each of them shaped similarly to the incident pulse (Fig 4, center and bottom).

This is where the situation stood regarding the acoustic excitation of the resonances of submerged elastic objects until the mid-1970s. Detailed explanations of the observed phenomena, both steady-state and transient, followed with the advent, in 1978, of the Resonance Scattering Theory (RST) of Flax, Dragonette and Überall[5], and of the experimental methods for observing the resonances: in 1983, the long pulse MIIR (Method of Isolation and Identification of Resonances) of Maze and Ripoche[6], and in 1984-86, the short-pulse method of Numrich et al.[7] and de Billy[8]. These developments are described most comprehensively in books[9,10] or review articles[11] up to end of the 1980s.

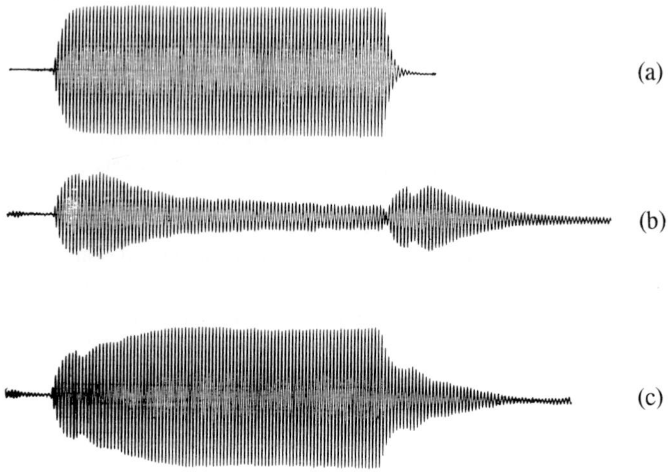

Fig. 3. (a) A 500-μm long incident pulse, and its reflection from a solid aluminum sphere in water at carrier frequency (b) ka = 20.78, and (b) ka = 21.21. From Ref. 2.

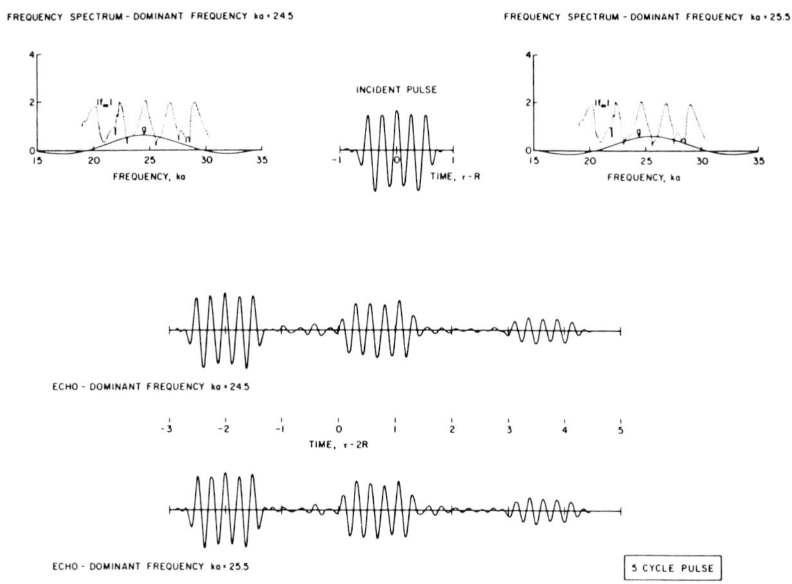

Fig. 4. Calculated echoes returned by an iron sphere in water for a short (5-cycle) incident pulse with center frequencies at ka = 24.5 and 25.5. Top also shows pulse spectrum overlaid on the form function (dotted curve). From Ref. 4.

2. Documented History of the Resonance Scattering Theory (RST)

In view of some confusing statements that have recently appeared in the literature regarding the exact history of the RST, it behooves us to present here a brief documentation of the historical development of this subject, and of the personalities involved in it, in order to clarify the roles of the actual originators of this theory, and of those having played only an auxiliary role (i.e., having participated strictly in the subsequent applications of the RST after it had already been established). The listing of the literature references reflecting this development of the theory, as it will be given below, should uniquely characterize the role of the participants and eliminate any possible doubts regarding their individual contributions.

The literature entries listed below fall in the three categories:
(a) Abstracts of presentations at periodic conferences, i.e. the semi-annual meetings of the Acoustical Society of America, as published in the Meeting Bulletins of the Journal of this society; these reflect the step-wise development of the theory leading up to the definitive *establishment* of the RST, and they further reflect the subsequent *applications* of the RST which will be clearly characterized as such;
(b) The reference of the one single journal publication establishing the RST itself, and of the journal papers that immediately preceded it and laid the groundwork for it; further, journal references of the first subsequent applications of the RST, again clearly characterized as being applications; and finally
(c) Presentations at a non-periodic conference (published in the Conference Proceedings) where an overview of the RST was given *after it had already been established*, as well as descriptions of the first applications of this theory that were ongoing at that time.

2.1. *Abstracts of Presentations at the Meetings of the Acoustical Society of America*

A brief statement of content follows each reference in square brackets, to be described in detail in Section 3 and thereafter.

2.1.1 Preparatory work towards the RST

(a) H. Überall, L. R. Dragonette and L. Flax, "Relation between creeping waves and normal modes of vibration of a curved body", 92nd ASA Meeting, San Diego, CA, 15-19 Nov. 1976 (abstract deadline: 17 August 1976), J. Acoust. Soc. Am. Vol. 60, Suppl. 1, Fall 1976), talk Y6 [Mode resonances at the eigenfrequencies].

(b) J. W. Dickey and H. Überall, "Amplitudes of transmitted and circumferential waves in sound scattering from an elastic cylinder," ibid. talk Y7 [Separation of "geometrical" background and "diffracted" resonating surface (creeping) waves].

2.1.2. The RST established

In this talk, the establishment of the Resonance Scattering Theory is described as it was accomplished by the three mentioned authors:

(c) L. R. Dragonette, L. Flax, and H. Überall, "Regge pole character of resonances in sound scattering from elastic bodies", 93nd ASA Meeting, Pennsylvania State University, 6-10 June 1977 (abstract deadline: March 8, 1977), J. Acoust. Soc. Am. Vol. 61, Suppl. 1, Spring 1977), talk T1 [Resonances in partial waves; relation to nuclear physics].

2.1.3. Applications of the RST

These presentations describe the first applications of the RST *as performed after it had been established*, listed in chronological order with the authors that had helped carry out each particular application.

(d) K. A. Sage, J. George and H. Überall, "Multipole resonances in sound scattering from gas bubbles in a liquid", 95th ASA Meeting, Providence, R.I., 16-19 May 1978 (abstract deadline: 3 Feb. 1978), J. Acoust. Soc. Am. Vol. 63, Suppl. 1, Spring 1978), talk D13 [Resonances of gas bubbles, described by nuclear physics techniques].

(e) A. J. Haug, S. G. Solomon, and H. Überall, "Scattering of elastic waves by a resonating fluid-filled cylindrical cavity", ibid. talk O1 [Resonance scattering from cylindrical cavities].

(f) J. D. Murphy, E. D. Breitenbach, and H. Überall, "Resonance scattering of acoustic waves from cylindrical shells", ibid, Talk O2.

(g) G. C. Gaunaurd and H. Überall, "Scattering of dilatational waves incident on a resonating fluid-filled spherical cavity inside a sound-absorbing material", ibid., talk O3 [Resonance scattering from spherical cavities].

2.2. Journal publications of the RST, and of its subsequent applications

2.2.1. Preparatory work towards the RST

(a) H. Überall, L. R. Dragonette and L. Flax, "Relation between creeping waves and normal modes of vibration of a curved body", *J. Acoust. Soc. Am.* **61**, No. 3 (March 1977) 711-715. (Paper received by the journal 22 June 1976) [Resonances in the form function are discussed and labeled].

(b) J. W. Dickey and H. Überall, "Surface wave resonances in sound scattering from elastic cylinders", *J. Acoust. Soc. Am.* **63**, No. 2 (Feb. 1978) 319-320 (Paper received by the journal 13 August 1977) [Resonances at the target eigenfrequencies are related to the phase matching (standing waves) of circumferential surface waves].

2.2.2. The RST established

(c) L. Flax, L. R. Dragonette, and H. Überall, "Theory of elastic resonance excitation by sound scattering", *J. Acoust. Soc. Am.* **63**, No. 3 (March 1978) 723-731. (Paper received by the journal 7 September 1977). [In this paper, and *only* here, the resonance algorithm of the RST has been formulated for the first time in acoustics together with the separation of the scattering amplitude into resonant parts and a nonresonant background. This was carried out exclusively by the three mentioned authors, who thereby established the RST in acoustics].

2.2.3. Applications of the RST

Only the first three journal publications on the earliest applications of the RST, as performed *after* it had been established, are listed here in chronological order with the authors that had helped carry out each particular application. For a more complete list of applications, see H. Überall (ed.), *Acoustic Resonance Scattering* (Gordon and Breach, New York, 1992), Ref. 9.

(d) A. J. Haug, S. G. Solomon, and H. Überall, "Resonance theory of elastic wave scattering from a cylindrical cavity", *J. Sound Vib.* **57** No. 1 (8 March 1978) 51-58 (Paper received by the journal 14 June 1977) [The first application of the RST concerned elastic-wave scattering from fluid-filled cylindrical cavities; note that the reception date precedes that of the RST formalism itself, ref. (c) of Sect. 2.2.2, since the RST had already been formulated by June 1977 but sent to the journal only in September 1977.] The reception date also precedes by far those of the following:

(e) G. C. Gaunaurd and H. Überall, "Theory of resonance scattering from spherical cavities in elastic and viscoelastic media", *J. Acoust. Soc. Am.* **63** No. 6 (June 1978) 1699-1712. (Paper received by the journal 1 Dec. 1977). [The second application, in chronological order of publication, of the RST concerned elastic-wave scattering from fluid-filled spherical cavities. Note that according to reception date by the journal, this is actually only the *third* application of the RST].

(f) J. D. Murphy, E. D. Breitenbach, and H. Überall, "Resonance scattering of acoustic waves from cylindrical shells", *J. Acoust. Soc. Am.* **64** No. 2 (Aug. 1978) 677-683. (Paper received by the journal 28 Nov. 1977). [The third application, in chronological order of publication, of the RST concerned acoustic-wave scattering from cylindrical elastic shells; note that the reception date precedes that of paper (e) above so that according to reception date, this is actually the *second* application of the RST].

2.3. *Special Conference Presentation of the RST, and of its Subsequent Applications*

Some ambiguities regarding the establishment of the RST have arisen in connection with two presentations given at the Symposium on Modern Problems in Acoustic Wave Propagation, held at Northwestern University, Evanston, IL, September 12-15, 1977. The Proceedings of the conference were published by Wiley Interscience in 1978 under the title *Proc. Internat. Union Theoret. Appl. Mech. (IUTAM) Symposium: Modern Problems in Elastic Wave Propagation*, ed. J. Miklowitz and J. Achenbach (John Wiley, New York, 1978). While the above-quoted abstract publications (Sec. *2.1*) and journal publications (Sec. *2.2*) are all clearly documented, including deadline and publication dates, in the easily-accessible literature, the Evanston IUTAM conference proceedings are not easily available and hence leave room for ambiguities. The two presentations in question in these Proceedings are:

(a) H. Überall, "Modal and surface wave resonances in acoustic wave scattering from elastic objects and in elastic wave scattering from cavities", Evanston IUTAM Proceedings pp. 239-263.

The contents of this 25-page invited paper are:

(i) a presentation of the RST formalism as it had been established in the original reference by Flax, Dragonette and Überall, *J. Acoust. Soc. Am.* **63** (1978) pp. 723-731, see Secs. 2.1.2 and 2.2.2 above;

(ii) a description of the subsequent applications which had been initiated by the time of the publication of the Evanston Proceedings, including those of the above Secs. 2.1.3 (e-g), and 2.2.3 (d-f).

(b) G. C. Gaunaurd, "Resonant theory of elastic and viscoelastic wave scattering and its applications to the spherical cavity in absorptive media", Evanston IUTAM Proceedings p. 550.

The contents of this 1/2 page note duplicate the specific application of the RST to elastic-wave scattering from a spherical cavity as contained in (a) above, and as also mentioned in the application sections 2.1.3 (g) and 2.2.3 (e). The note concerns one of the subsequent applications (the third) of the RST after the latter had already been established, the first application having been to elastic-wave scattering from a cylindrical cavity, cf. 2.1.3 (e) and 2.2.3 (d), the second to acoustic-wave scattering from cylindrical shells, cf. 2.1.3 (f) and 2.2.3 (f). The actual establishment of the RST, with its actual authors, is fully and unambiguously documented in the easily accessible literature listed in Sections *2.1* and *2.2* above, and the sequence of submission-deadline and receipt-by-the-journal dates bear this out.

3. The Formalism of the Resonance Scattering Theory

The echo returned in acoustic scattering from (submerged) elastic objects consists of two coherent parts: an amplitude which varies smoothly with frequency, and which would be present even if the object were impenetrable to sound (this will be referred to as the acoustic "background amplitude"); and a resonant amplitude which, plotted as a function of frequency, shows a spectrum of resonance peaks which coincide with the eigenfrequency spectrum of the vibrations of the scatterer. The Resonance Scattering Theory[5,9] (RST), correspondingly, carries out two functions:

(1) It mathematically separates in the total scattering amplitude the background amplitude from the resonant amplitude (such a first step had already been accomplished by Junger[12], but in the RST this step is carried out using the formalism of the resonance scattering theory of nuclear physics; see below; no resonance formalism was presented by Junger for the actual resonant part of the amplitude), and

(2) it mathematically expresses the resonant amplitude in a form where its resonance character is exhibited explicitly. This includes a determination of the target eigenfrequencies, as well as of the widths of the resonance peaks.

Steps (1) and especially (2) were accomplished in Ref. 5 by patterning the mathematical approach after the resonance scattering formalism previously developed in nuclear physics. Resonances in nuclear scattering were first discussed qualitatively by Niels Bohr (Nature Vol. 137, 1936, p. 344), and are generally associated with the names of Breit and Wigner[13,14]. A simplified, lucid derivation of nuclear resonance scattering theory was provided by Feshbach, Peaslee, and Weisskopf[15], which has been utilized in our corresponding derivation of acoustic resonance scattering.

The derivation of the resonance formula, by linearization of the denominator of the amplitude via a Taylor expansion; is also given in the book by Blatt and Weisskopf, *Theoretical Nuclear Physics*, John Wiley, New York; 1952, Chap. VIII, 7-9. A corresponding procedure was applied to the acoustic scattering amplitude. For normal incidence on a solid elastic cylinder, this amplitude is[5]

$$P_{sc} = \frac{1}{2} \sum (2 - \delta_{n0})\, i^n\, (S_n - 1) H_n^{(1)}(kr) \cos n\varphi, \tag{1}$$

the diagonal terms of the "S matrix" (Heisenberg), S_n, being given explicitly in quotient form N_n/D_n. Out of this quotient one may factor the rigid-cylinder scattering amplitude

$$S_n^r = -H_n^{(2)\prime}(x)/H_n^{(1)\prime}(x) \equiv e^{2i\xi_n}, \tag{2}$$

($x = ka$), which here plays the role of the acoustic background mentioned above. This leads to S_n being of the form

$$S_n = S_n^r \frac{L_n^{-1} - z_2^{-1}}{L_n^{-1} - z_1^{-1}}, \tag{3}$$

with

$$z_i^{-1} = H_n^{(i)}(x)/xH_n^{(i)\prime}(x) = \Delta_n \pm i s_n. \tag{4}$$

Expanding the known quantity

$$L_n^{-1}(x) \cong \Delta_n + \beta_n (x - x_n) + \ldots \tag{5}$$

about the resonance frequencies x_n, which are determined from the zeros of the denominator in Eq. (3), we get

$$S_n \equiv e^{2i\delta_n} = S_n^r \frac{x - x_n - (i/2)\Gamma_n}{x - x_n + (i/2)\Gamma_n}, \tag{6a}$$

This establishes the resonance form of S_n, and hence establishes the RST in acoustics, via its crucial step (2), see above.

The resonance width Γ_n is determined by the expression

$$\Gamma_n = -2s_n/\beta_n \tag{6b}$$

so that all expressions in Eq. (6a) are known.

In the scattering amplitude of Eq. (1), there enters the quantity $S_n - 1 = 2i \exp(i\delta_n) \sin \delta_n$, which we may thus write

$$S_n - 1 = 2ie^{2i\xi_n} \left\{ \sum_l \frac{\frac{1}{2}\Gamma_n}{x_{nl} - x - (i/2)\Gamma_{nl}} + e^{-i\xi_n} \sin \xi_n \right\}. \tag{7}$$

This expression, first used in acoustics by Flax, Dragonette and Überall[5], shows that each modal amplitude has resonant terms; but also a nonresonant "background" term, i.e., step (1), see above [second term in Eq. (7)], corresponding here to rigid-body scattering (called "potential scattering" in nuclear physics). From Eq. (7), it is evident that if we subtract the rigid-body scattering amplitude $S_n^r - 1 = 2i \exp(i\xi_n) \sin \xi_n$ from $S_n - 1$, then we are left with the purely resonant amplitude

$$S_n^{res} = 2ie^{2i\xi_n} \sum_l \frac{\frac{1}{2}\Gamma_{nl}}{x_{nl} - x - (i/2)\Gamma_{nl}}. \tag{8}$$

This completes both step (1) and (2) of the RST, as mentioned above.

In addition to this purely mathematical algorithm, the RST can also help us understand the physics behind the resonances. Franz[16] has shown, using the Watson transformation, that the diffracted field about an obstacle can be mathematically resolved into a coherent sum of what he called "creeping waves" (also later known as surface waves or circumferential waves) which propagate along the surface of the target. He did this for impenetrable objects where the surface waves are excited at tangential incidence and propagate solely in the ambient medium. For elastic targets of acoustic waves, elastic body-type surface waves may be excited also at the angle of incidence $\alpha = \sin^{-1}(c/c_\ell)$, where c = sound velocity in the ambient medium, c_ℓ = phase speed of the ℓ th elastic surface wave. These propagate below the surface of the target, and may be classified as a (leaky) Rayleigh-type wave, and a series of Whispering Gallery waves[17-19]; an additional Scholte-Stoneley wave also appears in the ambient medium.

The resonance representation of the scattered field can now be converted to a description in terms of surface waves[20,21], from which the origin of the resonances can be recognized as the condition that after each circumnavigation, the surface wave is in phase with itself. This "principle of phase matching" entails that a resonant reinforcement of surface waves will take place, generating the resonance at a specific resonance frequency (since the surface waves are dispersive). In this way, the physical origin of the resonances is understood. To sketch this derivation briefly, we represent the scattering amplitude of a cylinder in the form

$$f(\varphi) = \sum_{n=0}^{\infty} f_n(\varphi) \;, \tag{9}$$

f_n having the resonance form (see above)

$$f_n(\varphi, \omega) \cong A_n(\varphi) \frac{(\Gamma/2\pi)^{1/2}}{\omega - \omega(n) + \frac{1}{2} i\Gamma} , \tag{10}$$

now employing the frequency variable ω and the resonance frequencies $\omega(n)$. This amplitude has a pole at the complex frequency $\omega = \omega(n) - (1/2)\, i\Gamma$, located below the real axis of the ω plane.

For a fixed frequency ω, $f_n(\varphi,\omega)$ may be considered a function of n; thus, one may expand

$$\omega(n) \cong \omega(n_\omega) + (n - n_\omega)\, \omega'(n_\omega), \tag{11}$$

where[5] $\omega'(n_\omega) > 0$, and one may choose n_ω so that $\omega(n_\omega) = \omega$, i.e., the fixed incident frequency lies at the resonance peak when $n = n_\omega$. Then, the amplitude

$$f_n(\varphi, \omega) \cong \frac{-A_n(\varphi)}{[\omega'(n_\omega)]^{1/2}} \frac{(\hat{\Gamma}/2\pi)^{1/2}}{n - n_\omega - \frac{1}{2} i\hat{\Gamma}} \tag{12}$$

appears with a pole in the complex n plane, located at n = n + (1/2) i above the real n axis since

$$\hat{\Gamma} \equiv \Gamma/\omega'(n_\omega) > 0 . \tag{13}$$

In terms of the n variable, this is known as a Regge pole[22], and we see that one and the same pole of the scattering amplitude may appear as a resonance in the ω as well as in the n variable (although integer values of n correspond to physical values of the amplitude). If n is considered a parameter and ω the variable, the n dependence of the resonance frequency $\omega(n)$ in Eq. (10) causes a "Regge recurrence" of the one Regge pole under consideration in all successive partial waves at successively higher resonance frequencies.[5,20] In the following, however, we shall discuss the aspects of the scattering amplitude given by Eq. (12), as it provides a physical picture of the scattering process in terms of surface waves, as well as a physical explanation of the mechanism of resonance excitation.

Applying the Watson transformation to the normal-mode series for sound scattering from an elastic cylinder, the surface waves generated in the scattering process were shown to be given by[23]

$$p_{sc} = -2\pi \sum_l \frac{\cos\nu_l\varphi}{\sin\pi\nu_l} e^{-i\nu_l\pi/2} \frac{b_{\nu_l}}{\dot{D}_{\nu_l}} H^{(1)}_{\nu_l}(kr) , \tag{14}$$

where $\dot{D}_\nu = \partial D_\nu/\partial\nu$, ν is the modal number of the Rayleigh series turned into a complex variable, b_ν is a 3x3 determinant, D_ν is the secular determinant of the problem[23] and $\nu = \nu_\ell(\omega)$ are the complex roots of the latter,

$$\nu_l(\omega) = \mathrm{Re}\,\nu_l(\omega) + i\,\mathrm{Im}\,\nu_l(\omega) \tag{15}$$

($\ell = 1,2,3,...$). As a function of frequency, the ℓth root $\nu_\ell(\omega)$ will, at some resonance frequency $\omega = \omega_{n\ell}$, move past the integer n, so that $\mathrm{Re}\nu(\omega_{n\ell}) = n$. In Ref. 20, the resonance frequencies $\omega_{n\ell}$ were shown to coincide with the eigenfrequencies of the elastic vibrations of the cylinder. In the vicinity of the resonance frequency where $\nu_\ell = n + \delta\nu_\ell$, and considering the case of $\mathrm{Im}\nu_\ell \ll 1$, one has

$$\frac{1}{\sin\pi\nu_l} \cong \frac{1}{(-1)^n\pi\delta\nu_l} = \frac{-(-1)^n}{\pi}\,\frac{1}{n-\nu_l}\,. \tag{16}$$

If we now identify $\mathrm{Re}\,\nu_\ell = n_\omega$ and $\mathrm{Im}\,\nu_\ell = (1/2)\Gamma$, then Eq. (16) and hence the scattering amplitude of Eq. (14) assumes precisely the Regge pole form of Eq. (5) where we can consider n to be a continuous variable. This shows that as the incident frequency ω passes past the nth resonance frequency of the ℓth surface wave, $\omega_{n\ell}$, the corresponding surface wave goes through a resonance.

It is not hard to explain the physical origin of the frequency resonances with the help of the expansion

$$\frac{\cos\nu_l\varphi}{\sin\pi\nu_l} = -i\sum_{\lambda=\pm1}\sum_{m=0}^{\infty} e^{i\nu_l(\lambda\varphi+\pi+2m\pi)}\,. \tag{17}$$

If used in Eq. (14), and remembering the time factor $\exp(-i\omega t)$, Eq. (17) shows that for $m = 0$, two surface waves of type ℓ propagate in the counterclockwise ($\lambda = 1$) and in the clockwise ($\lambda = -1$) direction around the cylinder, joined by other such waves with $m > 0$ that already have encircled the cylinder m times previously (of course having undergone larger and larger attenuation since $\mathrm{Im}\,\nu_\ell \neq 0$), corresponding to the steady-state situation considered here. The wavelength of the ℓth surface wave is given by

$$\lambda_l = 2\pi a/\mathrm{Re}\,\nu_l(\omega)\ , \tag{18}$$

and at the resonance frequency $\omega = \omega_{n\ell}$ where $\mathrm{Re}\nu_\ell = n$, one sees that exactly n wavelengths of the surface wave fit the circumference of the target (which may be referred to as the "phase matching condition"), and hence lead to a resonant reinforcement of the circumferential wave in the course of its repeated circumnavigations.

The development presented above that leads from Eq. (8) to Eq. (18) shows how a physical explanation of the resonances can be obtained from the resonance expression of Eq. (8). As to the background term in this equation, which was taken there as corresponding to a rigid object, experiments have shown that for solid metal cylinders submerged in water, such an assumption is indeed justified; even for the case of a cylinder with a borehole of moderate diameter. It is clear, however, that when dealing with a very thin air-filled shell, the rigid-background assumption must be given up since then, the background will in the limit of vanishing shell thickness approach that of a completely soft object (or, an air bubble) which is also known. However, for a certain intermediate shell thickness neither of these background expressions applies and the exact form of the appropriate background remained unknown for a long time. This problem has recently been solved by Werby[24] who published his first article on the subject in Ref. 9. Related independent developments are due to Marston[25] and Veksler[26].

By considering the entrained mass of the ambient water, Werby obtained the spherical-shell scattering amplitude corresponding to the acoustic background contained in the amplitude coefficient b_n [the spherical equivalent of b_n in Eq. (14)]:

$$b_n = -\frac{\dfrac{3\rho}{\rho_s[1-(1-h)^3]}\,\mathrm{Im}\left(i\dfrac{h_n(ka)}{h_n'(ka)}\right)j_n(ka) - j_n'(ka)}{\dfrac{3\rho}{\rho_s[1-(1-h)^3]}\,\mathrm{Im}\left(i\dfrac{h_n(ka)}{h_n'(ka)}\right)h_n(ka) - h_n'(ka)} \tag{19}$$

where h is the ratio of the shell thickness to the shell radius, and ρ_s and ρ are the densities of shell material and water, respectively. The expression correctly tends towards the rigid-body expression in the limit $\rho/\rho_s \to 0$, and toward the soft-body expression for $\rho_s/\rho \to 0$. This remarkable results leads to our complete ability, now, to subtract the correct background from theoretical scattering amplitudes in order to obtain the pure resonance terms in the scattering amplitude.

4. Experimental Methods for the Study of Resonance Scattering

4.1. *The Long-Pulse Method*

Early resonance experiments had provided results for the backscattering amplitude such as shown in Fig. 1, where the resonant amplitude and the non-resonant

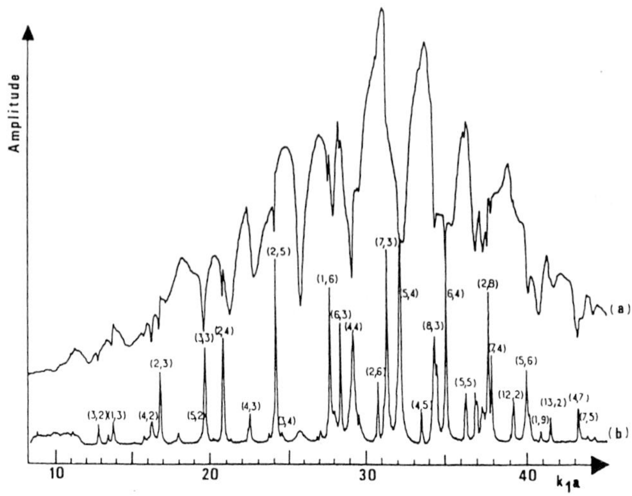

Fig. 5. Experimental spectra for aluminum cylinder in water: (a) form function, (b) isolated resonance spectrum. From Ref. 27.

background are not separated from each other, but interfere coherently so that the resonances manifest themselves only in the form of interference minima (for the Rayleigh-wave resonances), or other resonance features[5] (for the Whispering-Gallery waves). It was recognized by Maze and Ripoche[6] that the resonant amplitude, Eq. (8), could be measured separately (with the non-resonant background eliminated) by observing the ringing tail of a long-pulse echo such as seen in Fig. 3 (b) or (c); they termed this method the MIIR (Method of isolation and indentification of resonances). An example of their results[27], and of the precision of the measurements, is displayed in Fig. 5 which shows (for the case of a solid aluminum cylinder in water) in curve (a), top, the same unseparated form-factor curve as in Fig. 1 (here, depressed toward both sides by reduced detector sensitivity), but in curve (b), bottom, shows the resonance spectrum of the target obtained from an observation of the ringing tails analogous to those of Fig. 3 as functions of ka, evidently representing a very sensitive indicator of the resonances.

Long incident pulses, modulated with a carrier frequency ω_0, have a spectrum that can be made so narrow that it overlaps just one single complex-frequency pole. Accordingly, the echo response will contain the ringing of just this one particular resonance (cf. Fig. 3) whenever the carrier frequency coincides with its resonance frequency, so that much of the resonance spectrum of a given target can be acquired through a frequency sweep with sufficiently small frequency steps. This is the basis of the MIIR method of Maze and Ripoche[6], which will be described here.

Computer simulations of the acoustic scattering of modulated long pulses were carried out for elastic spheres by Howell, Numrich, and Überall[28,29], and for elastic cylinders by Numrich et al.[30]; the latter reference also contains experimental results. In these references, the ringing of a given resonance was explained by the coherent superposition of overlapping repeated circumferential wave trains; such wave trains had been considered earlier by de Lustrac and Carbo-Fite[31] and by Veksler[32]. However, the establishment of the MIIR method[6] had preceded the calculations of Howell et al.[28,29] and Numrich et al.[30] and it has given rise to numerous experimental studies at French laboratories (University of Le Havre, University of Paris VII, CNRS Marseilles, University of Lille), which up to 1986 are summarized in a comprehensive book[33] (GESPA, 1987) written by 24 French acousticians; for later work, see Ref. 9. Computer simulations of the scattering of long pulses have also been carried out for the electromagnetic case by Howell and Überall[34,35]; acoustic reflections of long pulses from plates have been measured and analyzed by Maze et al.[36]. It is also interesting to note that the transient elastic response of cylinders has been obtained by Burvingt et al. based entirely on the complex poles of the RST; cf. Ripoche and Maze, in Ref. 9.

A series of experimental studies on elastic cylinders by Ripoche et al.[6,37-40] has been devoted to the measurement of transients as they occur in the scattering of long incident pulses. Together with theoretical analysis by Derem et al.[41,42] these studies have developed the method of isolation and identification of resonances (MIIR), in which the transients trailing the main reflected pulse at a resonance frequency (i.e., the ringing), but absent off resonance have been employed as a highly sensitive means for obtaining the pure resonance spectrum of elastic cylinders or shells, thus contributing significantly to the emerging technology on the classification of submerged targets, cf. Numrich et al.[43]. By employing a long pulse with a spectrum narrower than the spacing between poles, it is possible to excite a given resonance selectively by choosing the carrier frequency to coincide with the resonance frequency; or not to excite it by choosing a carrier frequency sufficiently different from a resonance frequency. As a consequence, and as demonstratred by the Le Havre experimental work, it is possible to excite one resonance at a time, a procedure that may be employed for target identification purposes.[44]

We shall now describe the excitation of elastic-sphere resonances by long sinusoidal wave trains and investigate how the initial and final transients in the echo, which appear near a resonance frequency, can be interpreted by multiply circumnavigating circumferential wave trains. The relations of initial transients, and of the interference of reflected and circumnavigations pulses, to the acoustic form function of the target is established in this connection.

Theoretical expressions for the acoustic scattering amplitude from an elastic sphere have been given in the literature.[1] A plane wave

$$p_{\text{inc}} = e^{i(kz - \omega t)} \tag{20}$$

incident on an elastic sphere of radius a centered at the origin, with $k = \omega/c$, leads to a far-field backscattering amplitude

$$p_{sc} = (a/2r)e^{i(kr-\omega t)}f(x), \tag{21}$$

with a form function

$$f(x) = \sum_{n=1}^{\infty} f_n(x), \tag{22a}$$

$$f_n(x) = (4/ix)(-1)^n(n+\tfrac{1}{2})B_n, \tag{22b}$$

where $x = ka$ denotes the nondimensional frequency of sound in the ambient fluid. The coefficients B_n are given in the literature[1,29].

Form functions have been calculated by Howell et al.[29] as a function of real frequency for tungsten carbide (WC), aluminum, and lucite spheres. Results are displayed in Fig. 6 for WC, showing the form function modulus $|f(x)|$ plotted versus frequency. The complex-eigenfrequency patterns (poles of the scattering amplitude) are shown[45] in Fig. 7, and one notices the coincidence of the $n = 2,3\ldots$ Rayleigh-wave (R) poles with the sequence of deep interference minima in Fig. 6, and of the $n = 1,2\ldots$ first-Whispering Gallery (WG) wave ($\ell = 2$) poles with the narrower intermediate resonance features of that figure. The physical origin of these interferences can be attributed[5] to the destructive (180° out-of-phase) interference of the R-wave echoes with the specularly reflected echo, while the WG waves interfere with the specular echo at a smaller phase difference[5].

The calculation of transients proceeds as follows. The incident plane-wave pulse can be represented by the Fourier integral

$$p_{inc}(z, t) = \int G(k)e^{ik(z-ct)}\,dk/2\pi. \tag{23}$$

A sinusoidal wave train

$$p_{inc}(z=0, t) = \begin{cases} \sin\omega_0 t, & 0 \leqslant t \leqslant t_p, \\ 0 & \text{elsewhere,} \end{cases} \tag{24a}$$

with carrier frequency ω_0 and time duration t_p has a spectrum

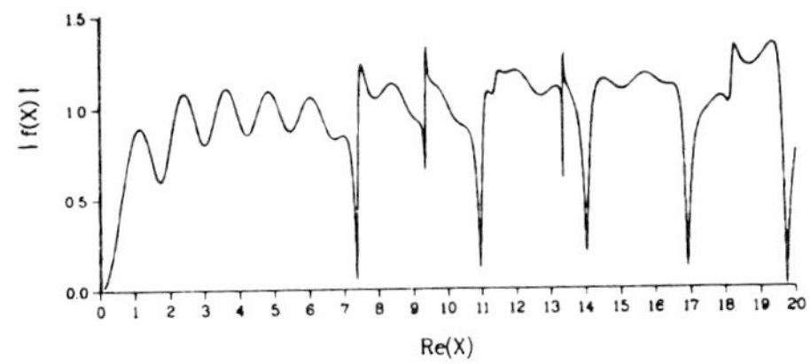

Fig. 6. Form function modulus for a tungsten carbide sphere in water, plotted vs. the frequency parameter ka. From Ref. 29.

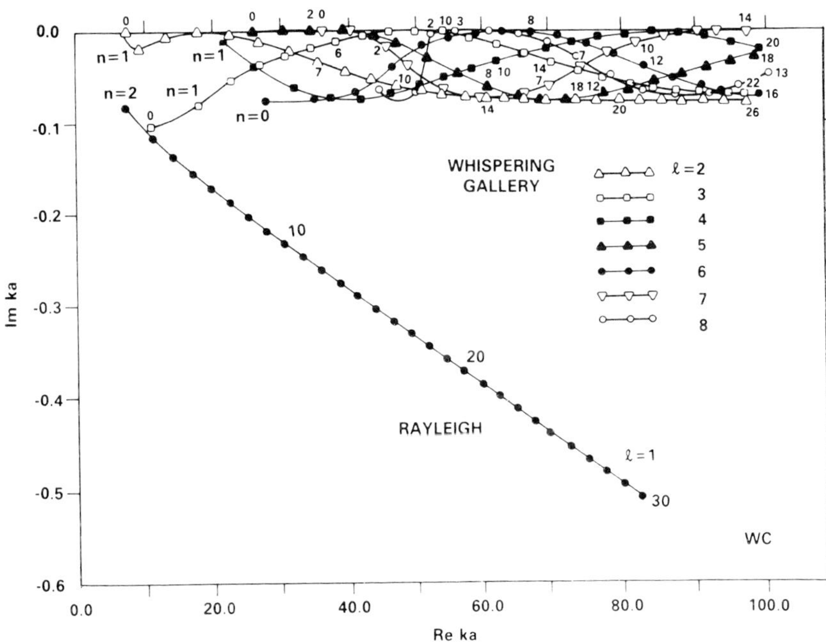

Fig. 7. Resonance frequencies of a tungsten carbide (WC) sphere in water, shown in the complex frequency plane ka. From Ref. 45.

$$\frac{1}{a}G(k) = \frac{1 - e^{i(x+x_0)\tau_p}}{2(x+x_0)} - \frac{1 - e^{i(x-x_0)\tau_p}}{2(x-x_0)}, \tag{24b}$$

where $\tau_p = ct_p/a$, $x_o = a\omega_o/c$. Introducing backscattered form functions f(x) and $f_n(x)$ by

$$\tilde{f}(x) = \sum_{n=1}^{\infty} \tilde{f}_n(x), \tag{25a}$$

$$\tilde{f}_n(x) = e^{i(kr-\omega t)} f_n(x), \tag{25b}$$

results in the expressions for the pulsed case:

$$\tilde{f}(\tau) = \sum_{n=1}^{\infty} \tilde{f}_n(\tau), \tag{26a}$$

$$\tilde{f}_n(\tau) = (-1)^n \frac{2n+1}{i\pi a} \int G(k) B_n(k) e^{-ix\tau} dx/x, \tag{26b}$$

where we have used the nondimensional time variable $\tau = (ct-r)/a$. In the calculations to follow, the expressions in Eqs. (26) were evaluated using numerical integration.

Figure 8 shows in the center part an incident pulse train, having a length $\tau_p = 25$ (i.e., covering a length of 25 radii) and a carrier frequency $x_o = 14.07$. The top portion of the figure depicts its spectrum G(x) superimposed on the form function $|f(x)|$ of a tungsten carbide sphere.

We choose x_o to coincide with the deep minimum of $|f(x)|$ caused by the x = 14.07 R zero of Fig. 6. The bottom portion of the figure shows the scattered pulse. The scattered pulse arrives at the observer at $\tau = -2$, due to the first reflection of the indicent pulse at the vertex of the sphere, while the origin of our time scale ($\tau = 0$) was chosen as the arrival of the incident pulse at the plane through the center of the sphere, z = 0. After an initial transient, the scattered pulse reaches a strongly constricted quasi-steady-state amplitude with the constriction being due to the fact that the steady-state form function $|f(x)|$ has a deep minimum at the selected center frequency x_o. After a duration of 25 units of τ (equal to that of the incident pulse), the scattered signal shoots up to its original value; and thereafter it consists of a slowly decaying final transient that can be attributed to the ringing of the R pole at x = 14.07, which has thus been selectively excited by our incident pulse. Note that in Hickling's[4]

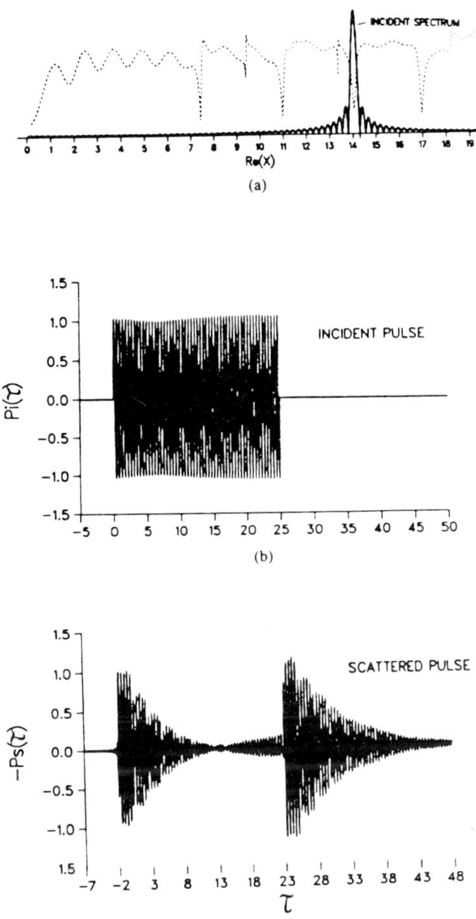

Fig.8. (a) Spectrum of incident wave train superimposed on the form function of a tungsten carbide sphere. (b) Incident wave train. (c) Scattered pulse. Carrier frequency coincides with the Rayleigh-wave interference dip at Rex = 14.07 in the form function. From Ref. 29.

study, the quasi-steady-state amplitude has been associated with interference features in the form function, but he did not discuss initial or final transients, clearly visible in his results.

Closer inspection of the tail in Fig. 8 reveals that it contains a series of descending steps. This is demonstrated in Fig. 9, which presents a blowup of the initial (top) and final (bottom) transient regions of Fig. 8 (with the τ-scale displaced again so that the

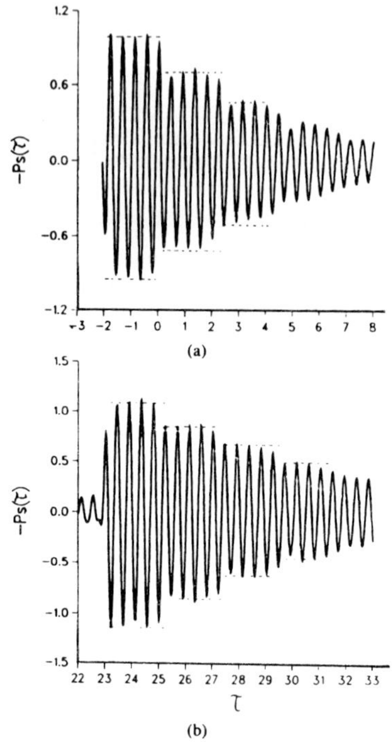

Fig. 9. (a) Enlarged portions of initial transient region and (b) final tail region of Fig. 8, showing the staircase effect. From Ref. 29.

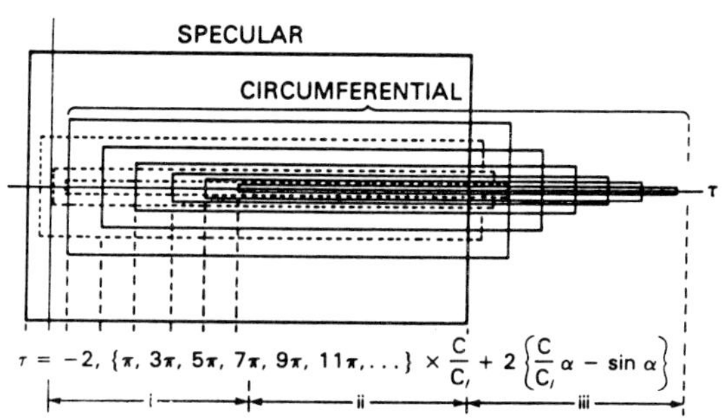

Fig. 10. Schematic view of superposition (without any coherent or incoherent eddditions) of specular (large solid rectangle), penetrating and multiply internally reflected (dotted rectangles), and circumferential wave trains (smaller solid rectangles), showing initial transient region (i), quasi-steady-state region (ii), and final transient region (iii). From Ref. 29.

pulse starts at $\tau = 0$). Note that de Lustrac and Carbo-Fite[31], and also Veksler[32] have found and discussed these steps previously, but they are also visible in earlier experimental long-pulse scattering results by Neubauer et al.[2], Fig. 3.

The physical interpretation is best discussed using the schematic graph of Fig. 10. The large rectangle denotes the envelope of a rectangular incident pulse (as in the center portion of Fig. 8), which essentially coincides with the shape of the specularly reflected pulse, due to minimal dispersion of the specularly reflected wave. As discussed earlier, the head of this specular pulse arrives at the observer at $\tau = -2$. The dotted rectangles schematically indicate further wave trains that penetrate the sphere, get reflected from its rear backside vertex, and reach the observer with short time delays (due to the high interior sound speeds) after a possible series of back-and-forth bounces with ever decreasing amplitude, as indicated. These interior reflections will not be our concern here: Numerical studies have shown them to be quite small[46], and no visible evidence exists for their presence in our computed results in Figs. 8 and 9, at least at the frequencies considered here. In addition, circumferentially traveling wave trains with rectangular shapes not too different from that of the incident pulse will (in spite of some dispersion[23] also be present, propagating with their own characteristic speed c_ℓ. They are launched at a location $x = -a \sin \alpha$ in front of the shadow boundary at the sphere's equator, where $\alpha = \cos^{-1}(c/c_\ell)$, and circumnavigate the sphere with successively decreasing amplitudes (due to radiation loss) as indicated in Fig. 10 (smaller solid rectangles). In backscattering, they reach the observer after having propagated the half-circumference plus two extra sectors, $\pi a + 2a\alpha$, with possibly an additional number of full circumnavigations. The successive arrival times are thus $\tau = [(2n - 1)\pi + 2\alpha]c/c_\ell - 2\sin\alpha$, as shown. For Franz-type waves (propagating in the surrounding fluid) where $c_\ell < c$, one has $\alpha = 0$.

Actually, there exists a large number of such surface waves (R, WG) that in general, after each circumnavigation, are out of phase with each other, so that the overlap of their rectangles in Fig. 10 will effectively cancel their contribution. However, if the carrier frequency coincides with an eigenfrequency of the sphere, which corresponds to a certain surface wave producing a phase match with itself after each circumnavigation[47], then all the successive circumferential wave trains in Fig. 10 corresponding to that surface wave (only) will add in phase. This leads not only to the stepwise decaying tail in region (iii) of Fig. 10, which represents the ringing of the resonance at the frequency in question, but also to the initial transient in region (i) of Fig. 10, stepped similarly as the final transient (tail), in which the successively arriving circumferential wave heads also add in phase. It now depends whether the resonance in question interferes with the specularly reflected wave destructively (the more usual case for metal targets) or constructively. In the first case, there will be a descending staircase in region (i) of the reflected signal down to the reduced steady-state amplitude of region (ii), as is clearly evident in the bottom portion of Fig. 8. In the second case, the initial transients (i) will form an ascending staircase leading to an augmented steady-state amplitude in region (ii), as, e.g., in the center portion of Ref. 48, Fig. 7.

Figure 10 shows that the width of the steps is given by

$$\Delta\tau = 2\pi c/c_l. \tag{27a}$$

Resonances originate if, for the case of the sphere, a half-integer number of wavelengths (Überall et al.[49])λ_ℓ of the ℓ th surface wave span the circumference of the sphere:

$$(n + \tfrac{1}{2})\lambda_l = 2\pi a. \tag{27b}$$

Since $\lambda_\ell = 2\pi c_\ell \omega$, this leads to

$$(n + \tfrac{1}{2})c_l/c = ka, \tag{27c}$$

showing that the ratio of surface wave speed to sound speed in the external fluid can be obtained from reading off the ka difference of adjacent resonance features caused by the ℓ th surface wave in the form function

$$c_l/c = \Delta(ka)_{nl}. \tag{27d}$$

With c = 1482m/s assumed in Fig. 6 this leads to an average speed of the Rayleigh wave in the vicinity of the ka = 14.07, resonance (note its dispersion!) of c_ℓ = 4366m/s, or the stepwidth of $\Delta\tau$ = 2.13, in perfect agreement with Fig. 9.

The conclusions of the preceding discussion may be summarized as follows:

(1) At a resonance frequency, the succession of multiply circumnavigating pulses builds up in phase but interferes with the preceding reflected pulse mainly destructively (typically, for metal targets). We obtain a picture of an initially square reflected pulse, which due to the initial transient builds down stepwise to a quasi-stationary reduced amplitude value. (If the interference with the reflected wave were constructive, this being the case for some Whispering Gallery waves in metals and more generally for lightweight targets such as lucite, a buildup to an augmented quasi-stationary amplitude value would occur. This was observed by Hickling see Fig. 9).

(2) Still at resonance, after the termination of the reflected pulse, a stepwise decaying tail of multiply circumnavigating constructively interfering surface waves is observed.

(3) Moving off resonance, the multiple surface wave trains get out of phase and begin to interfere with each other destructively, thus causing both the initial transients and the tail to disappear.

The MIIR method of Maze and Ripoche[6] uses the presence of the ringing tail in order to acquire the resonance spectrum of a given elastic target. The frequency is stepped up in sufficiently narrow intervals, and the free-vibration tail amplitude is measured,

(a) (b)

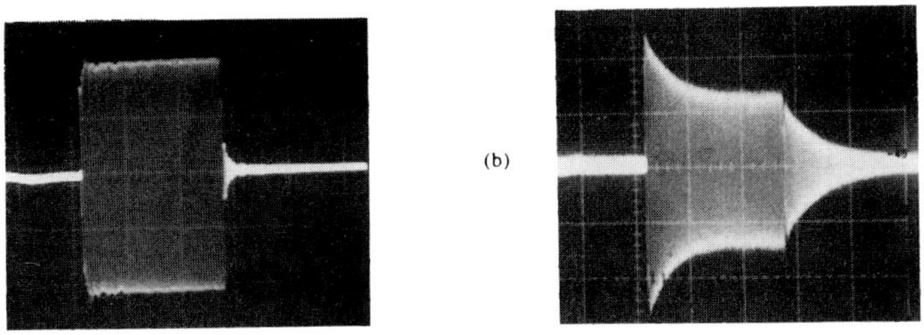

Fig. 11. Reflected pulse from an aluminum cylinder in water (a) off resonance, and (b) at a resonance frequency.

e.g., at $\tau = 33$ in Fig. 8. This amplitude plotted versus frequency constitutes the resonance spectrum. (If the amplitude of the forced vibration were measured as a function of frequency, e.g., at $\tau = 10$ in Fig. 8, the result would furnish the form function). Figure 11 shows the observed long pulses for an aluminum cylinder in water (a) off resonance, and (b) at a resonance frequency. Finally, in Fig: 5 (Maze et al.[50]) we see the form function (top) and the pure resonance spectrum (bottom) of a solid aluminum cylinder, measured in the described way. (Note that a bias in the measuring hydrophone causes a fall-off of these spectra towards both ends, which can be corrected for). In the previous theoretical work,[5,9] pure resonance spectra had been obtained by coherently subtracting out the nonresonant background amplitude (last term of Eq. 7), approximated by the rigid-scatter amplitude). The MIIR method as well as the Numrich-de Billy (NdB) method[7,8], to be described below, furnish an experimental approach for obtaining pure resonance spectra.

4.2. *The Short-Pulse Method*

The short-pulse (NdB) method[7,8] used for obtaining resonance spectra of submerged objects, employs acoustic pulses which have a duration that is short compared with their transit time of the target dimensions. Historically, the scattering of such pulses was observed in early experimental studies by Barnard and McKinney[51] and by Diercks et al.[52], where it was noted that a single incident pulse gave rise to a series of returned echo pulses, such as shown in Fig. 4. Subsequently, Horton et al.[53] interpreted this multiplicity of echo pulses as being due to repeated circumnavigations of circumferentially propagating surface-wave pulses, which radiated off one short pulse toward the observer upon each encirclement of the finite-sized target. Similar surface wave arguments were involved later on as part of the interpretation of the scattering of wide-bandwidth pulses from fluid-filled cavities in elastic solids.[54]

The connection between the pole layers of the complex-eigenfrequency patterns (Fig. 7), and the acoustic (and electromagnetic) scattering of wide-band pulses has first been established by Überall et al.[55,56] These original studies established the fact that each individual complex-frequency pole layer (solid curves in Fig. 7) was due to the successive resonances of one particular type of surface wave circumnavigating the target, and matching phases after each encirclement, so that successively n,n+1,n+2,...wavelengths spanned the circumference of the target (augmented, however, by a quarter-wavelength[57] each time the geodesical closed path touched a caustic[58], as it is the case, e.g., at the poles of a sphere). In addition, the studies showed to first order that the propagation speed of the pulses was given by the *group* velocity of the surface waves.

Experimental examples of short-pulse scattering include scattering experiments of wide-band pulses from evacuated, water-immersed cylindrical shells[59,60] in which more than seven circumnavigations of circumferential waves have been observed (Fig. 12).

As was mentioned before, the first arrival of short pulses returned, e.g., from a sphere (Fig. 4) corresponds to the specular reflection of the incident pulse from the vertex of the scatterer, with an arrival at the reduced time $\tau = -2$. It is clear that Fourier analysis of the return echo which contains the spectrum of the entire incident pulse, will furnish the total form function of the scattering process such as shown in Fig. 1. It was, however, noted independently by Numrich et al.[7] and by de Billy[8] that the return pulses could be gated, with the first arrival pulse left out of the Fourier integral. This can be seen, e.g., in Fig. 12, for the scattering of short pulses by a hollow aluminum cylindrical shell in water. Figure 12 represents the echo pulse series received in backscattering. With the first (specular) echo gated out, Fourier transformation of the return echoes (which are now due to the resonating circumferential waves) then furnishes[61] in Fig. 13 the pure spectrum of resonances in the scattering amplitudes, rather than the total form function such as shown in Fig. 1, consisting of a generally flat specular background spectrum with which the resonances interfere mainly destructively, had in the Fourier transform the specular echo in Fig. 12 not been gated out. It is seen that the pure resonance spectrum of the target can be obtained by this method of Numrich and de Billy (NdB) with the use of a single echo pulse measurement (rather than by a time consuming frequency sweep as used in the MIIR method), which represents the principal advantage of this approach, but probably renders its results less accurate than those of MIIR.

Comparing the short-pulse approach discussed here with the aforementioned long-pulse method, we may contrast the mathematical details as follows:

(1) Short pulses cover a wide band of complex-frequency poles; each pole contributes a decaying sinusoid to the scattering amplitude and the coherent sum of these decaying sinusoids synthesizes the radiation from multiply circumnavigating short circumferential pulses[55,56].

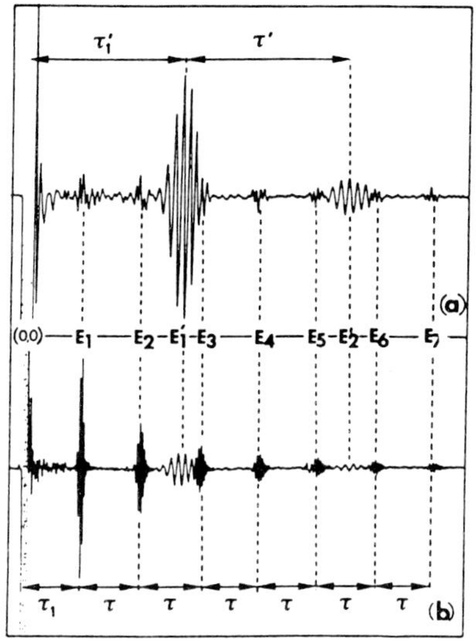

Fig.12.Echo pulse series of two different types of circumferential waves (E and E') returned from an empty duraluminum cylindrical shell in water. Center frequencies of incident pulses are (a) 1MHz, and (b) 2.25 MHz. From Ref. 60.

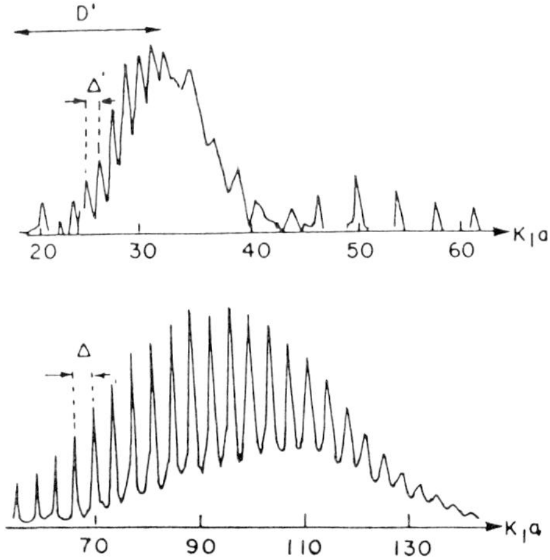

Fig.13.Resonance spectrum from a gated Fourier analysis of short pulses backscattered from a water-immersed, empty cylindrical aluminum shell 0.4 mm thick with inner-to-outer radius ratio of 0.96. From Ref. 61.

(2) Long pulses can be made to cover a sufficiently narrow frequency band so as to selectively excite just one single pole resonance, or else not to cover any resonance at all. In the former case, they will cause a ringing (final transient) of the resonance in the return pulse, which is synthesized by the coherent addition of multiply circumnavigating long circumferential wave trains, leading to a ringing tail in the echo that decays in a stepwise rather than an exponential fashion.

5. Experimental Results and their Analysis

In this chapter, we shall discuss some noteworthy earlier observations on the scattering from a variety of submerged elastic scattering targets, as well as some of the more recent experiments on such objects. At the same time, the analysis of these experimental results will be described, and the physical concepts and phenomena that arise from such an analysis will be introduced and discussed at a time as they refer to each of the object shapes under consideration.

5.1. Solid Elastic Spheres and the Time-Frequency Representation

The early experiments by Neubauer et al.[2] on the acoustic scattering from submerged solid metal spheres have been mentioned in Section 1, see Fig. 1 for steady-state, and Fig. 3 for transient scattering. Related experiments by this group[62] dealt with the scattering of pulses, and showed that by using shorter and shorter pulses, a transition from the case of Fig. 3 to the case of Fig. 12 took place, with the "staircase" in Fig. 3 (see also Fig. 10) thus stretching out and eventually decomposing into the individual pulse series of Fig. 12. The physical interpretation given in Sections *4.1* and *4.2* was, however, still missing at that time. Very precise experiments on the form function of tungsten carbide spheres were performed later on by Dragonette et al.[63], using the short-pulse method and obtaining the absolute value of the form function from Fourier-transforming the entire pulse return and dividing by the modulus of the specular-reflection echo. This method was proposed as a calibration method for return signals from arbitrary objects, due to its accuracy which is demonstrated in Fig. 14. Additional resonance scattering experiments on solid spheres are due to a team at the Mechanics and Acoustics Laboratory at Marseille[64,65].

These experiments were first carried out in the conventional fashion of the long-pulse method. Subsequently, however, this group introduced here a different method of analysis, the time-frequency representation, which they subsequently applied to spherical shells[66,67]. Applications of this method to acoustic scattering from solid spheres and spherical shells are due to Yen et al.[68]

In this latter study, it was shown that the various types of surface waves which, e.g., when employing the short-pulse technique appear in Fig. 12 in an interloping fashion, can be well separated from each other by the time-frequency analysis. Some of the circumferential wave types on an elastic sphere in water can be recognized in the pole pattern of Fig. 7. The various Whispering-Gallery waves, closing around the sphere,

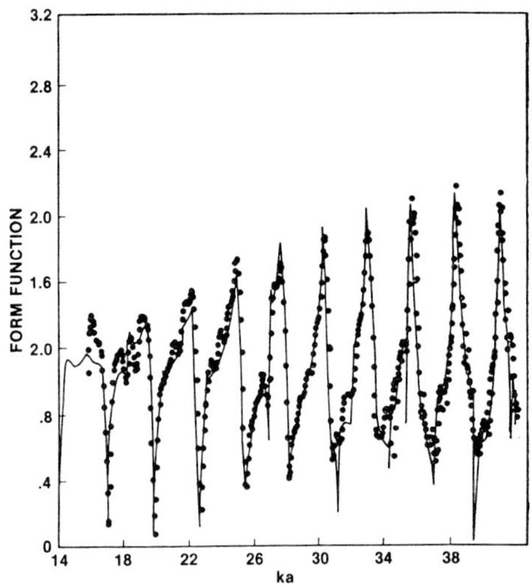

Fig.14.Comparison of experimentally obtained form function (points) and analytic computation (solid lines) for a tungsten carbide sphere. From Ref. 63.

form the pole series just below the real axis. The diagonally descending pole series is due to the (slower) Rayleigh wave. The so-called Franz-type, or creeping waves that circumnavigate the sphere in the surrounding fluid, with speeds below and comparable to the sound speed in water, would form further diagonally descending pole series below that of Rayleigh wave; they are not shown on the figure, nor is the Scholte-Stoneley wave which is similar to the Franz waves.

The time-frequency analysis often makes use of the Wigner-Ville distribution function. When applied to the form function f(ka) with its argument expressed by the radian frequency ω, the time-frequency distribution function is given by[68]

$$W(t,\omega) = \frac{1}{2\pi}\int_{-\infty}^{\infty} f_\infty\left(\omega + \frac{\nu}{2}\right) f_\infty^*\left(\omega - \frac{\nu}{2}\right) e^{j\nu t}\, d\nu. \tag{28}$$

this being a real-valued function. It is usually plotted as a contour graph above the (t,ω) plane as, e.g., shown in Ref. 11, Fig. 38. In Ref. 68, such 3-dimensional graphs are simplified to "vein diagrams", which indicate the locus of the crests of W(t,ω) in the (t,ω) plane.

Figure 15 shows the experimental form function of a stainless steel sphere[68] (top), and its corresponding vein diagram (bottom). The dominant crests that appear in the

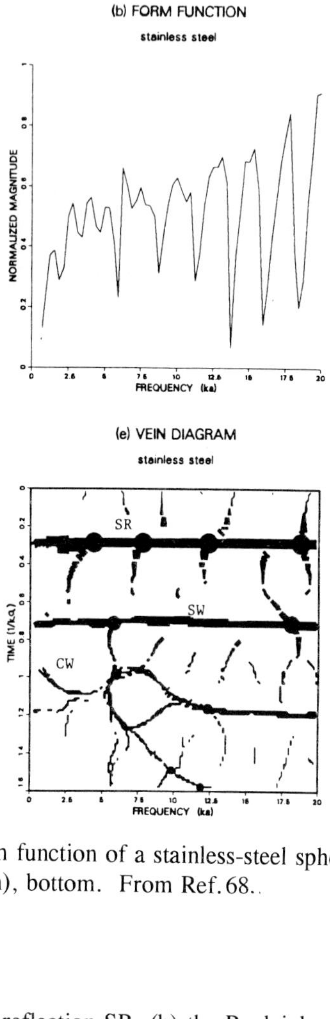

Fig. 15. Experimental form function of a stainless-steel sphere (top), and time frequency plot (vein diagram), bottom. From Ref. 68.

latter are (a) the specular reflection SR, (b) the Rayleigh wave, and (c) the fluid-borne creeping wave CW, indicating the dispersion of the latter. For the simple case of a solid sphere, the same separation could also be obtained from the short-pulse data, but for e.g. spherical shells (Ref. 68, or Ref. 11, Fig. 38) the Wigner-Ville separation is more significant, see below.

Other recent sphere-scattering results obtained at the Applied Research Laboratory, University of Texas[69] are noteworthy for having obtained a separation of specular and resonance echoes from a sphere by the use of frequency-modulated short pulses.

5.2. Spherical Shells and Similarities to Plate Waves

While for a solid sphere or a spherical shell of large wall thickness, the specular-reflection background [second term of the sphere-analog[5] of Eq. (7)] can be well represented by one corresponding to a rigid sphere as shown in Ref. 5, an extremely thin shell resembles an air bubble in water and its background will be represented by one corresponding to a soft sphere. For intermediate shell thickness, the correct background will be given by Eq. (19). This is demonstrated[24] in Fig. 16 for a steel shell of thickness 0.3%, part (a) showing the calculated form function and part (b) the form function after coherent subtraction of a soft-sphere background; it is seen that only the low-frequency part of the background is eliminated. In Fig. 16(c), however, the correct background of Eq. (19) is coherently subtracted and it is seen that the pure resonance response remains.

Observations of plane-wave scattering from spherical shells were carried out in Ref. 66 using both short and long pulses. The observed hill-shaped form function is similar to that of Fig. 16 (or of Fig. 13), and corresponds to a calculated resonance response[70], shown in Fig. 17 for a spherical aluminum shell of 2.5% thickness. One notices here two interpenetrating series of resonances denoted S_o (modes n = 8-20) and A (modes n = 40-74). These can be explained as the shell-analogs of Lamb waves as follows.

Lamb waves are known as the eigenvibrations of a flat plate (of thickness d) in vacuum. The dispersion curves of the phase velocities of the various Lamb wave modes plotted vs. fd (f being the frequency) are well-known, and are shown in Fig. 18 for an aluminum plate with bulk wave velocities c_L = 6.4mm/μs (compressional) and c_T = 3.1mm/μs (shear). They fall into two families, dots A_j (antisymmetric with respect to the center plane of the plate, i.e. shear or flexural) and circles S_j (symmetric, i.e. compressional). Only the S_o and A_o waves exist down to kd = 0, while the higher modes have low-frequency cutoffs. For ka $\to\infty$ they converge towards the Rayleigh-wave speed on a half-space, while the higher order modes converge toward c_T. The mentioned vibration types are, however, coupled by the boundary conditions so that the modes are not pure. Pure modes would exist on a fluid plate (compressional only, and c_T = 0), or on an artificial material that solely supports shear waves (flexural, and c_L = 0), as shown[71] in Fig. 19. One may test the character of the Lamb waves[71] by artificially reducing the coupling, cf. Fig. 20 with c_T reduced to 1mm/μs. Here, the dispersion curves are seen to form flat segments that tend to join up to form the compressional dispersion curve labeled c_T = 0 in Fig. 19, hence exhibiting compressional character. But at the ends of the flat segments, they get repelled by the preceding or following dispersion curves and tend, for ka $\to\infty$, toward the value of c_T as before; they obviously are of shear character here. One and the same Lamb waves hence changes its physical character along the dispersion curve, being of shear type on the steep portions, and compressional on the flat segment.

For plates that are fluid-loaded on one side only (the so called Grabowska problem[59,72]), relevant for comparison with our fluid-loaded shell, the above Lamb

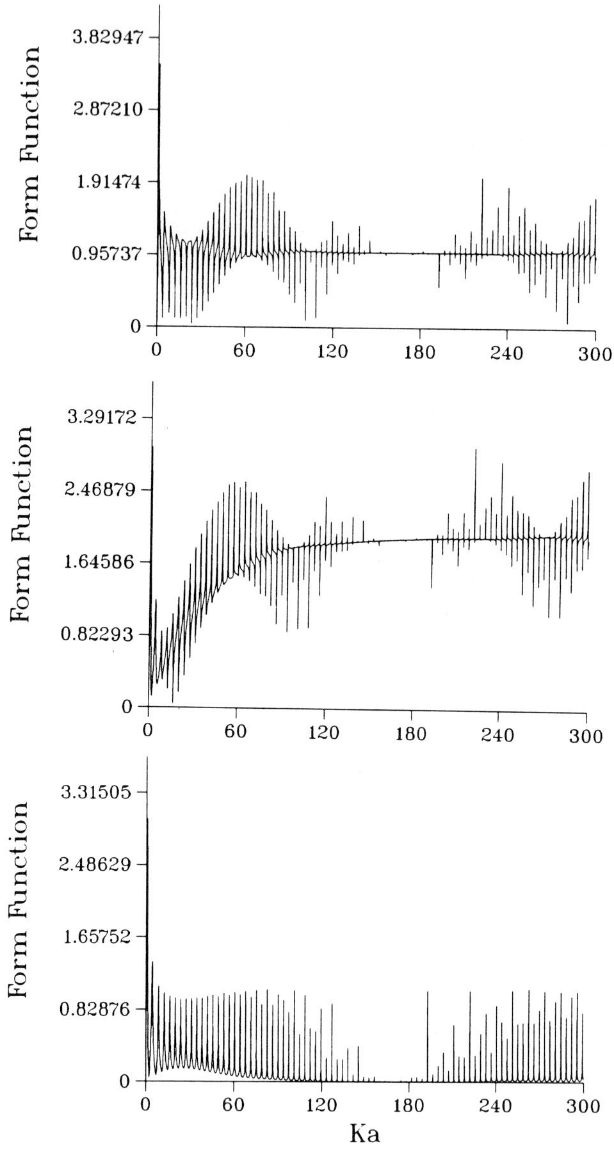

Fig. 16.(a) Calculated form function for a spherical steel shell of thickness 0.3%; (b) after coherent subtraction of soft-sphere background; (c) after coherent subtraction of background of Eq. (19). From Ref. 24.

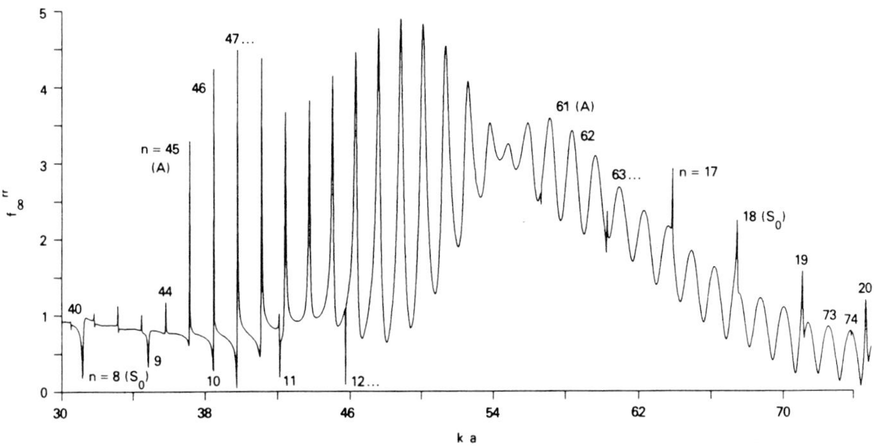

Fig. 17. Resonance response of a 2.5% aluminum shell. From Ref. 70.

curves change very little under water loading. However, a new fluid-borne wave gets added to the Lamb waves, known as the Scholte-Stoneley wave[73,74]. Its dispersion curve (labeled A) is shown[70] in Fig. 21 together with those of the A_0 and S_0 waves. However, just as in the case of the Lamb waves, a coupling phenomenon takes place between the A_0 and the A wave so that to the right of the region of repulsion between

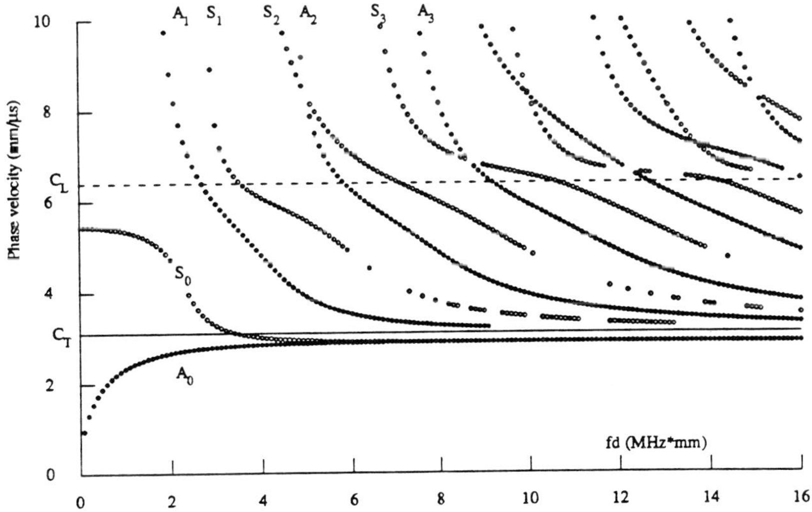

Fig. 18. Dispersion curves for Lamb waves in an aluminum plate. Dots: antisymmetric modes; circles: symmetric modes. Bulk speeds c_L, c_T are indicated.

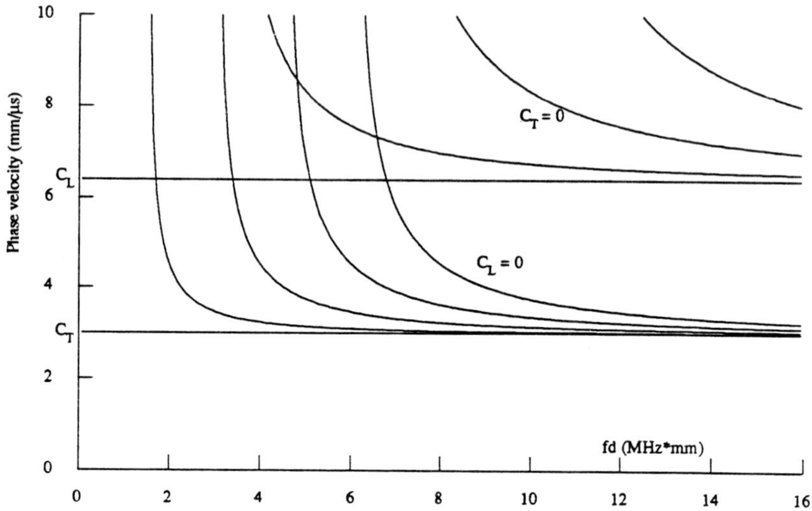

Fig. 19. Dispersion curves of a fluid plate in vacuum with c_L equal to that in aluminum and $c_T = 0$; and of an artificial plate supporting shear only, in vacuum with c_T equal to that in aluminum and $c_L = 0$. From Ref. 71.

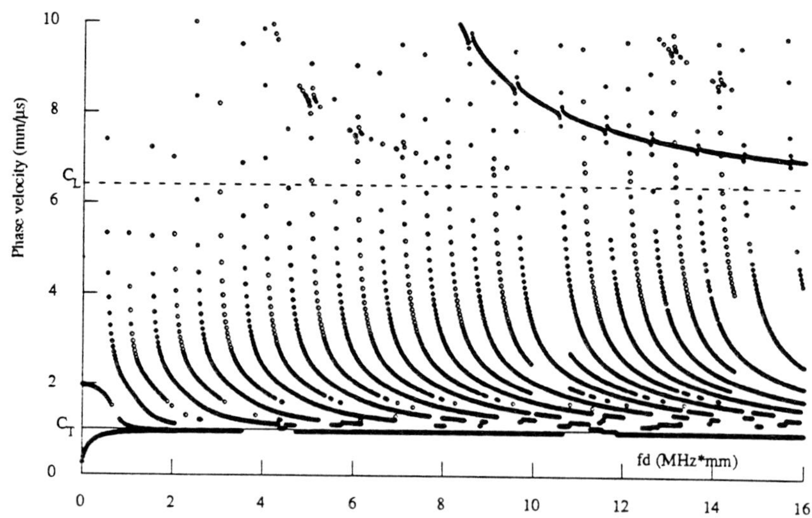

Fig. 20. Dispersion curves of an "aluminum" plate with reduced shear speed $c_L = 1$mm/μ. From Ref. 71.

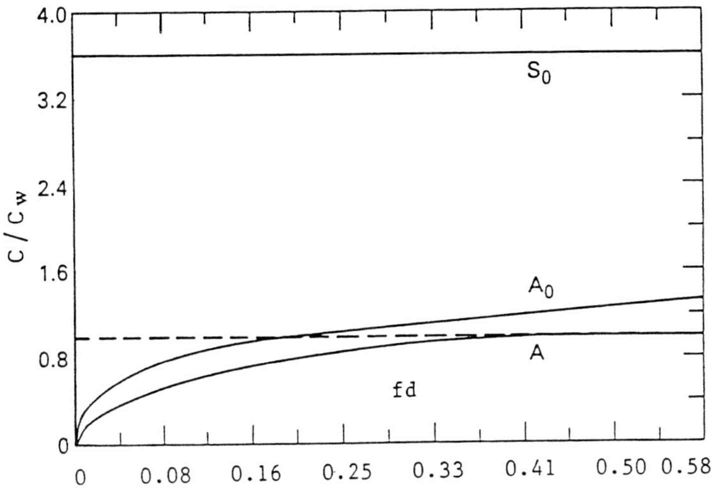

Fig.21. Dispersion curves S_0, A_0 and A waves for an aluminum plate bounded on side by water, on the other by vacuum. From Ref. 70.

the two curves (fd ≅ 0.3) the curve marked A_0 is Lamb type (plate-borne) and A is Stoneley type (fluid-borne), while for smaller fd values the upper curve is fluid-borne and the lower one is of Lamb type[71]. Indeed, with this identification the water-borne portions of the A_0 and A curves are such as to maintain the combined phase speed as close as possible to the sound speed in water, c_W (note that the curves are normalized to $c_W = 1$).

The circumferential waves that are supported by a fluid-loaded thin air-filled or evacuated elastic shell have a great similarity with the above-described Lamb and Stoneley waves. Figure 22 shows[75] the dispersion curves of the A_0 and A waves on a water-loaded, evacuated spherical shell of thickness 5%, as well as of the A_0 wave for the shell in vacuum where there is no A wave present. The same remarks apply as to Fig. 21; the main difference with the plate is that for the spherical shell, an upturn with ka→0 takes place (for reasons of the shell curvature) for the dispersion curve of the shell-borne wave (cf. the vacuum A_0 curve) but not for that of the water-borne wave. This proves that below the repulsion region, the water-borne A wave has converted to a shell-borne wave since it shows this upturn, while the A_0 curve tends to zero.

Returning now to Fig. 17 and counting the number of standing waves around the circumference of the sphere in order to find the resonance frequencies corresponding to mode number n, for a given wave of phase velocity c_ℓ, we use[70] Eq. (27c) and the calculated shell-wave phase velocities such as in Figs. 21 or 22 in order to interpret the resonance peak series in Fig. 17. This leads to the interpretation of the peaks labeled S_0, and their mode numbers n, as well as of the peaks labeled A, and their mode

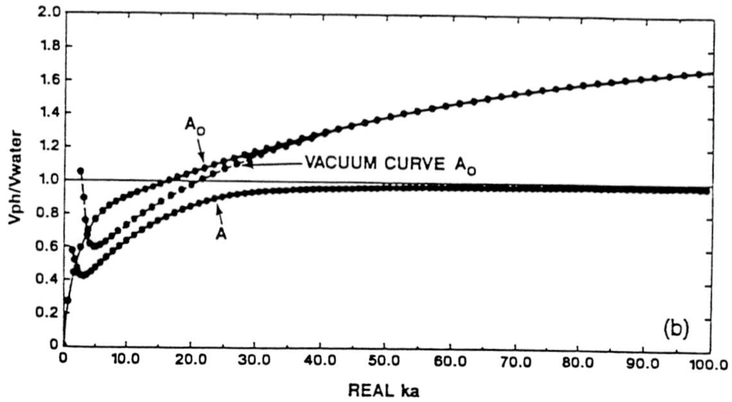

Fig.22. Dispersion curves of A_0 and A waves on a 5% thick water-loaded, evacuated spherical shell. From Ref. 75.

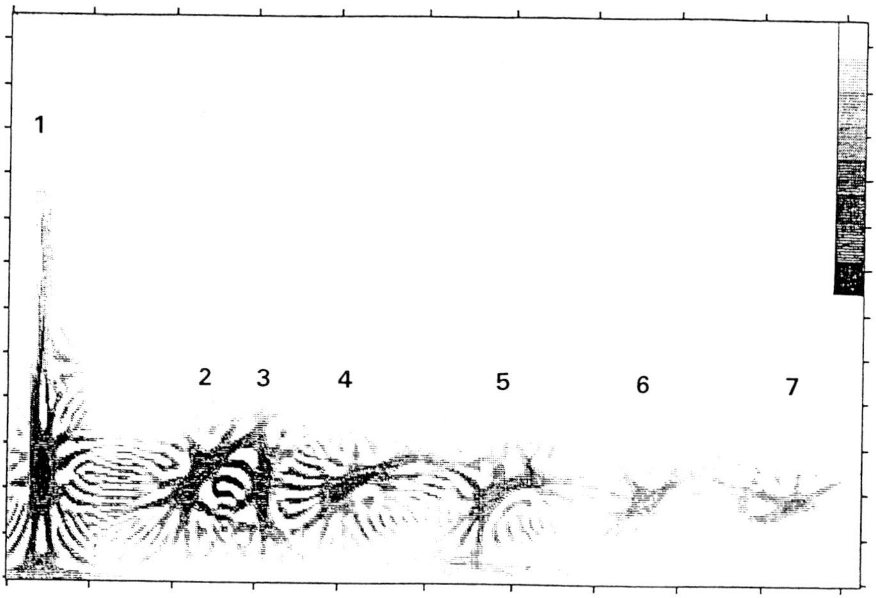

Fig.23. Wigner-Ville plot of the echo from spherical duraluminum shell of 10% thickness. Vertical axis: frequency; horizontal axis: time; 20μs/div., 200kHz/div., 3dB/gray level. From Ref. 76.

numbers n as given in this figure. The A_0 wave, however, cannot be identified by individual peaks: these have large amplitudes only in the "hill" region of Fig. 17, ka = 50-70, and their widths are large enough to make the individual peaks overlap and form one continuous hill. It will be shown later (section on cylindrical shells) that it is a common phenomenon of the A_0-wave to get excited only in a limited frequency region.

As mentioned, Wigner-Ville analyses have been carried out for the experimental echoes received from spherical shells; see, e.g., Ref. 76, or Ref. 11, Fig. 38 where the $\ell = 0$ and $\ell = 2$ waves are clearly separated. Figure 23 shows such a plot for a 10% thick duraluminum shell in water. Arrivals of energy packets are labeled 1-7. The first packet corresponds to the specular echo, the others to the successive arrivals of the repeatedly circumnavigating S_0 wave.

5.3. *Infinite Solid Cylinders and Guided Waves*

If a solid cylinder gets insonified by a plane wave incident in a direction normal to its axis, circumferential waves are excited that close around its circumference over a circular surface path. The locus of their excitation is a generatrix parallel-displaced from the geodesic to which the incident wave is tangent, by an angle α such that

$$\cos\alpha = c/c_\ell \tag{29}$$

where c_ℓ is the phase velocity of the circumferential wave. This is known as Kramer's coincidence condition. For the Rayleigh (R) and Whispering-Gallery (WG) type waves, c_ℓ is of the order of the elastic-wave speeds, i.e. $\lesssim$ 3000m/s but for the creeping and Scholte-Stoneley waves, one has $c_\ell \lesssim c$ = 1500m/s, in which case $\alpha = 0°$ and the locus of surface wave generation corresponds to tangential incidence.

This will influence the pulsed echoes as seen in Fig. 10, and especially the travel times of short pulses. Early cylinder-scattering experiments have been carried out by Maze and Ripoche[6], with their resonances analyzed using the MIIR method, the resonance frequencies agreed with those predicted by elasticity theory for a cylinder. Figure 5 shows the experimental resonance spectra, with the resonances labeled by (n,ℓ), ℓ being the type of wave (ℓ = 1...Rayleigh wave, ℓ = 2,3... Whispering Gallery waves) and n the number of standing wavelengths around the circumference. The value of n can be detemined by a bistatic measurement where the receiving transducer is rotated around the cylinder by varying the angle φ from the transmitter. Since the angular distribution in Eq. (1) is proportional to cos $n\varphi$, the resonantly enhanced nth partial wave will exhibit this angular dependence. Results are shown in Fig. 24, where the number of lobes in the "daisy pattern" equals 2n.

In later experiments on cylinders, however, a different type of resonance families was discovered[77], generated by waves termed "guided waves". These are surface waves, again of R, WG etc. types, that propagate along the cylinder parallel to its axis. Figure 25 shows[78] the dispersion curve of the lowest-order (Rayleigh-type) guided wave (solid

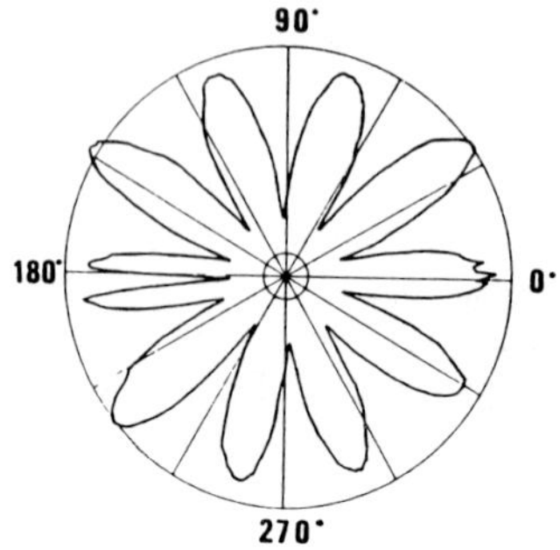

Fig.24. Distribution of pressure versus azimuthal angle around an aluminum cylinder at an n = 5 resonance. From Ref. 27.

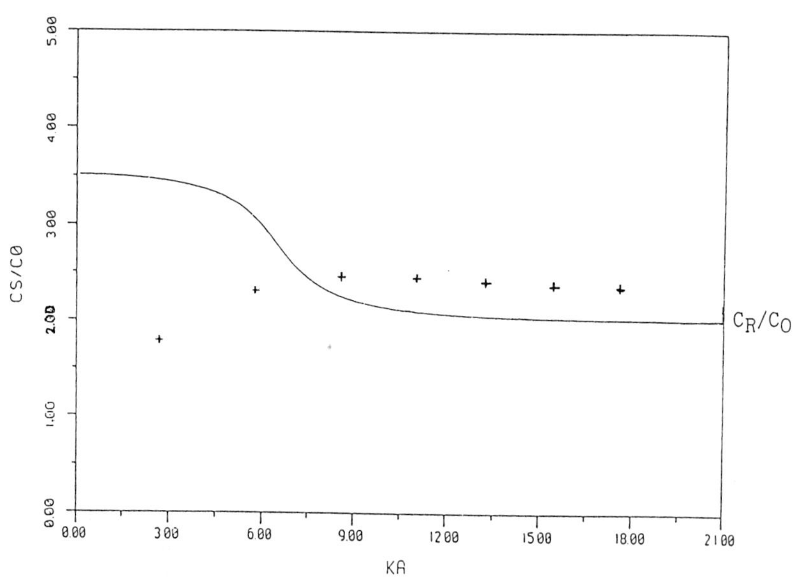

Fig.25. Dispersion relations of surface waves $\ell = 1$ on an infinite solid steel cylinder (solid curve) propagating along the axial direction, and on a solid steel sphere (crosses). From Ref. 78.

line) on a solid steel cylinder in water, normalized by the sound speed in water c; it is seen to converge toward the Rayleigh speed c_R. Also indicated by crosses is the corresponding dispersion curve on a steel sphere.

If the resonance frequencies of a series of resonances of given wave type ℓ are plotted vs. their mode numbers n, and the axes inverted, one obtains curves which for a given value of ℓ and known as Regge trajectories. Figure 26 shows these for a solid aluminum cylinder as obtained from the experimental results at Le Havre[27]. The trajecories belonging to the resonances of circumferential waves are labeled by ℓ , ℓ = 1 being the Rayleigh wave; those of the guided waves are labeled by p, p = 1 being the axial Rayleigh wave again. Crosses indicate the values of n that were identified bistatically.

5.4. *Cylindrical Shells, Helical Waves*

5.4.1. Normal Incidence

Early scattering experiments for air-filled, water-immersed infinite-cylindrical shells at normal incidence are due to Maze and Ripoche[6], investigating the circumferential waves. Guided waves on cylindrical shells were studied subsequently[79] as well as the Scholte-Stoneley wave. Figure 27 shows Regge trajectories for the experimental resonances of a 10% thick cylindrical shell, exhibiting (a) the first circumferential WG wave (ℓ = 2), (b) the Rayleigh-type first guided-wave (p = 1),

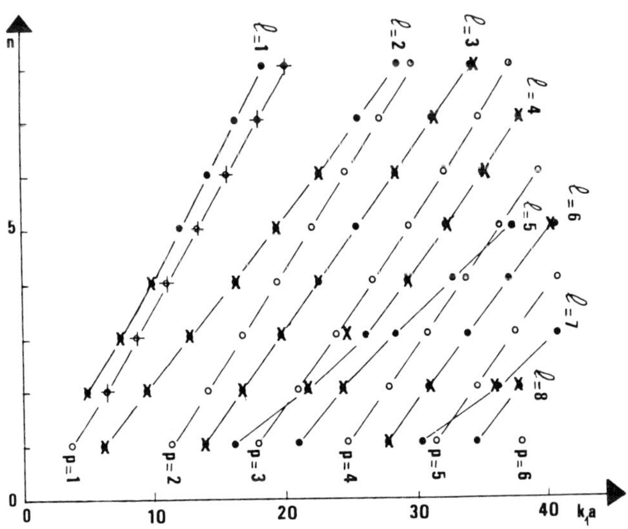

Fig.26. Regge trajectories for circumferential and guided-wave resonances of a solid aluminum cylinder in water From Ref. 27.

and (c) the Scholte-Stoneley wave (often termed $\ell = 0$). Experimental Scholte-wave studies are due to Refs. 61 and 80, and theoretical work is due to Ref. 81. The experimental resonance spectrum of the cylindrical shell is very similar to that of the spherical shell, Fig. 17, in which the $\ell = 0$ resonances are prominent in a limited ka-region. The limits of this region vary with shell thickness, as shown in Fig. 28 for the experimental $\ell = 0$ Regge trajectories of cylindrical aluminum shells with inner-to-outer radius ratios b/a = 0.67, 0.87, 0.90, 0.93, 0.95, 0.97 and 0.98.

5.4.2. Oblique incidence

Experiments on surface wave resonances created by the oblique incidence of a plane wave on an infinite cylindrical shell were carried out by Léon[82]. Figure 29 shows the geometry of his experimental arrangement of emitter (E) and receiver (R), with beams directed at the cylinder at an angle α from the normal, and Fig. 30 shows an example of the simultaneous detection of a circumferential and a guided wave (at 9° incidence) by the MIIR method.

For the case of oblique incidence both circumferential and guided waves can form helical waves on the cylinder [83,84], the helix being a geodesic surface path. References 83 and 84 discuss the refraction effects occurring in the excitation of the helical surface wave: its propagation direction, after its creation, is not a straight-ahead

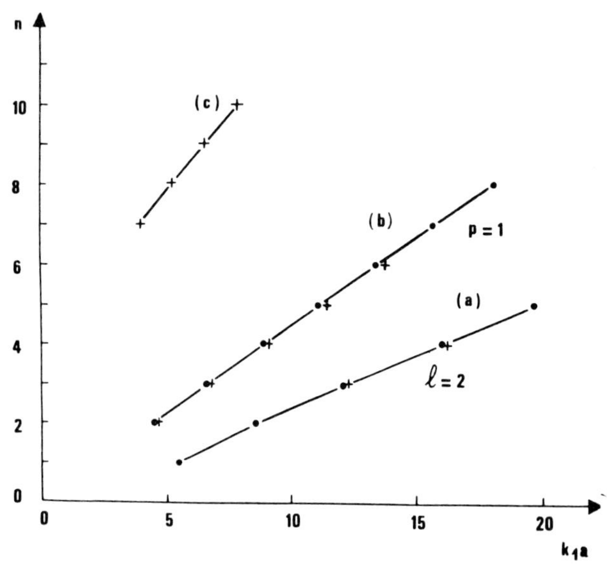

Fig.27. Regge trajectories of (a) first WG wave ($\ell = 2$), (b) Rayleigh type guided wave (p = 1), and (c) Scholte wave ($\ell = 0$) on a 10% thick cylindrical aluminum shell. From Ref. 79.

continuation of the direction of the incident wave that created it. The occurrence of the resonances of helical waves has been most lucidly analyzed by Conoir et al.[85]. Figure 31 shows point M on the (helical) wavefront of a helical surface wave. Its trajectory proceeds over a length L along the helical propagation path to point N, where the helical wave front interferes with itself-generating a resonance if the interference is constructive. This will be the case if

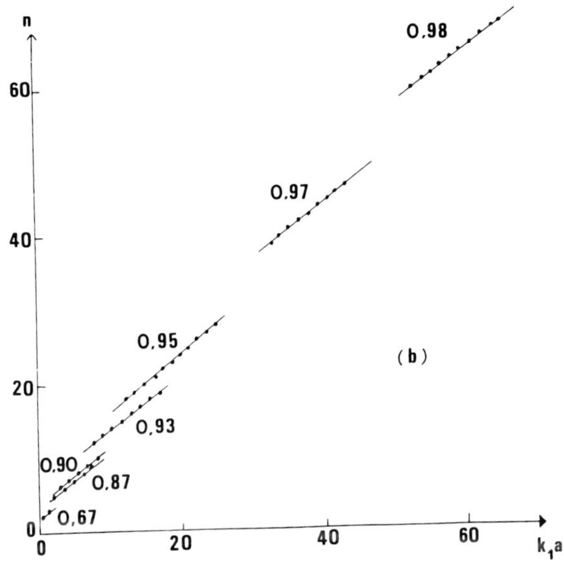

Fig.28 Regge trajectories of the Scholte wave on cylindrical aluminum shells of various wall thickness. From Ref. 27.

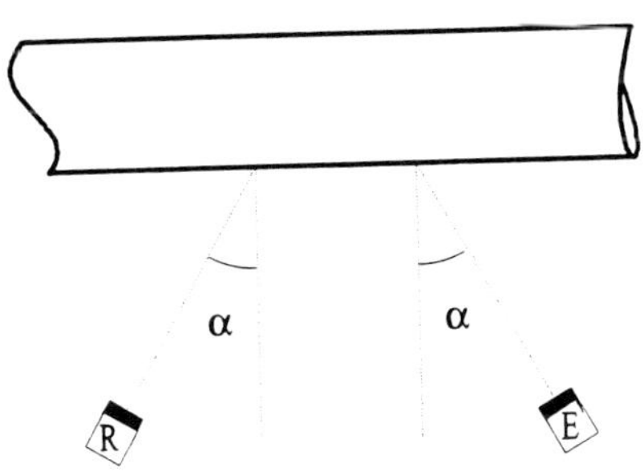

Fig.29. Experimental arrangement at oblique incidence on a cylindrical shell. From Ref. 82.

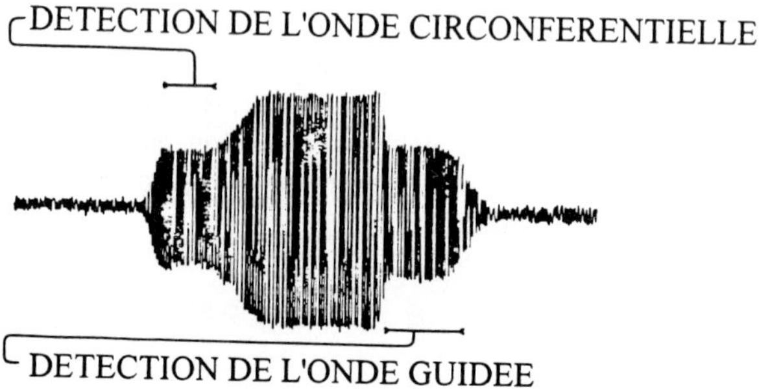

Fig.30. Simultaneous observation of a circumferential and a guided wave by the MIIR. From Ref. 82.

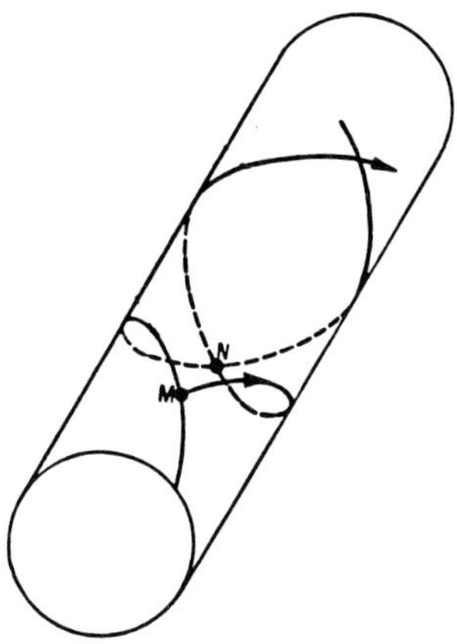

Fig. 31. Geometrical interpretation of the phase matching of a helical surface wave. From Ref. 85.

$$L \equiv 2\pi a \cos\gamma_\ell = n\lambda_\ell, \tag{30}$$

λ_ℓ being the wavelength of the ℓ th surface wave and γ_ℓ the refraction angle. In Ref. 85, it was shown that if $x = ka$ (a = cylinder radius) and $\upsilon_\ell(x)$ the azimuthal propagation constant of the (dispersive) surface wave, the resonance or phase matching conditions of Eq. (30) may be written as

$$\mathrm{Re}[\upsilon_\ell(x_{n\ell})] = n \tag{31}$$

where $x_{n\ell}$ is the resonance frequency at the (n,ℓ) resonance, which has exactly the same form as for normal incidence leading to circularly propagating surface waves[20].

Figure 32 shows[82] the form function (top) and resonance spectrum (bottom) at α = 40 of the cylindrical aluminum shell with b/a = 0.6, indicating both circumferential (ℓ = 2,3) and guided (p = 1,2) helical wave resonances. Determination of n can be done by bistatic measurements as in the normal-incidence case. The theoretical resonance frequency curves for $x_{n\ell}$ and $x_{p\ell}$ are shown in Fig. 33 as a function of α, with experimental helical guided-wave (p = 1) resonance frequencies indicated by circles and circumferential (ℓ = 2) ones by squares; note that cutoff values α_ℓ and α_T are given by the critical shear-wave angle $\alpha_T = \sin^{-1}(c/c_T)$ which forms a cutoff for the guided waves, and by the "plate-wave" critical angle $\alpha_\ell = \sin^{-1}(c/c_P)$ where c_P is the limiting velocity of the S_0 wave on a plate for fd→0, cf. Figs. 18 or 21. The latter angle is not related to the critical compressional-wave angle $\alpha_L = \sin^{-1}(c/c_L)$ which is also indicated in Fig. 33. Good agreement with the theory is evident.

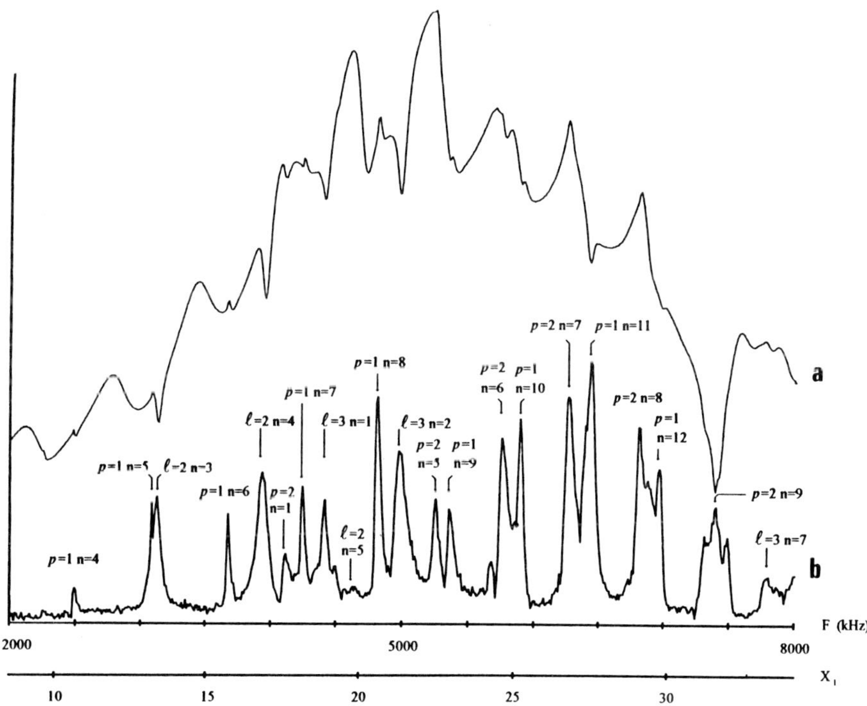

Fig.32 Form function (top) and resonance spectrum (bottom) for a cylindrical aluminum shell (b/a = 0.6) at α = 4°. From Ref. 82.

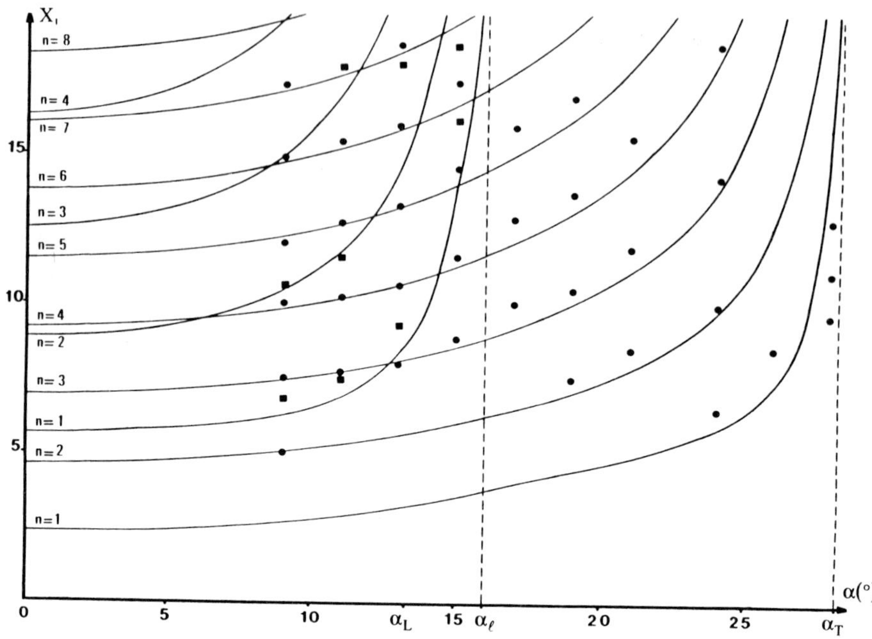

Fig.33.Resonance frequencies vs. α for a cylindrical aluminum shell (b/a = 0.6). From Ref. 82.

5.4.3. Repulsion of phase velocity curves

Calculations of acoustic scattering from infinite-cylindrical, air-filled metal shells at broadside incidence were performed in Ref. 61, and more recently in Ref. 86. Using Eq. (27c) without 1/2, for the cylinder case to obtain the phase velocities of the surface waves at the resonance values of ka (the latter having been calculated from exact cylindrical-shell theory), the phase velocity dispersion curves of Fig. 34 were obtained, with the thickness of the water-loaded steel shell being determined by the inner-to-outer radius ratio b/a = 0.8.

These correspond to two waves, termed A_{0+} and A_{0-} whose dispersion curves repel each other at ka ~ 5. Without fluid loading, however, one finds the dashed curve which corresponds to the Lamb wave A_0, and which follows A_{0+} above the repulsion point but A_{0-} below. The latter has the character of the fluid-borne Scholte-Stoneley wave, with phase velocity always below the sound velocity c in water.

It could be shown[86] that although the A_{0+}, A_{0-} curves are well separated, they

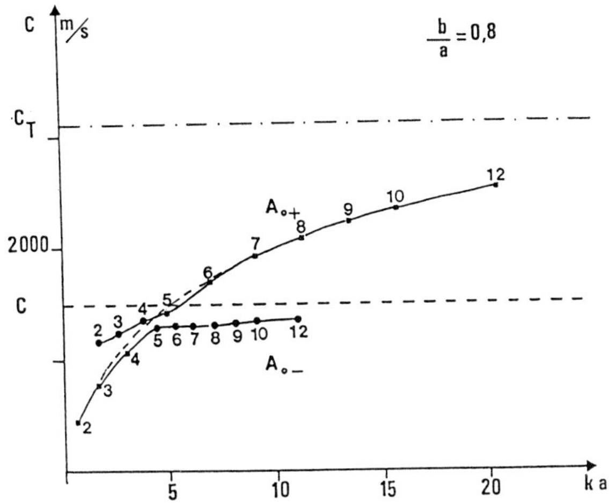

Fig.34. Phase velocity dispersion curves for Scholte-Stoneley and A_0 Lamb waves on cylindrical steel shell in water, with b/a = 0.8. From Ref. 86.

don't each possess uniformly the same character. The portions of the curve that follow the dashed line represent the A_0 Lamb wave, and those that follow the horizontal dashed line at c represent the Scholte-Stoneley wave. This switch-over of the nature of the waves at a repulsion point has been known previously for the shift of atomic energy levels during molecular formation[87]. It is similar to the case of the plate waves, Fig. 21, or the corresponding waves on a spherical shell, Fig. 22.

5.4.4. Infinite cylindrical shells with internal attachments

Solutions to the scattering problem for this case have been pioneered by Guo[88], and by Klauson and Metsaveer[89].

Guo considered the scattering problem for an evacuated infinite-cylindrical shell at normal sound incidence, the shell having a mass diametrically attached to its interior by two equal springs from each side. The backscattered amplitude plotted vs. ka is shown in Fig. 35. While the empty shell amplitude is given by the dotted curve, the internally loaded shell (solid curve, at incidence angles $\theta_0 = 0°$ parallel to the shell diameter along the mass-spring system, and at 45° to it) shows an additional resonant fine structure due to the interaction between structural waves and the attachment; in fact the fine structure is not correlated to the resonance frequencies of either the empty shell or the internal system, but can be shown to be due to the excitation of flexural waves in the shell.

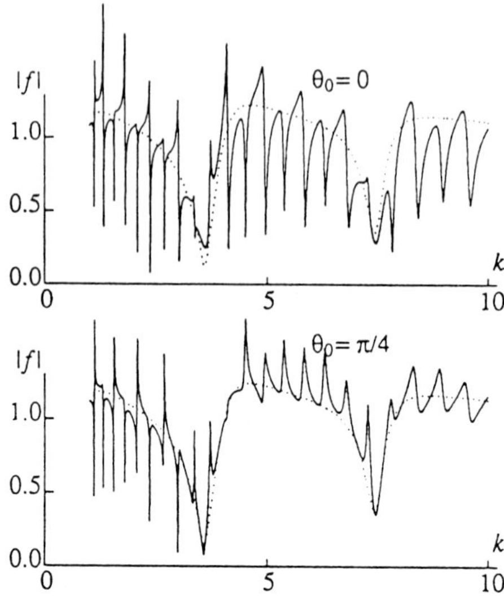

Fig.35.Backscattered amplitude at different incident angles for a cylindrical steel shell of 1% thickness in water, with internally attached mass-spring system of 1/3 the shell mass (solid curve); without internal loading (dotted). From Ref. 88.

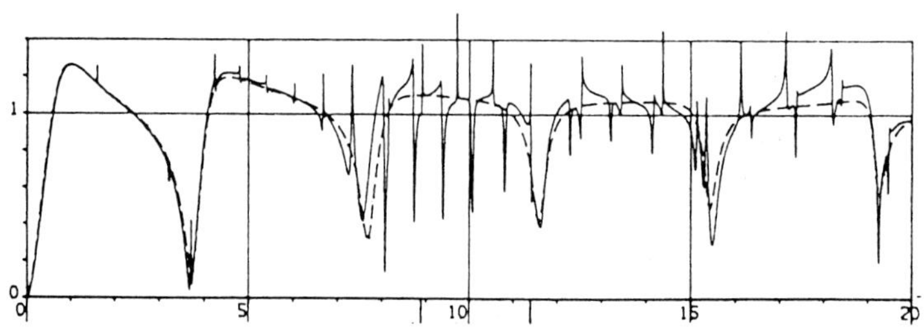

Fig.36.Backscattered amplitude vs. frequency for a 2% thick aluminum shell in water, reinforced by a length-wise stringer (solid curve); without stringer (dashed). From Ref. 89.

Similar observations apply to the example considered in Ref. 89, the infinite-cylindrical shell here being reinforced by a "stringer", i.e. two longitudinal plates normally to the shell interior in a diametrically opposite fashion. Figure 36 shows (dashed) the backscattering amplitude for the unreinforced shell, and (solid curve) the effect of the stringer.

5.4.5. Finite-length cylindrical shells

Sound scattering from finite-length cylindrical shells, with flat disks closing the ends, was studied by Lecroq et al.[90-92]. (The first reference also includes some results on the scattering from solid cylinders with flat ends). Further numerical studies are due to Ref. 93.

In this problem, new resonances arise due to the finite length of the shell. Basically, they are caused by standing waves forming in the axial direction, the length of the shell L being spanned by an integer number m of half-wavelengths[84]. This condition can be written as

$$L = m\lambda_z/2 \tag{32}$$

where λ_z is the axial projection of the wavelength of the helical wave. The experiments were performed[90,92] at normal and at oblique incidence, with a geometry as in Fig. 29 but without the geometrical gap. The theory of this scattering process was likewise presented in these references. In Fig. 37, the theoretical resonance frequencies $x_{n\ell} \equiv (ka)_{n\ell}$ of an infinite aluminum shell with b/a = 0.89, as functions of α are shown as solid curves for the p = 1 guided wave, and are combined with the resonance condition of Eq. (32) shown as dashed curves; this leads to theoretical discrete families of resonance values at the intersection of the curve families. The asterisks show the experimental results that all lie on the solid curves. Some observed deviations from the dashed lines may well stem from a phase shift of the longitudinal

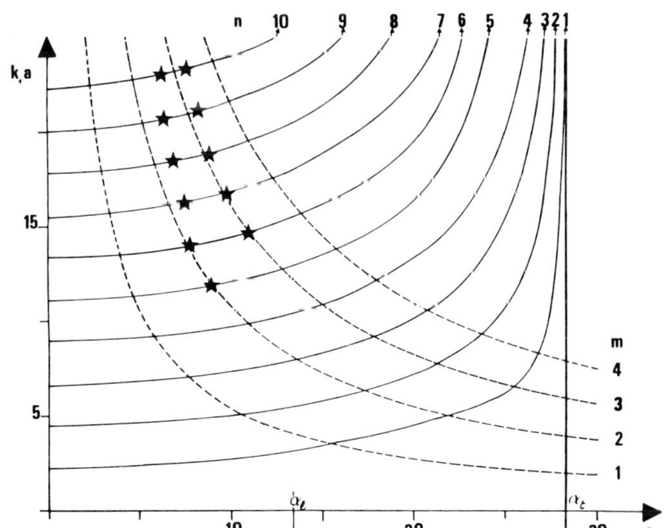

Fig.37. Resonance frequency as a function of angle of incidence from the p = 1 guided wave on a b/a = 0.89 cylindrical aluminum shell. From Ref. 27.

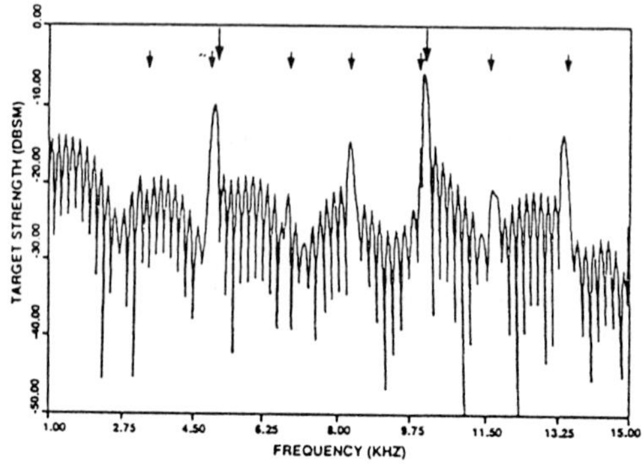

Fig.38. Total target strength of a soft finite-length cylinder with 30 circular ribs vs. frequency at aspect angle 40°. Arrows indicate predicted Bragg interference maxima. From Ref. 94.

wave components at the cylinder ends which will modify Eq. (32); such phase shifts should be due to the radiation from the cylinder ends of the axial waves into the fluid.

The case of a cylindrical shell of finite length with a series of circular internal ribs spaced almost uniformly was treated in Ref. 94. The shell scattering was obtained by the Geometrical Theory of Diffraction, and the scattering from the ribs by curvature-corrected results of the theory for a flat plate carrying a rib. Figure 38 shows the target strength vs. frequency of a soft cylinder (radius = 10.5 in, length = 230in) with five times six ribs, at aspect angle 40°. Rib scattering effects are clearly visible, and can be predicted by the Bragg law both for the entire shell (short arrows), and for the six-rib sub-assemblies (long arrows).

5.5. *Hemispherically-Endcapped Cylinders and the Phase Matching Condition*

The shape of this scattering object is shown in Fig. 39. In the case where the incident wave propagates parallel to the cylinder axis, a surface wave will be created that propagates along a meridian. It will resonate if n + 1/2 wavelengths span the length of a meridian, i.e.

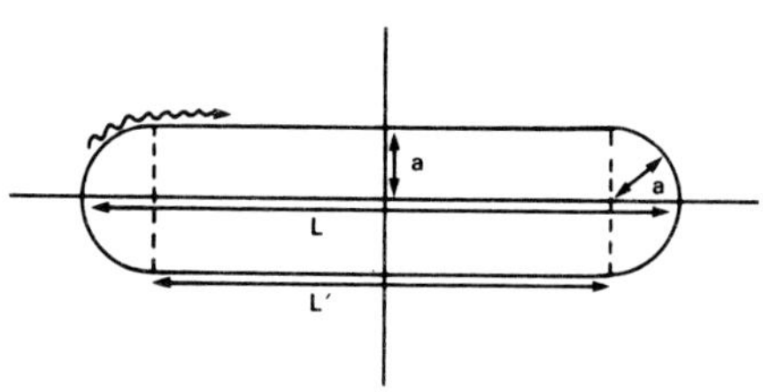

Fig.39. Surface wave propagation along the meridian of a cylinder with hemispherical endcaps.

$$\oint ds/\lambda_\ell = n + 1/2, \qquad n = 1,2,3... \tag{33}$$

where λ_ℓ is the wavelength of the ℓ th surface wave. With $k_\ell = \omega/v_\ell$ (v_ℓ being its phase velocity), this also reads

$$\oint k_\ell \, ds = 2\pi(n + 1/2). \tag{34}$$

The term 1/2 added to the number n of wavelengths spanning the path originates from the fact that at the two vertices of the cylinder, focal points arise at which a $\lambda/4$ phase jump each takes place; this is true for axial incidence but the 1/2 is absent for broadside incidence. In order to solve Eq. (34) for the (complex) resonance frequencies, the phase velocities of the surface waves must be known. Equation (34) was rigorously proved for the case of an axially symmetric shell[57].

If the cylinder were treated as an elastic bar, then according to the "bar wave" approximation[95], the longitudinal resonance condition is

$$2\int_0^L k_\ell{}^b \, dx = 2\pi m, \qquad m = 1,2,3... \tag{35}$$

where $k_\ell{}^b$ is the wave number of the ℓ th longitudinal axisymmetrical wave for an infinite cylinder, and m is the number of half wavelengths along the whole length L.

At the resonant values of the frequency, the relation $n = m+1$ is obtained immediately, as shown by figures of displacements[95]. The resonance frequencies can be estimated by using Eq. (34) or Eq. (35) depending on the situation of the objects, including aspect ratio and the frequency region considered.

In the resonance conditions as shown in Eq. (34) and Eq. (35), the wave numbers k_ℓ and $k_\ell{}^b$ are dispersive and vary with position around the surface or along the axis of targets. The wave numbers usually are unknown for a general finite elastic cylinder, and can only be estimated approximately. In Reference 96, a simple "tangent sphere" model has been developed, where the known wavenumber on a sphere is used locally, the sphere being tangent to the object's surface along the propagation direction. This model is especially suitable for objects with low aspect ratio, or in the higher frequency region. But because only the local curvature along the propagation direction is matched by a tangent sphere, errors are brought in while the effects of transverse curvature can not be ignored. A typical example is the cylindrical portion of the finite cylinder with hemispherical endcaps. For an $\ell = 1$ type surface wave along the axial direction, the constant phase velocity of the Rayleigh wave was often used in terms of the "tangent sphere" model. But in this case, it is more reasonable to estimate the wavenumbers (or phase velocities) by using the "infinite cylinder" model, where the known wavenumber on a infinite cylinder is used locally. An example of dispersion curves for an infinite solid steel cylinder is shown in Fig. 25, tending to the constant Rayleigh speed in the higher-ka region. In Eq. (34), the "infinite cylinder" model rather than the approximation of the Rayleigh speed limit should be applied to the cylindrical portion of endcapped cylinders for estimating the wavenumber of surface waves, but the "tangent sphere" model is still applied to the terminal portions.

5.5.1. Solid cylinders with hemispherical endcaps

For a solid cylinder with low aspect ratio L/D = 2(D = 2a), there are results from the University of Le Havre for a tungsten carbide (WC) cylinder. For estimating the resonance frequencies, the two phase matching results differ in using the "Rayleigh-limit" model (PM - RL) or the "infinite-cylinder" model (PM-IC) on the cylindrical portion, respectively. The other method is based on the longitudinal "bar wave" approximation (LBW).

Figure 40 shows the Regge trajectory of resonance frequencies of the ℓ = 1 type

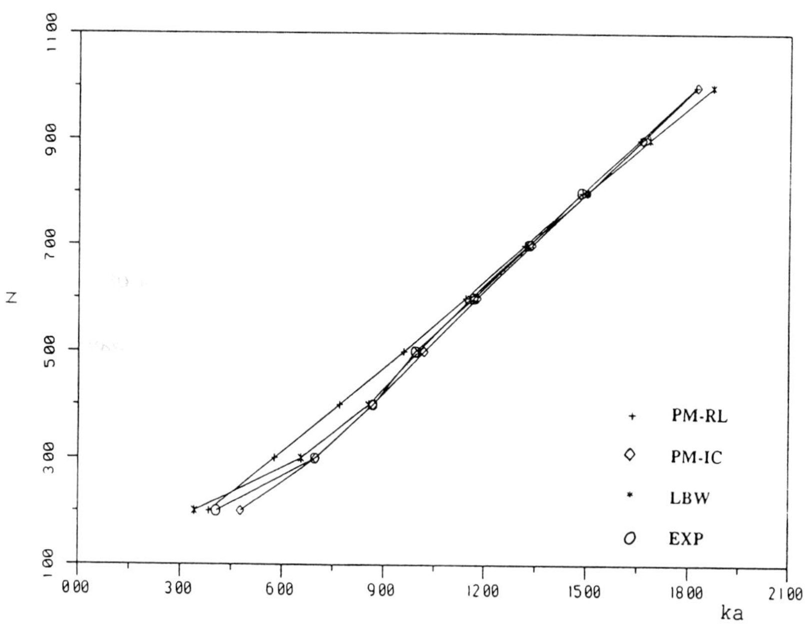

Fig.40. Regge trajectory for a solid WC cylinder with hemispherical endcaps, of aspect ratio L/D = 2, for the case of end-on incidence.

wave for a solid WC cylinder with hemispherical endcaps, of aspect ratio L/D = 2, for the case of end-on incidence. Experimental resonance response curves[97,98] are shown in Fig. 41, with phase-match predictions (PM-IC) entered as vertical bars. The experimental resonance values are also entered in Fig. 40 as circles. The comparisons show that the phase match model is better than the LBW model in the high-frequency region. The PM-IC model is better than the PM-RL model for low frequencies, but they coincide at higher frequencies.

5.6. *Solid Spheroids*

Experiments on the scattering of short sound pulses from steel spheroids were carried out in a water tank at Catholic University. The incident signal was a 3-cycle pulse of 600 kHz. Aspect ratios of the spheroids were a:b = 1:5, 1:2 (prolate) and 2:1, 5:1 (oblate); 180° backscattering was measured. Figure 42 shows a complete set of echo time series for various aspect angles β on a 1:2 spheroid (β = 0°...end-on, β = 90°...broadside incidence). The time series show an initial return pulse due to specular reflection, followed by a sequence of delayed pulses due to the radiation from circumnavigating surface-wace pulses (directly transmitted and internally reflected echoes are imperceptible small[46]).

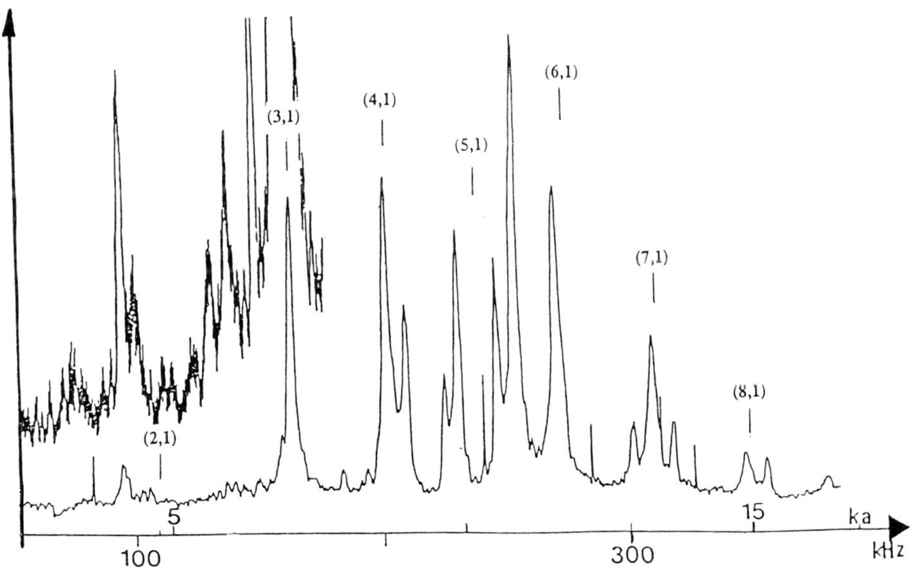

Fig.41. Resonance response obtained from experiments performed at the University of Le Havre[97,98], for a solid WC cylinder with hemispherical endcaps and aspect ration L/D = 2, case of end-on incidence.

Analysis of the specular pulses furnishes the target strength of the spheroids, defined by TS = 20 log p(β), p being the pressure amplitude of the echo. Figure 43 shows the experimental values as dashed curves, compared to the solid lines given by a theoretical formula based on simple geometry.[99]

The arrival times of the first surface wave pulses (arrows in Fig.42) can be estimated theoretically if one knows the geodesics along which the surface waves propagate, examples being shown in Fig. 44 for β = 30° and 60° (top: side view; bottom: top view). Here, the small ellipse on the left constitutes the locus of entry at which the incident wave generates the surface wave according to Kramer's coincidence condition[100], cf. Eq. (29). Entry at point A entails a backward return from point B, by symmetry. In addition, one needs to know the group velocities of the surface waves which will mainly depend on the radius of curvature along the propagation direction. These group velocities can be estimated from those on a steel sphere, and it is found that these depend very little on the sphere radius (i.e., the radius of curvature) so that they could be assumed as nearly constant for our purposes. Under these assumptions, the measured arrival times of Fig. 45 (crosses for 1:5 spheroids, diamonds for 1:2 spheroids) are seen to be well described by our theory of propagation along geodesics.

6. Conclusions

The Resonance Scattering Theory (RST) of Flax, Dragonette, and Überall[5,9,10,33] forms the theoretical underpinning on which the present experiments on acoustic scattering from submerged elastic objects, using either long (Maze and

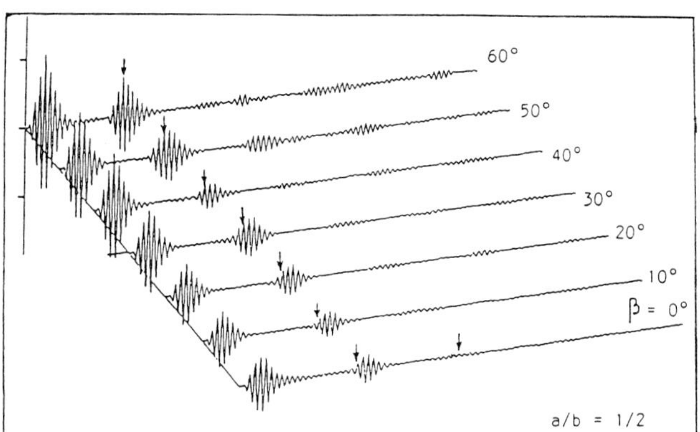

Fig.42. Experimental echo time series vs. incident angles for a 2:1 prolate steel spheroid.

Ripoche[6]) or short sound pulses (Numrich et al.[7], de Billy[8]) are based. The past decade has witnessed enormous experimental progress being made in this field largely at French institutions, predominantly the University of Le Havre's Laboratory of Acoustics, Ultrasonics and Electronics (LAUE) under Prof. J. Ripoche, with Prof. G. Maze and 25 additional faculty, research personnel and graduate students working on their PhD degrees; also at the University of Paris VII's acoustics institute (GPS) under Prof. G. Quentin, with Drs. de Billy, Talmant and others; at the University of Bordeaux I's Laboratory of Physical Mechanics under Prof. A. Gérard, with Drs. Hosten, Deschamps, and others, and the CNRS's Laboratory of Mechanics and Acoustics, Marseille, with Drs. Gazanhes, Sessarego, Sageoli and others. All this is due to the keen appreciation of the importance of this subject by the French organizations for research funding, both scientific (CNRS) and military (DRET, with notably Dr. B. Poirée who is actively engaged in acoustics research himself).

In the U.S., one academic laboratory continuously performs sound scattering experiments (at Washington State University, Pullman, WA, under Prof. P. L. Marston) with the support of the U.S. National Science Foundation. Profoundly significant theoretical research is being carried out at the Naval Research Laboratory, Stennis, MS by the group of Dr. M. F. Werby, with considerable activity also at the Department of Physics, Catholic University, Washington, DC (Prof. H. Überall). The great potentialities for target recognition by resonant sound scattering have so far been explored to a very small degree only, and it is hoped that this will be the direction in which the field will develop in the coming decade.

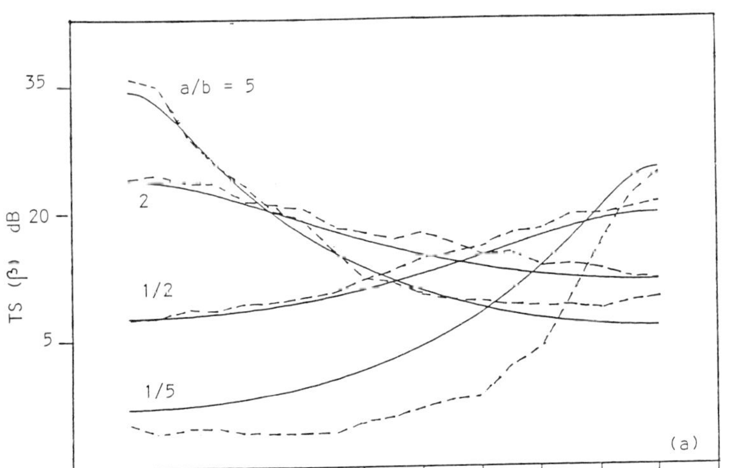

Fig.43. Target strength of steel spheroids (aspect ratios 5, 2, 1/2, 1/5) vs. angle of incidence. Solid lines: formulas of geometrical acoustics. Dashed lines: from specular reflection pulses.

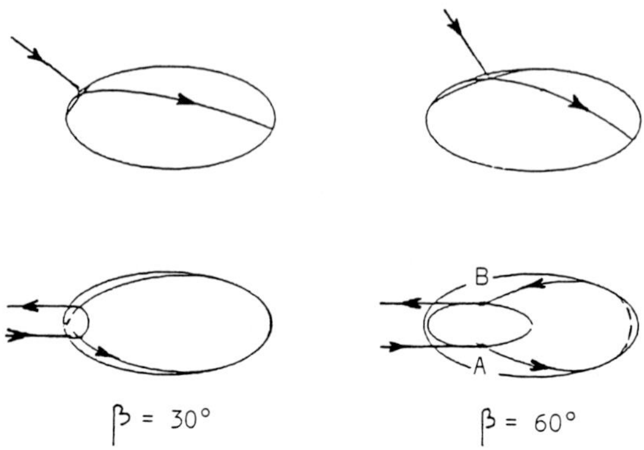

Fig.44. Examples of geodesic paths of surface waves on spheroids.

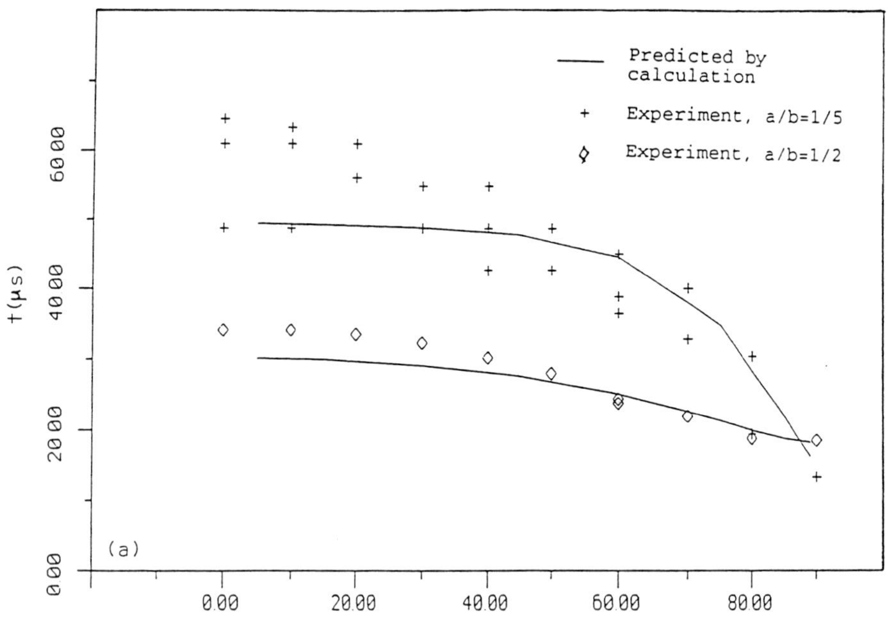

Fig.45. Arrival time of the first elastic echo vs. angles of incidence, for 5:1 and 2:1 prolate spheroids.

Acknowledgements

The authors are profoundly grateful to Prof. X. L. Bao for his invaluable collaborative effort at Catholic University, and for the use of his as yet unpublished figures,40, 42, and 43-45, and to Dr. M. F. Werby for his essential contributions. Further thanks go to J. D. Alemar, R. Carbo-Fite, P. P. Delsanto, R. Hickling, P. L. Marston, A. Nagl, W. G. Neubauer, J. Niemiec, S. K. Numrich, and E. Rosario, for their interest and contributions, and especially to L. R. Dragonette, and (posthumously) L. Flax, for their collaboration on the establishment of the Resonance Scattering Theory. We are also greatly indebted to the researchers at the Institute LAUE, University of Le Havre, France, notably Prof. J. L. Izbicki, and J. M. Conoir, D. Décultot, F. Lecroq, O. Lenoir, F. Léon, P. Pareige, P. Rembert and N. D. Veksler, and at other French laboratories, notably A. Derem, M. de Billy, M. Deschamps, A. Gérard, B. Hosten, G. Quentin, J. L. Rousselot, J. Sageloli, J.-P. Sessarego, and M. Talmant for the use of their results in the present review. One of us (H.Ü.) wishes to acknowledge many years of support by the Office of Naval Research, U.S. Navy; the Naval Research Laboratory, the David Taylor Research Center, and the Naval Surface Warfare Center. The others (J. R. and G.M.) acknowledge the continuing support of LAUE by the University of Le Havre, the CNRS, the DRET, and the Regional Council of Haute Normandie.

7. References

1. J. J. Faran, *J. Acoust. Soc. Am.* **23** (1951) 405.
2. W. G. Neubauer, R. H. Vogt, and L. R. Dragonette, *J. Acoust. Soc. Am.* **55** (1974) 1123.
3. K. J. Diercks and R. Hickling, *J. Acoust. Soc. Am.* **41** (1967) 380.
4. R. Hickling, *J. Acoust. Soc. Am.* **34** (1962) 1582; ibid **36** 1124 (64).
5. L. Flax, L. R. Dragonette, and H. Überall, *J. Acoust. Soc. Am.* **63** (1978) 723.
6. G. Maze and J. Ripoche, *Rev. Phys. Appl.* **18** (1983) 319.
7. S. K. Numrich, N. H. Dale, and L. R. Dragonette, in *Advances in Fluid-Structure Interaction 1984*, ed. G. C. Everstine and M. K. Au-Young, PVP Vol. 78/AMD Vol. 64 (Amer. Soc. of Mech. Engineers, New York 1984), p. 59.
8. M. de Billy, *J. Acoust. Soc. Am.* **79** (1986) 219.
9. H. Überall (ed.), *Acoustic Resonance Scattering* (Gordon and Breach, Philadelphia, 1992).
10. N. D. Veksler, *Resonance Acoustic Spectroscopy* (Springer, Berlin and Heidelberg, 1993).
11. S. K. Numrich and H. Überall, *Phys. Acoustics* **21** (Academic Press, New York, 1992) p. 235.
12. M. C. Junger, *J. Acoust. Soc. Am.* **24** (1952) 366.
13. G. Breit and E. P. Wigner, *Phys. Rev.* **49** (1936).
14. E. P. Wigner, *Phys. Rev.* **70** (1946) 15.
15. H. Feshbach, D. G. Peaslee, and V. F. Weisskopf, *Phys. Rev.* **71** (1947) 145.
16. W. Franz, *Z. Naturfosch.* **A9** (1954) 705.
17. G. V. Frisk, J. W. Dickey and H. Überall, *J. Acoust.Soc.Am.* **58** (1975) 996.
18. G. V. Frisk and H. Überall, *J. Acoust. Soc. Am.* **59** (1976) 46.
19. J. W. Dickey, G. V. Frisk, and H. Überall, *J. Acoust. Soc. Am.* **59**, (1976) 1339.
20. H. Überall, L. R. Dragonette, and L. Flax, *J. Acoust. Soc. Am* **61** (1977) 711.
21. J. W. Dickey and H. Überall, *J. Acoust. Soc. Am.* **63** (1978) 319.
22. See, e.g., R. G. Newton, *The Complex j-plane* (Benjamin, 1964).
23. R. D. Doolittle, H. Überall, and P. Ugincius, *J. Acoust. Soc. Am.* **43** (1968) 1.
24. M. F. Werby, "Recent developments in scattering from submerged elastic and rigid targets," in Ref. 9.
25. S. G. Kargl and P. L. Marston, *J. Acoust. Soc. Am.* **88** (1990) 1114.
26. N. D. Veksler, *Acustica* **76** (1992) 1.
27. J. Ripoche and G. Maze, " A new acoustic spectroscopy", in Ref. 9.
28. W. E. Howell, S. K. Numrich, and H. Überall, *J. Acoust. Soc. Am.* **78** (1985) 1125.
29. W. E. Howell, S. K. Numrich, and H. Überall, *IEEE Trans. Ultras., Ferroel., Freq. Contr.* **UFFC-34** (1987) 22.

30. S. K. Numrich, W. E. Howell, J. V. Subrahmanyam, and H. Überall, *J. Acoust. Soc. Am.* **80** (1986) 1161.
31. J. de Lustrac and R. Carbó-Fite, *Acustica* **20** (1986) 21.
32. N. D. Veksler, *Information Analysis in Hydroelasticity* (Acad. Sci. Estonian SSR, Tallinn, 1982).
33. N. GESPA, *La Diffusion Acoustique*, ed. B. Poirée (CEDOCAR, Paris, 1987).
34. W. E. Howell and H. Überall, *Proc. IEEE* **73** (1985) 1138.
35. W. E. Howell and H. Überall, *IEEE Trans. Antennas Propag.* **AP-38** (1990) 293.
36. G. Maze, J. L. Izbicki, J. Ripoche, A. Nagl, H. Überall, and K. B. Yoo, *J. Acoust. Soc. Am.* **80** (1986) 295.
37. G. Maze, B. Taconet, and J. Ripoche, *Phys. Lett.* **84a** (1981) 309.
38. G. Maze, B. Taconet, and J. Ripoche, *Revue du CETHEDEC* **72** (1982) 103.
39. G. Maze and J. Ripoche, *J. Acoust. Soc. Am.* **73** (1983) 41.
40. J. Ripoche, G.Maze, and J. L. Izbicki, *J. Nondestr. Eval.* **5** (1981) 69.
41. A. Derem, *Revue du CETHEDEC* **58** (1979) 43.
42. A Derem, J. L. Rousselot, G. Maze, J. Ripoche, and A. Faure, *Acustica* **50** (1982) 39.
43. S. K. Numrich, L. R. Dragonette, and L. Flax, in *Elastic Wave Scattering and Propagation,* eds. V. K. Varadan and V. V. Varadan (Ann Arbor Science, Ann Arbor MI, 1982).
44. H. Überall, *Traitement du Signal* **2**, No. 5sp. (1985) 479.
45. M. F. Werby, Y. J. Stoyanov, J. W. Dickey, M. Keskin, J. M. D'Archangelo, A. Nagl, N. J. Stoyanov, J. J. Castillo, and H. Überall, *J. Acoust.* **3** (1990) 201.
46. K. L. Williams and P. L. Marston, *J. Acoust. Soc. Am.* **79** (1986) 1702.
47. H. Überall, L. R. Dragonette, and L. Flax, *J. Acoust. Soc. Am.* **61** (1977) 711.
48. R. Hickling, "Early work on calculating and measuring the elastic response of underwater acoustic targets", in Ref. 9.
49. H. Überall, J. George, A. R. Farhan, G. Mezzorani, A. Nagl, and K. A. Sage, *J. Acoust. Soc. Am.* **66** (1979) 1161.
50. G. Maze, J. Ripoche, A. Derem, and J. L. Rousselot, *Acustica* **55** (1984) 69.
51. G. R. Barnard and C. M. McKinney, *J. Acoust. Soc. Am.* **33** (1961) 226.
52. K. J. Diercks, W. R. King, and C. W. Horton, *J. Acoust. Soc. Am.* **33** (1961) 1652.
53. C. W. Horton, W. R. King, and K. J. Diercks, *J. Acoust. Soc. Am.* **34** (1962) 1929.
54. Y. H. Pao and W. Sachse, *J. Acoust. Soc. Am.* **56** (1974) 1478.
55. H. Überall and G. C. Gaunaurd, *Appl. Phys. Lett.* **39** (1981) 362.
56. H. Überall, G. C. Gaunaurd, and J. D. Murphy, *J. Acoust. Soc. Am.* **72** (1982) 1014.
57. D. G. Vasil'ev, *Sov. Phys. Dokl.* **24** (1979) 720.

58. B. L. Merchant, A. Nagl, and H. Überall, *IEEE Trans. Antenn. Propag.* **AP-37** (1989) 629.
59. M. Talmant, PhD thesis, University of Paris 7, 1987.
60. M. Talmant and G. Quentin, *J. Appl. Phys.* **63** (1988) 1857.
61. M. Talmant, G. Quentin, J. L. Rousselot, J. V. Subrahmanyam, and H. Überall, *J. Acoust. Soc. Am.* **84** (1988) 681.
62. L. R. Dragonette, R. H. Vogt, L. Flax, and W. G. Neubauer, *J. Acoust. Soc. Am.* **55** (1974) 1130.
63. L. R. Dragonette, S. K. Numrich, and L. J. Frank, *J. Acoust. Soc. Am.* **69** (1981) 1186.
64. C. Gazanhes, J.-P. Sessarego, J.-P. Hérault, and J. Léandre, *Acustica* **52** (1983) 265.
65. J.-P. Sessarego, J. Sageloli, C. Gazanhes, and J. Léandre, in N. GESPA, *La Difusion Acoustique*, ed. B. Poirée (CEDOCAR, Paris, 1987).
66. M. Zakharia, P. Flandrin, J. Sageloli, and J.-P. Sessarego, *J. Acoust.* **1** (1988) 185.
67. P. Flandrin and J.-P. Sessarego, *J. Phys.* **51** Suppl., Colloque No. 2 (1990) 707.
68. N.-C. Yen, L. R. Dragonette, and S. K. Numrich, *J. Acoust. Soc. Am.* **87** (1990) 2359.
69. P. D. Thorne, T. J. Brudner, and K. R. Waters, *Proc. Inst. Acoust.* **15** Part 3 (1993) 597.
70. M. Talmant, H. Überall, R. D. Miller, M. F. Werby, and J. W. Dickey, *J. Acoust. Soc. Am.* **86** (1989) 278.
71. H. Überall, B. Hosten, M. Deschamps, and A. Gérard, *J. Acoust. Soc. Am.* (in press).
72. A. Grabowska, *Arch. Acoust.* **4** (1979) 57.
73. R. Stoneley, *Proc. Roy. Soc. London* **A106** (1924) 416.
74. J. C. Scholte, *Proc. Kon. Akad. Von Wetensch., Amsterdam* **45** (1942) part I, 20; Part II, 159.
75. G. S. Sammelmann, D. H. Trivett and R. H. Hackman, *J. Acoust. Soc. Am.* **85** (1989) 114.
76. J. P. Sessarego, J. Sageloli, P. Flandrin, and M. Zakharia, in *Wavelets. Time-Frequency Methods and Phase Space*, ed. J. M. Combes, A. Grossmann, and P. Tchamitchian (Springer, Berlin, 1989).
77. G. Maze, J. L. Izbicki, and J. Ripoche, *J. Acoust. Soc. Am.* **77** (1985) 1352.
78. X. L. Bao, H. Überall, and J. Niemiec, *Appl. Mech. Rev.* **44** part 2 (1991) S16.
79. J. L. Izbicki, G. Maze, and J. Ripoche, *J. Acoust. Soc. Am.* **80** (1986) 1215.
80. J. L. Izbicki, G. Maze, and J. Ripoche, *Acustica* **61** (1986) 137.
81. J. L. Rousselot, *Acustica* **58** (1985) 291.

82. F. Léon, PhD thesis, Univeristy of Le Havre, 1994; F. Léon, F. Lecroq, D. Décultot, and G. Maze, *J. Acoust. Soc. Am.* **91** (1992) 1388.
83. A. Nagl, H. Überall, P.P. Delsanto, J.D. Alemar, and E. Rosario, *Wave Motion* **5** (1983)235.
84. H. Überall; P. J. Moser, B. L. Merchant, A. Nagl, K. B. Yoo, S. H. Brown, J. W. Dickey and J. M. D' Archangelo, *J. Appl. Phys.* **58** (1985) 2109.
85. J. M. Conoir, P. Rembert, O. Lenoir, and J. L. Izbicki, *J. Acoust. Soc. Am.* **93** (1993) 1300.
86. G. Maze, F. Léon, J. Ripoche, A. Klauson, J. Metsaveer, and H. Überall, *Acustica* (in press).
87. See, e.g., P. H. E. Meijer and E. Bauer, *Group Theory* (North Holland, Amsterdam, 1965).
88. Y. P. Guo, *J. Acoust. Soc. Am.* **91** (1992) 926.
89. A. Klauson and J. Metsaveer, *J. Acoust. Soc. Am.* **91** (1992) 1834.
90. F. Lecroq, PhD thesis, University of Le Havre, 1990.
91. G. Maze, J. L. Izbicki, and F. Lecroq, "Acoustic scattering from finite cylinders", in Ref. 9.
92. F. Lecroq, F. Léon, D. Décultot, and G. Maze, *Acustica* **74** (1991) 2185.
93. R. D. Miller, E. T. Moyer, H. Huang, and H. Überall, *J. Acoust. Am.* **89** (1991) 2185.
94. P. J. Moser, H. Überall, and J. R. Yuan, *J. Acoust. Soc. Am.* **94** (1993) 3342.
95. R. Hackman, G. Sammelmann, K. Williams, and D. Trivett, *J. Acoust. Soc. Am.* **83** (1988) 1255.
96. H. Überall, Y. Stoyanov, A. Nagl, M. Werby, S. Brown, J. Dickey, S. Numrich, and J. D'Archangelo, *J. Acoust. Soc. Am.* **81** (1987) 312.
97. G. Maze, F. Lecroq, D. Décultot, J. Ripoche, S. Numrich, and H. Überall, *J. Acoust. Soc. Am.* **90** (1991) 3271.
98. G. Maze, D. Décultot, F. Lecroq, J. Ripoche, X. L. Bao, and H. Überall, *J. Acoust. Soc. Am. (in press).*
99. R. J. Urick, *Principles of Underwater Sound* (McGraw Hill, New York, 1975).
100. L. M. Brekhovskikh, *Waves in Layered Media* (Academic Press, New York, 1960).

Wave Motion, Intelligent Structures and Nonlinear Mechanics, pp. 92–111
edited by A. Guran and D. J. Inman
Series on Stability, Vibration and Control of Structures: Vol. 1

THEORY OF FIBER ACOUSTIC WAVEGUIDES

AHMAD SAFAAI-JAZI

Bradley Department of Electrical Engineering
Virginia Polytechnic Institute and State University
Blacksburg, Virginia 24061, USA

ABSTRACT

Propagation of elastic waves in fiber-optic-type acoustic waveguides is addressed. A simplified theory based on small differences between the acoustic parameters of the core and cladding regions is presented. The simplified theory reveals the important features and provides considerable insight into the transmission properties of fiber acoustic waveguides. Two types of modes, guided shear-type and leaky longitudinal type, are examined. Eigenvalues, phase and group velocities, cutoff conditions, and field distributions for shear and longitudinal modes are analyzed, and numerical results for lower order modes are presented. Similarities and differences between acoustical and optical modes in glass fiber waveguides are pointed out. Highlights of the analysis for liquid-core fiber acoustic waveguides are also presented.

1. Introduction

Fiber acoustic waveguides are optical-fiber-type structures for use as elastic wave transmission media. Interest in such waveguides stems from their potential applications, particularly in signal processing, sensing, monitoring health and integrity of structures, and nondestructive evaluation of materials. They consist of a core surrounded by a cladding layer which serves to confine the elastic waves to predominantly the core region. As in optical fibers, the parameters of the core and cladding in fiber acoustic waveguides differ from each other only slightly. A small difference between the parameters of the core and cladding becomes mandatory in single-mode acoustic fibers and in situations where signal distortions need be kept low. They may be made of glass materials such as pure and doped fused silica, using optical fiber fabrication techniques. Fiber acoustic waveguides thus inherit many advantages of optical fibers, including small size, light weight, flexibility, isolation from mechanical supports, and low cost. The compatibility of fiber acoustic waveguides with fiber optic technology and the greater simplicity in signal processing due to propagation speeds much slower than the velocity of light enhance the prospect of fiber acoustic sensors. One notable exception, however, is the acoustic attenuation which is much higher than the optical loss, but this should not cause much concern because most applications do not require very long fiber acoustic waveguides. Other materials such as metals may also be considered for the fabrication of acoustic fibers; however, at higher frequencies non-metallic materials such as silica glass offer lower transmission losses and are thus preferred.

Fiber acoustic waveguides belong to the broader family of circularly cylindrical acoustic waveguides which have been studied during the past three decades[1-10]. The exact analysis of circular acoustic waveguides for cases of no cladding (rod), finite cladding, and infinite cladding are available[1,4,9]. These exact analyses, with the possible exception of application to torsional modes, lead to complicated solutions for fiber acoustic waveguides which cannot be utilized conveniently for engineering design applications and also provide little physical insight into their propagation properties. On the other hand, in fiber-optic-type acoustic waveguides, the core and cladding materials are doped and pure fused silica, respectively, whose acoustic parameters, for instance shear and longitudinal velocities and densities, differ less than a few percent. The small differences between the parameters of the core and cladding permit a much more simplified analysis of fiber acoustic waveguides[11-14]. This chapter is largely devoted to the development and discussion of this simplified analysis.

Approximate solutions may be obtained from the exact analysis[11], or directly from the governing wave equations[14]. The theory presented here is based on the latter approach. It is known that solutions of the scalar and vector wave equations for potential functions are, generally, coupled at the core cladding boundary. However, when the acoustic parameters of the core and cladding differ by a small amount, these solutions become nearly uncoupled. The uncoupled solutions for the scalar and vector potential functions form the basis for the simplified theory of fiber acoustic waveguides.

The simplified theory reveals the important features of fiber acoustic waveguides, provides considerable insight into their transmission characteristics and field behaviors, and elucidates the analogies and differences between fiber optic and fiber acoustic waveguides. This theory establishes that: guided modes with real propagation constants (assuming lossless materials) are, to a first approximation, independent of longitudinal velocities of the core and cladding materials and the axial component of the displacement vector is much smaller than the transverse components, thus guided modes are predominantly shear type. The shear-type modes are classified as torsional, radial, and flexural with one, two, and three non-zero displacement components, respectively [11].

It is further revealed that aside from shear-type modes, there exists another set of modes with complex propagation constants. The imaginary part of the propagation constant, which is much smaller than the real part, implies loss of mode power as it propagates along the waveguide. These modes are referred to as leaky and are predominantly longitudinal with the axial component of the displacement vector much larger than the transverse components. The loss of power occurs through radiation into the cladding region. Under the ideal conditions that core and cladding have equal shear velocities and densities, longitudinal modes become true guided modes with real propagation constants[12]. Of course, in practical situations these ideal conditions are not realized, and longitudinal modes become leaky.

In the following sections, fundamentals of elastic wave propagation in circularly cylindrical structures are reviewed and solutions of the scalar and vector wave equations and boundary conditions are summarized. A simplified theory is presented, shear-type and longitudinal-type modes are addressed, and phase and group velocities and cutoff frequencies for all modes are examined. Similarities and differences of acoustical and optical modes in glass fiber waveguides are discussed. The analysis of liquid-core fiber acoustic waveguide is also presented with particular attention paid to the transmission properties of flexural and Stoneley-type modes. Finally, highlights of elastic wave propagation in fiber acoustic waveguides are summarized.

2. Formulation and Basic Equations

Consider a circularly cylindrical structure consisting of two layers; namely, a core of radius a and a cladding extending to infinity in the radial directions. This structure serves as a model for an actual fiber acoustic waveguide with finite cladding. When the strength of elastic waves, which decays in an exponential manner away from the core, becomes negligible at the outer wall of the cladding, the infinite cladding used in this model is justified. The core and cladding are assumed to be homogeneous, isotropic, and lossless. The ith medium, with i=1 and 2 referring to the core and cladding regions, respectively, is characterized by a shear velocity V_{si}, a longitudinal velocity V_{Li}, and a density ρ_i. A cylindrical coordinate system (r,φ,z) in which the z-axis coincides with the waveguide axis is chosen. Figure 1 illustrates the geometry and coordinates used in the analysis of fiber acoustic waveguide. Only time harmonic waves propagating in the positive z-direction are considered. Therefore, the z and time dependencies of the acoustic fields take the form $exp[j(\omega t-\beta z)]$, where ω is the angular frequency and β is the axial propagation constant. The particle displacement vector $\vec{u}$ in an isotropic and homogeneous medium with density ρ and Lame constants μ and λ is the solution of the Navier equation

$$\mu\nabla^2\vec{u}+(\lambda+\mu)\nabla(\nabla\cdot\vec{u})=\rho\partial^2\vec{u}/\partial t^2\,. \tag{1}$$

The Lame constants are related to the shear and longitudinal velocities and density according to $V_s=(\mu/\rho)^{1/2}$ and $V_L=[(\lambda+2\mu)/\rho]^{1/2}$.

The standard procedure for solving Eq. (1) is to write $\vec{u}=\nabla\phi+\nabla\times\vec{\psi}$, then ϕ and $\vec{\psi}$ satisfy the scalar and vector wave equations, respectively[1]. Alternatively, $\vec{u}$ may be expressed as $\vec{u}=\vec{u}_a+\vec{u}_b$ such that $\nabla\times\vec{u}_a=\nabla\cdot\vec{u}_b=0$. Obviously, $\vec{u}_a=\nabla\phi$ and $\vec{u}_b=\nabla\times\vec{\psi}$. Substituting for $\vec{u}$ in terms of ϕ and $\vec{u}_b$ in Eq. (1) and noting that for the assumed z and time dependencies $\partial^2/\partial z^2=-\beta^2$ and $\partial^2/\partial t^2=-\omega^2$, yields

$$\nabla_t^2\phi+h^2\phi=0 \tag{2}$$

$$\nabla_t^2\vec{u}_{tb}+k^2\vec{u}_{tb}=0 \tag{3}$$

where

$$h^2=(\omega^2/V_L^2)-\beta^2 \tag{4}$$

$$k^2=(\omega^2/V_s^2)-\beta^2. \tag{5}$$

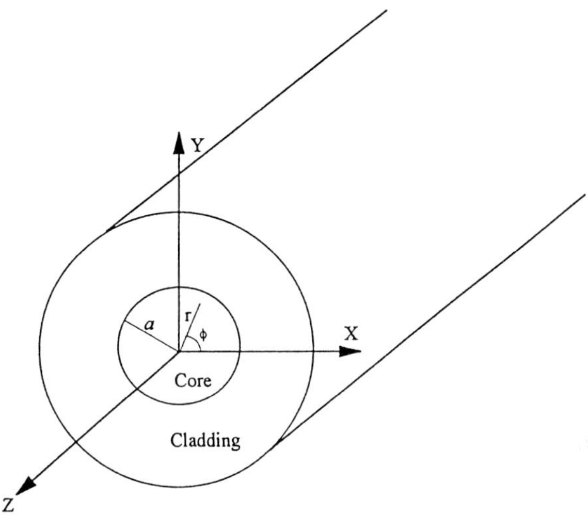

Fig. 1. Geometry and coordinates for fiber acoustic waveguide

In Eq. (3), $\vec{u}_{tb}$ is the transverse component of $\vec{u}_b$, and ∇_t^2 represents the transverse Laplacian operator of scalar type in Eq. (2) and of vector type in Eq. (3). Using the relation $\nabla \cdot \vec{u}_b = 0$, the axial component of $\vec{u}_b$ is obtained from

$$u_{zb} = (-j/\beta)\,\nabla_t \cdot \vec{u}_{tb}. \tag{6}$$

It is evident that the vector $\vec{u}_b$ can be completely specified by solving Eq. (3) and then using Eq. (6). In fiber acoustic waveguides, Eqs. (2) and (3) need be solved once for the core region with V_{L1} and V_{s1} replacing V_L and V_s and once for the cladding region with V_{L2} and V_{s2} used in these equations.

Generally, solutions of Eqs. (2) and (3) are coupled at the core-cladding boundary. In certain special cases, however, they are uncoupled; that is, each solution satisfies the boundary conditions independently. Torsional modes are an example of uncoupled solutions. Also, in the absence of any boundary when the core and cladding become the same, uncoupled solutions exist. These solutions are shear and longitudinal uniform plane waves. An interesting case of uncoupled solutions occurs when core and cladding regions have equal shear velocities and densities. This case which forms the basis for the analysis of longitudinal modes will be discussed in more details in Sec. 4. In fiber-optic-type acoustic waveguides, the core and cladding are nearly the same, thus making the solutions of Eqs. (2) and (3) nearly uncoupled. The uncoupled solution of Eq. (3) can describe the behavior of shear-type modes satisfactorily. For longitudinal-type modes, while Eq. (2) alone can adequately describe some properties such as phase and group velocities, but coupled solutions should be used to account for the leaky nature of these modes[12-14].

3. Shear-Type Modes

The displacement vector for core-guided modes in fiber acoustic waveguides is, to a first approximation, the solution of Eq. (3) subject to boundary conditions at the core-cladding interface. Expressing $\vec{u}_{tb} = u_r\hat{a}_r + u_\varphi\hat{a}_\varphi$, Eq. (3) is resolved into two coupled differential equations in terms of u_r and u_φ. These equations are solved simultaneously following the method employed in Ref. 15. Boundary conditions are then applied in order to determine the amplitude coefficients in terms of one coefficient chosen as an independent coefficient, and also obtain a dispersion relation from which the eigenvalue β can be calculated.

3.1. Boundary Conditions

At the core-cladding boundary, $r = a$, the displacement vector $\vec{u}$ and the stress components T_{rr}, $T_{r\varphi}$, and T_{rz}, given by Eqs. (A1) to (A3) in the Appendix, must be continuous. However, the continuity of u_r, u_φ, u_z, T_{rr}, $T_{r\varphi}$, and T_{rz} at $r = a$ is equivalent to the continuity of u_r, u_φ, $\partial u_r/\partial r$, and $\partial u_\varphi/\partial r$ at $r = a$. In other words, boundary conditions for the approximate shear-type modes amount to continuity of the transverse components of the displacement vector and their normal derivatives at the boundary $r = a$.

3.2. Field and Dispersion Relations

Carrying out the steps outlined above, the following results for displacement components and dispersion equations are obtained.

$$u_r = \begin{cases} A_n J_{n\pm 1}(k_1 r)Q_1, & r < a \\ \mp A_n \zeta_n K_{n\pm 1}(k_2 r)Q_1, & r > a \end{cases} \tag{7}$$

$$u_\varphi = \begin{cases} \mp A_n J_{n\pm 1}(k_1 r)Q_2, & r < a \\ A_n \zeta_n K_{n\pm 1}(k_2 r)Q_2, & r > a \end{cases} \tag{8}$$

$$u_z = \begin{cases} jA_n(k_1/\beta)J_n(k_1 r)Q_1, & r < a \\ jA_n(k_2/\beta)\zeta_n K_n(k_2 r)Q_1, & r > a \end{cases} \tag{9}$$

$$\pm\frac{J_{n\pm 1}(k_1 a)}{(k_1 a)J_n(k_1 a)} + \frac{K_{n\pm 1}(k_2 a)}{(k_2 a)K_n(k_2 a)} = 0 \tag{10}$$

where

$$k_i = |(\omega^2/V_{si}^2) - \beta^2|^{\frac{1}{2}}, \quad i = 1,2 \tag{11}$$

$$\zeta_n = [k_1 J_n(k_1 a)] \;/\; [k_2 K_n(k_2 a)] \tag{12}$$

$$Q_1 = exp[j(\omega t - \beta z)] \cdot \begin{cases} cos\ (n\varphi) \\ sin\ (n\varphi) \end{cases} \tag{13}$$

$$Q_2 = exp[j(\omega t - \beta z)] \cdot \begin{cases} -\,sin\ (n\phi) \\ cos\ (n\phi) \end{cases} \tag{14}$$

In Eqs. (7) to (9), A_n is a constant amplitude coefficient, J_n and K_n are, respectively, the Bessel function of the first kind and the modified Bessel function of the second kind, and n is an integer number known as the circumferential order or azimuthal number.

Equation (10), referred to as the dispersion equation, is a function of ω, β, n, and waveguide parameters a, V_{s1}, and V_{s2}. Solving this equation numerically, the propagation constant β is obtained. Different values of β correspond to different modes. The shear-type modes are classified as torsional, radial and flexural modes.

3.3. Torsional, Radial, and Flexural Modes

3.3.1. Torsional Modes

When $n = 0$, it is evident from Eqs. (13) and (14) that either $Q_1 = Q_0$ and $Q_2 = 0$, or $Q_1 = 0$ and $Q_2 = Q_0$, where $Q_0 = exp[j(\omega t - \beta z)]$. The two cases correspond to the upper and lower lines in Eqs. (13) and (14). For the latter case, putting $Q_1 = n = 0$ in Eqs. (7) to (9) results in

$$u_r = u_z = 0 \tag{15}$$

$$u_\varphi = \begin{cases} -\,A_0 J_1(k_1 r) Q_0, & r < a \\ A_0 \zeta_0 K_1(k_2 r) Q_0, & r > a. \end{cases} \tag{16}$$

In this case, the only non-zero component of the displacement vector is u_φ and the modes are referred to as torsional. The corresponding dispersion relation is obtained from Eq. (10) with $n = 0$. The result is

$$\frac{J_1(k_1 a)}{(k_1 a) J_0(k_1 a)} + \frac{K_1(k_2 a)}{(k_2 a) K_0(k_2 a)} = 0. \tag{17}$$

Torsional modes are designated as T_{om}, where the first subscript is the azimuthal number n which is zero here, while the second subscript $m = 1,2,3...$ indicates the order of the mode.

3.3.2. Radial Modes

When $Q_2 = n = 0$ and $Q_1 = Q_0$, the field expressions in Eqs. (7) to (9) reduce to

$$u_\varphi = 0 \tag{18}$$

$$u_r = \begin{cases} A_0 J_1(k_1 r) Q_0, & r < a \\ -A_0 \zeta_0 K_1(k_2 r) Q_0, & r > a \end{cases} \tag{19}$$

$$u_z = \begin{cases} jA_0(k_1/\beta) J_0(k_1 r) Q_0, & r < a \\ jA_0(k_2/\beta)\zeta_0 K_0(k_2 r) Q_0, & r > a. \end{cases} \tag{20}$$

Although both radial and axial components of the particle displacement vector are non-zero, but, as discussed below, the axial component is generally much smaller than the radial component, and the modes are predominantly radial. The dispersion relation for radial modes is obtained from Eq.(10) with $n = 0$ used in it. The result is identical to Eq. (17) which is the dispersion relation for torsional modes. This means that torsional and radial modes of the same order have (approximately) the same propagation constant and thus are (nearly) degenerate. Radial modes are designated as R_{om}, where the subscripts follow the same definitions given for torsional modes.

3.3.3. Flexural Modes

When $n \geq 1$, all three components of the particle displacement vector exist and the modes are referred to as flexural. Two groups of flexural modes corresponding to the upper and lower signs in Eqs. (7) to (10) exist. The group corresponding to the lower signs represents a family of flexural modes designated as $F_{n,(2m-1)}$, $m = 1,2,3,...$, while the group with the upper signs represents another family designated as $F_{n,2m}$, $m = 1,2,3....$ The lowest order mode, usually referred to as the fundamental or dominant mode, is the F_{11} mode. This mode, as discussed later, has no cutoff and can exist at all frequencies.

3.4. Particular Features of Shear-Type Modes

An examination of Eqs. (9) and (20) reveals that the u_z component for both flexural and radial modes contains k_1/β or k_2/β term, which upon using $\beta = \omega/V_p$, V_p being the phase velocity, and Eq. (11), reduces to $[(V_p/V_{s1})^2 - 1]^{1/2}$ or $[1-(V_p/V_{s2})^2]^{1/2}$, respectively. Furthermore, for guided modes the phase velocity is in the range $V_{s1} < V_p < V_{s2}$ and in the light of the fact that $(V_{s2} - V_{s1})/V_{s1} \ll 1$, it becomes clear that both terms k_1/β and $k_2/\beta \ll 1$. Therefore, the axial component u_z is generally much smaller than the transverse components u_r and u_φ, except only in the neighborhood of the possible zeros of $J_{n\pm1}(k_1 r)$, $r < a$, and also zeros of

$cos(n\phi)$ or $sin(n\phi)$, where u_r and u_φ are themselves small. This implies that all guided modes in fiber acoustic waveguides are predominantly shear. It is further observed that field and dispersion relations are independent of the longitudinal velocities of the core and cladding. This property arises from the nearly uncoupled solutions of the scalar and vector wave equations in Eqs. (2) and (3).

It was pointed out earlier that T_{0m} and R_{0m} modes are degenerate. Further examination of dispersion relations for flexural modes reveals that $F_{2,(2m-1)}$ modes are also degenerate with torsional and radial modes. Degeneracy also occurs between $F_{n,2m}$ and $F_{n+2,(2m-1)}$, $n \geq 1$ modes. For example, (T_{01}, R_{01}, F_{21}), (T_{02}, R_{02}, F_{22}), (F_{12}, F_{31}), (F_{22}, F_{41}), etc., are groups of degenerate modes. The $F_{1,(2m-1)}$ modes are not degenerate with any other modes. Thus, the fundamental F_{11} mode which belongs to this family is a non-degenerate mode[11-15]. The degeneracy of different modes allows the dispersion relations of all shear-type modes to be cast into a single equation expressed as

$$\frac{J_\nu(k_1 a)}{(k_1 a)J_{\nu-1}(k_1 a)} + \frac{K_\nu(k_2 a)}{(k_2 a)K_{\nu-1}(k_2 a)} = 0, \tag{21}$$

where

$$\nu = \begin{cases} n+1, & \text{for } F_{n,2m} \text{ modes} \\ n-1, & \text{for } F_{n,(2m-1)} \text{ modes} \\ 1, & \text{for } T_{0m},\ R_{0m}, \text{ and } F_{2,(2m-1)} \text{ modes} \end{cases}$$

3.5. *Phase and Group Velocities*

Phase and group velocities, denoted as V_p and V_g, respectively, are determined using the following relations

$$V_p = \omega/\beta \tag{22}$$

$$V_g = (d\beta/d\omega)^{-1}, \tag{23}$$

where β is now the solution of Eq. (21). Variations of β versus ω are dependent upon waveguide parameters. Thus, for the same mode, different $\beta-\omega$ plots for different waveguides are needed. In order to obtain characteristic curves which would be universal and thus applicable to different fiber acoustic waveguides with different parameters, we introduce normalized frequency Ω and normalized propagation constant b as defined below

$$\Omega = \omega a\left(\frac{1}{V^2_{s1}} - \frac{1}{V^2_{s2}}\right)^{\frac{1}{2}}, \tag{24}$$

$$b = \frac{\frac{1}{V^2_p} - \frac{1}{V^2_{s2}}}{\frac{1}{V^2_{s1}} - \frac{1}{V^2_{s2}}} \simeq \frac{V_{s2} - V_p}{V_{s2} - V_{s1}}. \tag{25}$$

It can be easily shown that in terms of Ω and b, $k_1 a = \Omega\sqrt{1-b}$ and $k_2 a = \Omega\sqrt{b}$. Hence, Eq. (21) becomes solely a function of Ω and b, allowing universal dispersion characteristics for shear-type modes. Once b is known, V_p and V_g are readily obtained from

$$V_p = V_{s2} - (V_{s2} - V_{s1})\, b \tag{26}$$

$$V_g = V_{s2} - (V_{s2} - V_{s1})\, \frac{d(b\Omega)}{d\Omega}. \tag{27}$$

Variations of b and $d(b\Omega)/d\Omega$ versus Ω for several lower order modes are shown in Figs. 2 and 3. Using these plots in conjunction with Eqs. (26) and (27), the phase and group velocities for a given mode in a given waveguide are readily obtained. The frequencies corresponding to $b = 0$ or $V_p = V_{s2}$ are referred to as the cutoff frequencies below which the exponential decay of fields away from the core is lifted, the propagation constants become complex, and modes become of radiation type. Cutoff frequencies may be obtained from the dispersion relation in the limit of $V_p \rightarrow V_{s2}$, or equivalently $k_2 \rightarrow 0$. Upon letting $k_2 \rightarrow 0$ in Eq. (21) and using the small argument approximations for the modified Bessel functions K_ν given by Eqs. (A-7) and (A-8) in the Appendix, the following result is obtained

$$J_{\nu-1}(\Omega_c) = 0, \tag{28}$$

where Ω_c is the normalized cutoff frequency and ν follows the same definition given in Eq. (21) for different modes. Accordingly, the normalized cutoff frequencies for the group of T_{0m}, R_{0m}, and $F_{2,(2m-1)}$ modes are the roots of $J_0(\Omega_c) = 0$. The first root of this equation is $\Omega_c = 2.405$ which defines the frequency range for single-mode operation. The cutoffs of $F_{n,(2m-1)}$ modes are obtained from $J_{n-2}(\Omega_c) = 0$. Except for n=2, $\Omega_c = 0$ is always a root; however, a closer examination of the dispersion relation in Eq. (21) indicates that $\Omega = 0, b = 0$ satisfies this equation only when $n = 1$. Thus, only the F_{11} mode possesses a zero cutoff frequency and is the sole guided mode in a single-mode fiber acoustic waveguide. Finally, the cutoff frequencies of $F_{n,2m}$ modes are determined from $J_n(\Omega_c) = 0$. A more detailed analysis of cutoffs in cylindrical acoustic waveguides is elsewhere[16].

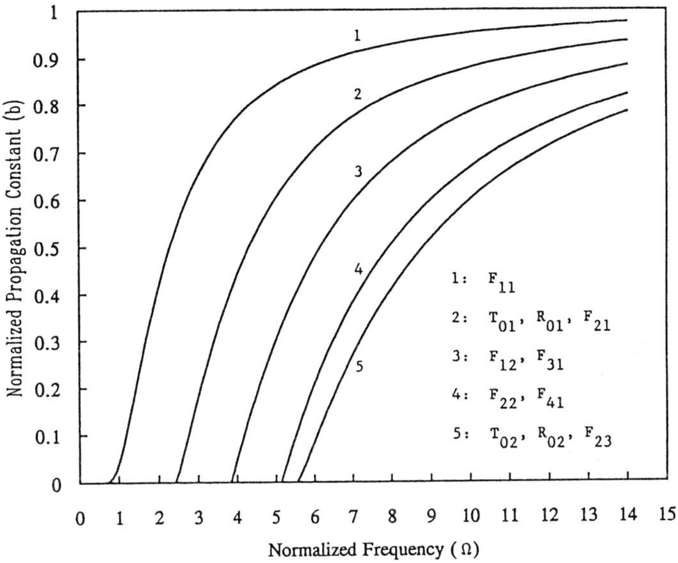

Fig. 2. Variations of normalized propagation constant b versus normalized frequency Ω for several lower order shear-type modes of fiber acoustic waveguide

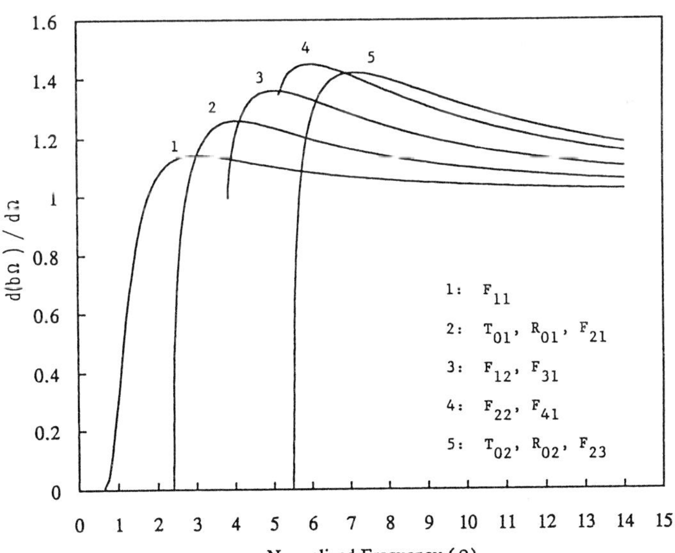

Fig. 3. Variations of $d(b\Omega)/d\Omega$ versus Ω for several lower order shear-type modes of fiber acoustic waveguide. $d(b\Omega)/d\Omega$ is used in Eq. (27) to calculate group velocity.

4. Leaky Longitudinal Modes

It was pointed out in Sec. 2 that solutions of Eqs. (2) and (3) in fiber acoustic waveguides are approximately uncoupled. Under the peculiar conditions of $V_{s1} = V_{s2}$ and $\rho_1 = \rho_2$; that is, when core and cladding have the same shear velocities and densities, exact uncoupled solutions exist for Eqs. (2) and (3). In fact, under these conditions the displacement vector $\vec{u}_b$ does not see any boundary, and the solution of Eq. (3) is simply a uniform shear plane wave. This solution may also be viewed as a transformation of guided shear-type modes into a uniform shear plane wave in the process of reducing the difference V_{s2}–V_{s1} to zero. As this difference is reduced, cutoff frequencies of all modes, except that of the F_{11} mode which is always zero, increase and eventually become infinity when $V_{s1} = V_{s2}$. In this process, the F_{11} mode transforms into a uniform shear plane wave and all other modes become permanently cutoff.

The solution of Eq. (2) leads to field expressions which represent a new type of modes referred to longitudinal modes. The propagation constants of such modes are, strictly speaking, real only when $V_{s1} = V_{s2}$ and $\rho_1 = \rho_2$. In practical fiber acoustic waveguides, these conditions are not ideally realized, with the implication that the propagation constant becomes complex, but with the imaginary part much smaller than the real part. Thus, longitudinal modes are leaky and decay as they propagate along the fiber.

4.1. Displacement Vector

Solving Eq. (2) and using $\vec{u} = \nabla\phi$, the components of the displacement vector for leaky longitudinal modes are summarized as

$$u_r = \begin{cases} B_n(h_1/\beta)J'_n(h_1r)Q_1, & r < a \\ B_n(h_2/\beta)\eta_n K'_n(h_2r)Q_1, & r > a \end{cases} \tag{29}$$

$$u_\varphi = \begin{cases} B_n(h_1/\beta)[(n/h_1r)J_n(h_1r)]Q_2, & r < a \\ B_n(h_2/\beta)\eta_n[(n/h_2r)K_n(h_2r)]Q_2, & r > a \end{cases} \tag{30}$$

$$u_z = \begin{cases} -jB_nJ_n(h_1r)Q_1, & r < a \\ -jB_n\eta_nK_n(h_2r)Q_1, & r > a \end{cases} \tag{31}$$

where

$$h_1 = [(\omega/V_{L1})^2 - \beta^2]^{1/2}, \tag{32}$$

$$h_2 = [\beta^2 - (\omega/V_{L2})^2]^{1/2}, \tag{33}$$

$$\eta_n = J_n(h_1a)/K_n(h_2a), \tag{34}$$

B_n is a constant amplitude coefficient, and Q_1 and Q_2 follow the same definitions as those in Eqs. (13) and (14). Boundary conditions require that the displacement components u_r, u_φ, and u_z and the stress components T_{rr}, $T_{r\varphi}$, and T_{rz} be continuous at $r = a$. Continuity of u_r results in

$$\frac{J_n(h_1 a)}{(h_1 a)J_{n-1}(h_1 a)} + \frac{K_n(h_2 a)}{(h_2 a)K_{n-1}(h_2 a)} = 0, \tag{35}$$

which is the dispersion equation for longitudinal modes. It can be shown that with Eq. (35) all remaining boundary conditions are exactly satisfied for the ideal case of equal shear velocities and densities for the core and cladding[12]. Equation (35) has real roots only when $V_{L1} < V_p < V_{L2}$, where $V_p = \omega/\beta$ is the phase velocity. When $n = 0$, the component u_φ vanishes, indicating that all modes are non-torsional. Generally, the modes are radial-axial for $n = 0$ and flexural for $n \geq 1$. However, in fiber acoustic waveguides where the difference between longitudinal velocities of the core and cladding is small, the transverse components u_r and u_φ are small compared with the axial component u_z and the modes are predominantly longitudinal. This may be seen more clearly if u_r and u_φ are expressed as

$$u_r = \tfrac{1}{2}B_n(h_1/\beta)[J_{n-1}(h_1 r) - J_{n+1}(h_1 r)]Q_1, \quad r < a \tag{36}$$

$$u_\varphi = \tfrac{1}{2}B_n(h_1/\beta)[J_{n-1}(h_1 r) + J_{n+1}(h_1 r)]Q_2, \quad r < a. \tag{37}$$

For $r > a$, similar expressions involving K_n function may be written. It is noted that u_r and u_φ both contain the term h_1/β (or h_2/β in the cladding) which is much smaller than unity if $(V_{L2} - V_{L1})/V_{L1} \ll 1$. Thus, except in the neighborhood of certain points where $cos(n\varphi)$, $sin(n\varphi)$ or $J_n(h_1 r)$ is zero and u_z is small, the transverse components are much smaller than the axial component and the modes are essentially longitudinal. Longitudinal modes are designated as L_{nm} with $n = 0, 1, 2, 3...$ and $m = 1, 2, 3...$.

4.2. Phase and Group Velocities

Phase and group velocities for longitudinal modes are obtained from the same relations used for shear-type modes; namely Eqs. (22) and (23), provided that β is the solution of dispersion relation given by Eq. (35). Comparison of Eqs. (35) and (21) reveals that they are of the same mathematical form and thus should have the same normalized solutions. The normalized frequency and normalized propagation constant for longitudinal modes are defined as

$$\Omega = \omega a\left(\frac{1}{V_{L1}^2} - \frac{1}{V_{L2}^2}\right)^{1/2} \tag{38}$$

$$b = \frac{\frac{1}{V_p^2} - \frac{1}{V_{L2}^2}}{\frac{1}{V_{L1}^2} - \frac{1}{V_{L2}^2}} \simeq \frac{V_{L2} - V_p}{V_{L2} - V_{L1}} . \tag{39}$$

The dispersion relation in Eq. (35) is readily expressed in terms of Ω and b if we note that $h_1 a = \Omega\sqrt{1-b}$ and $h_2 a = \Omega\sqrt{b}$. Having determined b, the phase and group velocities are obtained from

$$V_p = V_{L2} - (V_{L2} - V_{L1})\, b \tag{40}$$

$$V_g = V_{L2} - (V_{L2} - V_{L1}) \frac{d(b\Omega)}{d\Omega}, \tag{41}$$

where variations of b and $d(b\Omega)/d\Omega$ versus Ω are identical to those for shear-type modes as described by Figs. 2 and 3. In using these figures, one should note that the curves for F_{11}, (R_{01}, T_{01}, F_{21}), (F_{12}, F_{31}), etc., correspond to L_{01}, L_{11}, L_{21}, etc., respectively.

Cutoff frequencies correspond to $b = 0$ or $V_p = V_{L2}$ and are obtained from Eq. (35) in the limit of $h_2 \to 0$. The result is

$$J_{n-1}(\Omega_c) = 0, \quad \text{for } L_{nm} \text{ modes.} \tag{42}$$

The first mode with $n = 0$, the L_{01} mode, has a zero cutoff frequency and is the fundamental longitudinal mode. It is emphasized that $\Omega_c = 0$ is a valid cutoff frequency only for $n = 0$ in Eq. (42). This is because $b = 0$ and $\Omega = 0$ satisfy Eq. (35) only when $n = 0$.

4.3. Leaky Nature of Longitudinal Modes

In real fiber acoustic waveguides, the conditions $V_{s1} = V_{s2}$ and $\rho_1 = \rho_2$ which give rise to ideal longitudinal modes do not hold exactly. Therefore, the longitudinal modes discussed above are only partially guided and their energy leaks into the cladding as they propagate along the fiber. The rigorous analysis of leaky longitudinal modes requires the coupled solutions of wave Eqs. (2) and (3). However, since the differences between shear velocities and densities in fiber acoustic waveguides are small, leaky modes differ from ideal non-leaky modes only slightly. In other words, the propagation properties of leaky modes are those of ideal modes subject to a small perturbation.

To understand the leaky nature of longitudinal modes, let us examine a component of the displacement vector, say u_r in the cladding region. From the rigorous analysis of an acoustic waveguide with infinitely thick cladding, the expression for u_r, when $V_{L1} < V_p < V_{L2}$, is written as[12],

$$u_r = [A h_2 K_n(h_2 r) + B H_n'^{(2)}(k_2 r) + C(n/r) H_n^{(2)}(k_2 r)] Q_1, \qquad r > a \tag{43}$$

where A, B, and C are constant coefficients. The presence of the Hankel function $H_n^{(2)}$ in Eq. (43) implies that the dispersion relation and hence the propagation constant β are complex. Moreover, for $k_2 r \gg 1$, the function $H_n^{(2)}(k_2 r)$ behaves as $(1/\sqrt{k_2 r}).exp(-jk_2 r)$ which clearly represents an outgoing wave traveling in the radial direction. When $V_{s1} = V_{s2}$ and $\rho_1 = \rho_2$, the coefficients B and C vanish and radiation in the radial direction does not take place. For $|V_{s2} - V_{s1}|/V_{s1} \ll 1$ and $|\rho_2 - \rho_1|/\rho_1 \ll 1$, however, these coefficients are no longer zero but very small, and β is a complex quantity with an imaginary part much smaller than the real part. The imaginary part of the propagation constant accounts for the attenuation of longitudinal modes due to radiation. In summary, the solutions for leaky longitudinal modes are well approximated by those of ideal modes provided that β in $exp(-j\beta z)$ is replaced with $\beta - j\alpha$, where α is the attenuation coefficient.

The calculation of complex propagation constant for leaky longitudinal modes may be carried out using a complex root search extended to the dispersion relation resulting from coupled solutions, a perturbation method applied to this dispersion relation, or a Taylor series approximation method. The perturbation and Taylor series methods have been used and it has been shown that the numerical results of the two methods are in excellent agreement[15]. Details of the perturbation and Taylor series methods are quite involved and are not presented here. Instead, the numerical results for an example case which have been obtained using the perturbation analysis are reported and discussed.

A fiber acoustic waveguide whose core and cladding are Corning glasses 7971($V_{s1} = 3625$ m/s, $V_{L1} = 5736$ m/s, $\rho_1 = 2213$ kg/m^3) and 7940($V_{s2} = 3764$ m/s, $V_{L2} = 5933$ m/s, $\rho_2 = 2202$ kg/m^3), respectively, has been fabricated and studied for delay line application[8]. This fiber is a suitable candidate for numerical analysis of leaky modes. Figures 4 and 5 show variations of $\Delta V_p/V_p$ and the normalized attenuation coefficient, defined as $\bar{\alpha} = -20log_{10}[exp(-\alpha a)]$, versus normalized frequency for L_{01} and L_{11} modes. ΔV_p is regarded as the error incurred in the phase velocity of an assumed ideal mode due to non-ideal conditions; that is, non-zero differences between shear velocities and densities of the core and cladding. It is observed from Figs. 4(a) and 4(b) that, for the example acoustic fiber under consideration, $\Delta V_p/V_p$ is at worst not more than 0.12%, indicating that for all practical purposes the phase velocity of a leaky mode and hence the real part of its propagation constant can be accurately approximated with that of the corresponding ideal mode obtained from Eq. (35).

Variations of $\bar{\alpha}$ versus frequency for L_{01} and L_{11} modes are shown in Figs. 5(a) and 5(b), respectively. To obtain the actual attenuation in dB for a given length of the fiber, the value of $\bar{\alpha}$ should be multiplied by l/a, where a is the core radius and l is the fiber length. Both $\Delta V_p/V_p$ and $\bar{\alpha}$ exhibit fluctuations with frequency. Comparison of Figs. 4(a) and 5(a) and also Figs. 4(b) and 5(b) indicates that fluctuations of $\Delta V_p/V_p$ and $\bar{\alpha}$ have the same periods and thus are expected to have the same origin. From a mathematical point of view, these fluctuations are due to functions $J_n(k_1 a)$, $J_n(k_2 a)$, and $Y_n(k_2 a)$ which behave nearly sinusoidally for $k_1 a$ and $k_2 a$ much larger than unity. These functions appear in the coupled solutions (for example see Eq. (43) and note that $H_n^{(2)} = J_n - jY_n$) of the displacement vector and hence in the dispersion relation. At higher frequencies, $\Delta V_p/V_p$ and the amplitude of fluctuations in $\bar{\alpha}$ decrease with frequency, ultimately vanishing at very high frequencies. In summary, it is the attenuation coefficient which strongly affects the propagation of longitudinal modes and has a determining role in the selection of the operating frequency. Furthermore, wideband signals suffer severe amplitude distortions, thus longitudinal modes are useful for narrow-band applications.

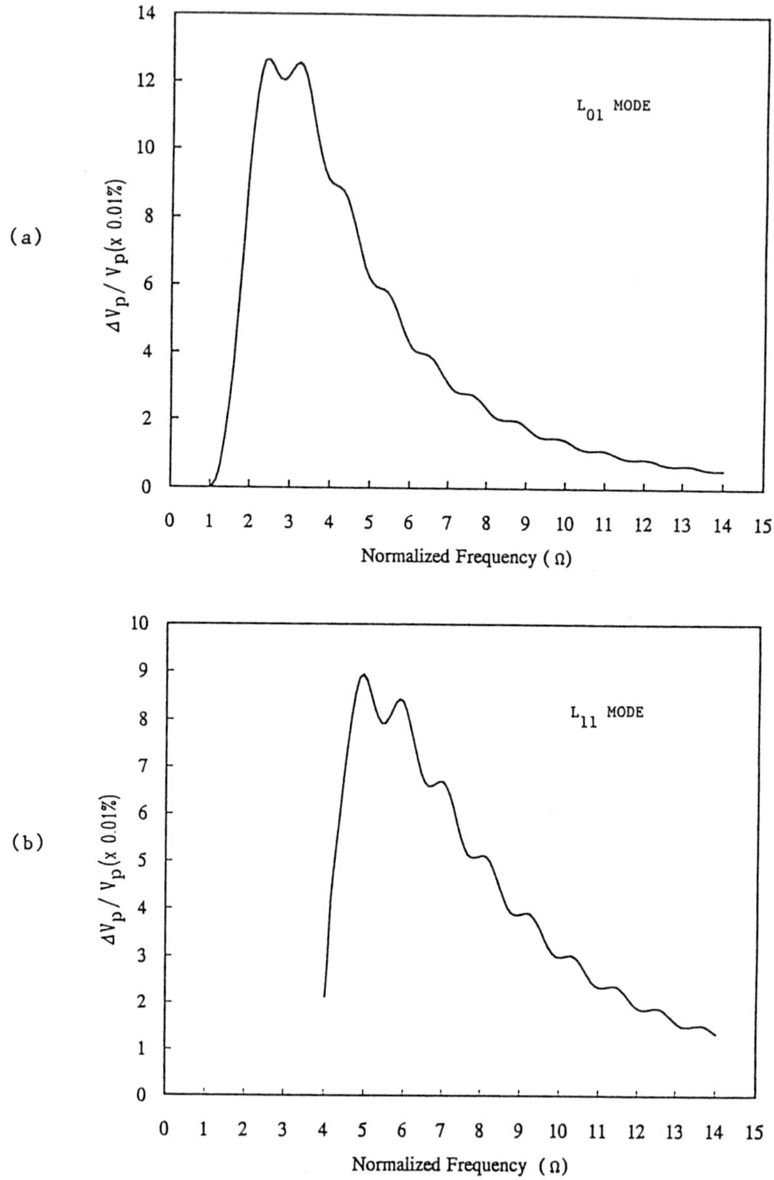

Fig. 4. Variations of phase velocity error $\Delta V_p/V_p$ versus normalized frequency Ω for (a) L_{01} mode, (b) L_{11} mode. The parameters of the waveguide are: $V_{s1} = 3625$ m/s, $V_{L1} = 5736$ m/s, $\rho_1 = 2213$ kg/m^3, $V_{s2} = 3764$ m/s, $V_{L2} = 5933$ m/s, $\rho_2 = 2202$ kg/m^3.

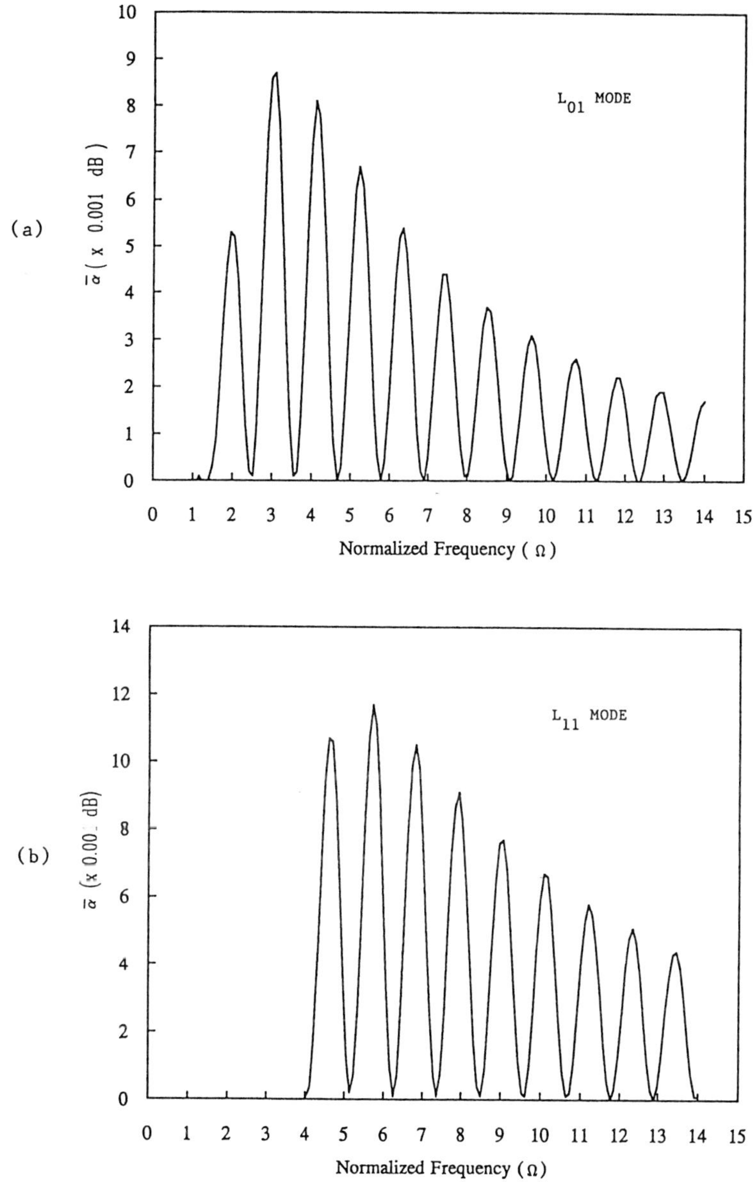

Fig. 5. Variations of normalized attenuation coefficient $\overline{\alpha}$ versus normalized frequency Ω for (a) L_{01}, mode, (b) L_{11} mode. The parameters of the waveguide are the same as those in Fig. 4.

5. Liquid – Core Waveguides

Liquid-core acoustic waveguides are of interest in temperature sensing and may also be used as dispersive delay lines. A simplified theory similar to that discussed for solid-core fiber acoustic waveguides is not generally applicable, because the parameters of the liquid core and solid cladding are usually considerably different. An exact analysis of liquid-core cylindrical acoustic waveguides is available[17]. Here, highlights of this analysis and important features of the guided modes are presented. In the core region, the shear velocity is zero and thus only the scalar potential function contributes to the particle displacement vector. In the cladding region, however, both scalar and vector potential functions contribute to the particle displacement vector. The components of the particle displacement vector are summarized as[17].

$$u_r = \begin{cases} \widehat{A}_n h_1 Z_n(h_1 r) Q_1, & r < a \\ [A_n h_2 K'_n(h_2 r) + B_n \beta K'_n(k_2 r) + C_n (n/r) K_n(k_2 r)] Q_1, & r > a \end{cases} \tag{44}$$

$$u_\varphi = \begin{cases} -\widehat{A}_n (n/r) Z_n(h_1 r) Q_2, & r < a \\ [A_n (n/r) K_n(h_2 r) + B_n (n\beta/k_2 r) K_n(k_2 r) + C_n k_2 K'_n(k_2 r)] Q_2, & r > a \end{cases} \tag{45}$$

$$u_z = \begin{cases} -j\widehat{A}_n \beta Z_n(h_1 r) Q_1, & r < a \\ -j[A_n \beta K_n(h_2 r) - B_n k_2 K_n(k_2 r)] Q_1, & r > a \end{cases} \tag{46}$$

where $\widehat{A}_n$, A_n, B_n, and C_n are constant coefficients, and

$$Z_n = \begin{cases} J_n, & V_p > V_{L1} \\ I_n, & V_p < V_{L1}. \end{cases} \tag{47}$$

In Eqs. (44) to (46), Q_1, Q_2, h_1, h_2, and k_2 follow the same definitions given by Eqs. (13), (14), (32), (33), and (11), respectively. At the liquid-solid interface, $r = a$, the normal components of displacement and stress vectors must be continuous and the tangential components of stress in the cladding must vanish. That is, $u_{r1} = u_{r2}$, $T_{rr1} = T_{rr2}, T_{r\varphi 2} = T_{rz2} = 0$ at $r = a$, where subscripts 1 and 2 refer to the core and cladding regions, respectively. The imposition of these conditions results in a system of four equations. Setting the determinant of coefficients in this system of equations equal to zero results in a dispersion relation from which β for guided modes can be determined. Examination of the dispersion relation reveals that the modes are radial-axial for $n = 0$ and flexural for $n \geq 1$. Thus, torsional modes are not supported by liquid-core acoustic waveguides. Cutoff frequencies correspond to $V_p = V_{s2}$ (or $k_2 = 0$). In solid-core fiber acoustic waveguides, as discussed earlier, only one mode with zero cutoff frequency exists. The analysis of cutoffs in liquid-core waveguides shows that there are two modes without cutoff; namely the first modes with $n=0$ and $n=1$[17].

The behavior of modes for the case $V_p < V_{L1}$, corresponding to $Z_n = I_n$ in Eqs. (44) to (46), is worth more attention. The radial distribution of particle displacement vector is described by the modified Bessel function $I_n(h_1 r)$ which for $h_1 r \gg 1$ increases in an exponential manner with r. However, if the radial dimension is measured from the core-cladding boundary, one may consider $I_n(h_1 r)$ as an exponentially decaying function of the distance measured from $r=a$ towards the axis of the waveguide. Thus, for $V_p < V_{L1}$ and $h_1 r \gg 1$ the fields decay exponentially away from the boundary $r=a$ in both core and cladding regions. This behavior is the characteristic of "Stoneley" or "interface" type waves. Hence, the modes with $V_p < V_{L1}$ are referred to as Stoneley-type modes. Stoneley-type modes are rare in solid-core acoustic waveguides.

6. Comparison of Fiber Optic and Fiber Acoustic Waveguides

Comparison of elastic wave propagation in fiber acoustic waveguides and electromagnetic wave propagation in optical fibers is quite revealing. The approximate dispersion relation of shear-type acoustic modes given by Eq. (21) is identical to that of electromagnetic modes in optical fibers[18]. This allows us to draw an analogy between the acoustic and electromagnetic modes. Assuming that the particle displacement vector is analogous to the electric field vector, it can be stated that torsional (T_{om}), radial (R_{om}), odd flexural ($F_{n,(2m-1)}$), and even flexural ($F_{n,2m}$) modes are, respectively, analogous to transverse electric (TE_{om}), transverse magnetic (TM_{om}), *HE* hybrid (HE_{nm}), and *EH* hybrid (EH_{nm}) modes. A comprehensive account of mode designation in cylindrical dielectric waveguides has been presented before[19].

The analogous modes have, within the accuracy of approximate analyses for both fiber optic and fiber acoustic waveguides, the same normalized propagation constants, the same transverse field distributions, and the same normalized cutoff frequencies. This analogy, however, is limited to shear-type modes only. Leaky longitudinal modes are purely an acoustic phenomenon and have no analogous optical counterpart. More details on similarities and differences between optical and acoustic fibers are given elsewhere[20].

7. Summary

The theory of elastic wave propagation in fiber acoustic waveguides was presented. This theory, while approximate, can predict with sufficient accuracy important propagation properties such as phase and group velocities, cutoff frequencies, field distributions, etc. The theory also reveals some important features of the modes of propagation which could not be readily grasped from the exact solutions. A summary of important results is given below.

- Universal normalized propagation characteristics applicable to both shear-type and longitudinal-type modes were obtained.
- For shear-type modes, the axial component of the displacement vector is

generally much smaller than the transverse components.

- Propagation properties of shear-type modes, such as phase and group velocities, cutoff frequencies, and power flow, are essentially independent of longitudinal velocities.
- The fundamental shear-type mode (F_{11}) has a zero cutoff frequency and exhibits a plane-wave like transverse distribution.
- Shear-type modes do not lose power due to radiation. Thus, if waveguide materials are assumed to be lossless, the propagation constants of these modes are real.
- For longitudinal-type modes, the axial component of the displacement vector is generally much larger than the transverse components.
- Propagation properties of longitudinal-type modes such as phase and group velocities, cutoff frequencies, and power flow are essentially independent of the shear velocities. The attenuation coefficient associated with the radiation of energy into the cladding, however, depends strongly on the differences between shear velocities and densities of the core and cladding materials.
- Longitudinal modes lose power due to radiation. Thus, even for the ideal case of lossless materials, the propagation constants of these modes are complex and they are referred to as leaky modes.
- The fundamental longitudinal mode is independent of φ, its displacement vector has no φ component, and has no cutoff frequency.
- In liquid-core acoustic waveguides, two modes with zero cutoff frequency exist, modes cannot be of torsional type, and Stoneley-types modes are supported.

8. Appendix

The stress components used in boundary conditions are:

$$T_{rr} = \lambda \nabla^2 \phi + 2\mu S_{rr} \tag{A-1}$$

$$T_{r\varphi} = 2\mu S_{r\varphi} \tag{A-2}$$

$$T_{rz} = 2\mu S_{rz} \tag{A-3}$$

where S_{rr}, $S_{r\varphi}$, and S_{rz} are strain components given by

$$S_{rr} = \partial u_r / \partial \varphi \tag{A-4}$$

$$S_{r\varphi} = \tfrac{1}{2}[r\partial(u_\varphi / r)/\partial r + (1/r)\partial u_r / \partial \varphi] \tag{A-5}$$

$$S_{rz} = \tfrac{1}{2}(\partial u_r / \partial z + \partial u_z / \partial r) \tag{A-6}$$

The small argument approximations for the modified Bessel function K_n are:

$$K_0(x) = ln(2/\gamma x), \quad x \ll 1 \tag{A-7}$$

$$K_n(x) = (1/2)(n\text{-}1)!/(x/2)^n, \text{ for } n>0, \quad x \ll 1 \tag{A-8}$$

where $\gamma = 1.781$ is the Euler's constant.

9. References

1. T. R. Meeker and A.H. Meitzler, in *Physical Acoustics,* ed. W. P. Mason, (Academic Press, New York, 1964), Vol. 1, pt. A.
2. I. L. Gelles, *J. Acoust. Soc. Amer.,* **39**, (1966) 1111.
3. R. A. Waldron, *IEEE Trans., Microwave Theory and Techniques,* **17**, (1969) 893.
4. J. L. Lai, E. H. Dowell, and T. R. Tauchert, *J. Acoust. Soc. Amer.,* **49**, (1971) 220.
5. B. A. Auld, *Acoustic Fields and Waves in Solids,* (Wiley, New York, 1973), Vol. II, Ch. 10.
6. G. D. Boyd, L. A. Coldren, and R. N. Thurston, *Appl. Phys. Letts.* **26**, (1975) 31.
7. R. N. Thurston, *IEEE Trans., Sonics and Ultrasonics,* **23**, (1976) 154.
8. G. D. Boyd, L. A. Coldren, and R. N. Thurston, *IEEE Trans. Sonics and Ultrasonics,* **24**, (1977) 246.
9. R. N. Thurston, *J. Acoust. Soc. Amer.,* **64**, (1978) 1.
10. J. B. Keller, in *Modern Problems in Elastic Wave Propagation,* eds. J. Miklowitz and J. D. Achenbach, (Wiley, New York, 1978).
11. A. Safaai-Jazi, C. K. Jen, and G. W. Farnell, *IEEE Trans. Ultrasonics, Ferroelectrics, and Frequency Control,* **33**, (1986) 59.
12. C. K. Jen, A. Safaai-Jazi, and G. W. Farnell, *IEEE Trans. Ultrasonics, Ferroelectrics, and Frequency Control,* **33**, (1986) 634.
13. A. Safaai-Jazi, C. K. Jen, and G. W. Farnell, *Elect. Letts.* **23**, (1987) 350.
14. A. Safaai-Jazi, and R. O. Claus, *IEEE Trans. Ultrasonics, Ferroelectrics, and Frequency Control,* **35**, (1988) 619.
15. A. Safaai-Jazi, C. K. Jen, and G. W. Farnell, *Fiber Acoustic Waveguides,* Final Report, (Department of Electrical Engineering, McGill University, Montreal, Canada, 1986).
16. A. Safaai-Jazi, C. K. Jen, and G. W. Farnell, *IEEE Trans. Ultrasonics, Ferroelectrics, and Frequency Control,* **33**, (1986) 69.
17. A. Safaai-Jazi, C. K. Jen, G. W. Farnell, and J. D. N. Cheeke, *J. Acoust. Soc. Amer.* **81**, (1987) 1273.
18. A. W. Snyder and J. D. Love, *Optical Waveguide Theory,* (Chapman and Hall, London, 1973), p. 318.
19. A. Safaai-Jazi and G. L. Yip, *Radio Science, Special Issue on Fiber and Integrated Optics,* **12**, (1977) 603.
20. C. K. Jen, *Proc. IEEE Ultrasonics Symposium,* San Francisco, CA, (1986) 1128.

Wave Motion, Intelligent Structures and Nonlinear Mechanics, pp. 112–134
edited by A. Guran and D. J. Inman
Series on Stability, Vibration and Control of Structures: Vol. 1

STRUCTURAL VIBRATION SUPPRESSION VIA PARAMETRIC CONTROL ACTIONS -- PIEZOELECTRIC MATERIALS WITH REAL-TIME SEMI-ACTIVE NETWORKS

K. W. WANG
Mechanical Engineering Department
The Pennsylvania State University,
University Park, PA 16802
USA

ABSTRACT

A new structural vibration control concept, using piezoelectric materials with real-time semi-active electrical networks, is investigated. With an adjustable circuit, a variable resistance and the changing rate of a variable inductance are used as control inputs. Based on energy analysis, parametric control schemes are discussed for different case studies. The major idea is to reduce the total system energy while constraining the energy flowing into the main structure. The concept is examined and illustrated on an adaptive beam example.

1. Background

Vibration in today's increasingly high speed and light weight complex structures severely limits their performance. Excessive vibration can cause catastrophic failures or diminished life. To attain international competitiveness, the needs for effective vibration controls are rapidly increasing in various areas of engineering.

Structural vibration suppression techniques can be categorized into three areas, namely, passive, active, and semi-active (also known as adaptive-passive) controls. In a passive system, dissipation and reactive mechanisms are introduced, such as viscous dampers, frictional dampers, composite damping materials, and dynamic absorbers. The control parameters are synthesized through off-line design techniques and no on-line feedback actions are used. The advantages of this approach are that the devices are usually relatively simple, and the system will always be stable. However, since it produces fixed designs, the control will not be optimal when the system or the operating condition changes.

Active control has been of popular interest in recent years. Force or torque inputs from actuators are usually used to suppress vibration amplitudes based on on-line measurements from sensors. The advantage of an active system is that the control can adapt for system changes, and thus can be much more effective than passive devices. The major difficulties in this kind of approach are that the power requirements are usually relatively high, and the system could be de-stabilized due to uncertainties from model parameter errors, actuator/sensor dynamics, and measurement noise. Active structural controls could also be further complicated by interactions between the controller and the uncontrolled modes (residual modes), the so-called spillover problem[1,2].

Comparing the above two methods, one can see the merit of combining the feedback concept in active systems with adaptable reactive and energy dissipation devices whose parameters can be varied according to the control commands. This semi-active approach normally requires less power than the fully-active systems. Also, since energy is almost always being dissipated, this method is less sensitive to spillover and is more stable than the active approach. In other words, it has the advantages of both the passive (stability, low power requirement) and active (feedback/feedforward actions) systems. With the recent development of adaptive materials, on-line parameter variations are physically achievable.

In the past few years, researchers have found that piezoelectric ceramics can be effective active vibration control actuators. Also because of high piezoelectric coupling, these materials have been explored for their ability to damp vibration passively. The general approach involves shunting the terminals of the piezoelectric element across some kind of electrical circuit. In the simplest case, the shunting circuit is simply a resistor; in the more general case, the shunting circuit may include inductors as well as resistors.

In the resistive shunting situation (Figure 1a), damping is a direct result of the electromechanical coupling inherent in piezoelectric materials[3,4,5]. When the base structure deforms, the piezoelectric element is strained, producing a voltage across its electrical terminals. When these terminals are shunted by a resistor, energy will be dissipated via Joule heating as current flows through the resistor.

Resonant shunting circuits can be used to reduce resonant structural response (Figure 1b). The circuit includes inductors as well as resistors[3,6,7]. With this configuration, the coupling of a second order RLC electrical circuit with a second order structure gives rise to behavior analogous to that of vibration absorber systems.

These original passive configurations can be modified to become semi-active devices. In concept, it involves the use of a variable resistance and/or a variable inductance in the network. Since the time constants for varying parameters in electrical circuits are normally much shorter than those experienced in mechanical systems, this type of actuators is very promising for on-line semi-active applications. A comprehensive study on real-time control of this class of adaptive structures has yet to be performed.

(a)

(b)

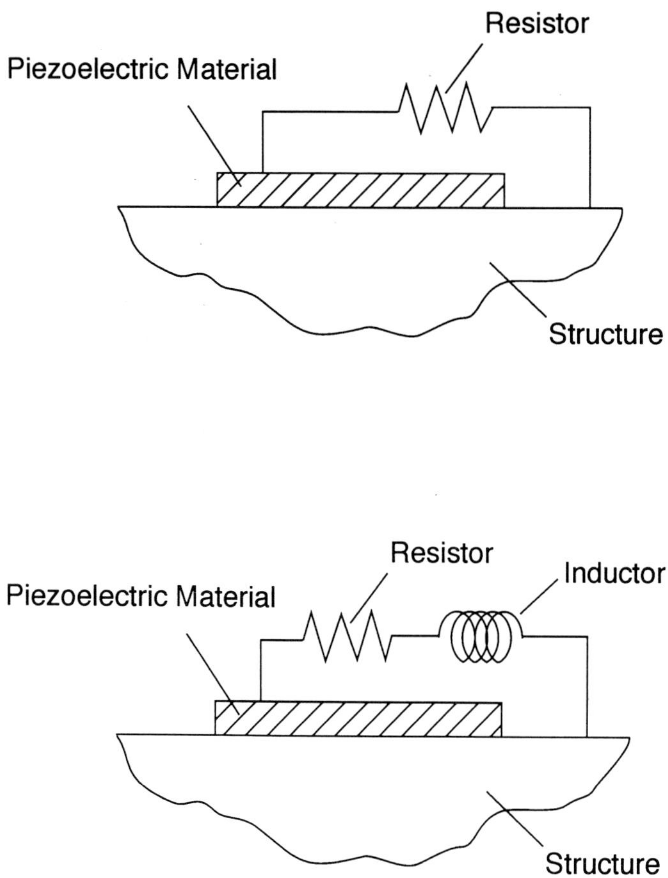

Figure 1. Schematic of (a) resistively-shunted and (b) resonant-shunted piezoelectric-based structures

2. Problem Statement and Objective

While the concept of structural vibration control using semi-active piezoelectric circuits is promising, more research work is needed to realize this idea. Novel methodologies need to be developed to address the unique characteristics of such structures. The major control issues are the system nonlinearities due to state-dependent parameters and actuator constraints (eg. positive inductance and resistance), and the balance between suppressing the main structure vibration versus stabilizing the sub-system (electric circuit) dynamics. The overall goals of this study are to address the above issues and create effective parametric control laws for such systems.

3. System Description

For the purpose of discussion and without losing generality, a cantilever beam with surface-bonded piezoelectric materials (PZT) is used as an example to illustrate the concept (Figure 2). The PZT layers are shunted with an external RL circuit. Here, the variable resistance (R) and inductance (L) are used as control inputs.

The system model is derived based on the following assumptions:

(a) The piezoelectric materials (PZTs) are thin and short comparing to the beam thickness and length.
(b) The inertial effects of the PZTs are negligible.
(c) The poling direction of the PZTs is in the positive w-direction.
(d) Only uniaxial loading of the PZT in the x-direction is considered.
(e) The two pieces of PZTs are connected with the external network in a parallel configuration as shown in Figure 2.
(f) The beam axial motion is negligible comparing to its transverse motion.

The model is derived using the Hamilton's Principle. The derivation is outlined in the Appendix. The equations of motion of the structure can be presented as:

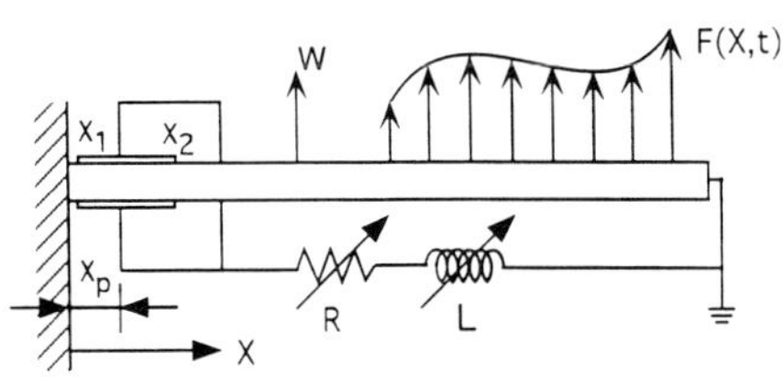

Figure 2. **Schematic of cantilever beam with surface-bonded piezoelectric material and adaptive electrical circuit**

$$m_b \frac{\partial^2 w}{\partial t^2} + c_b \frac{\partial w}{\partial t} + E_b I_b \frac{\partial^4 w}{\partial x^4} + 2E_s I_s \frac{\partial^4 w}{\partial x^4}(H(x-x_1)-H(x-x_2)) +$$
$$4E_s I_s \frac{\partial^3 w}{\partial x^3}(\delta(x-x_1)-\delta(x-x_2)) + [2E_s I_s \frac{\partial^2 w}{\partial x^2} - \frac{M_s h_{31} b_s}{A_s \beta_{33}(x_2-x_1)}(2b_s(x_2-x_1)(RI+\frac{d(LI)}{dt}) + \tag{1}$$
$$2b_s h_{31} M_s \frac{\partial w}{\partial x}\Big|_{x_1}^{x_2})] \frac{d(\delta(x-x_1)-\delta(x-x_2))}{dx} = F(x,t), \; 0<x<L_b$$

the boundary conditions

$$w(0,t) = \frac{\partial w(0,t)}{\partial x} = \frac{\partial^2 w(L_b,t)}{\partial x^2} = \frac{\partial^3 w(L_b,t)}{\partial x^3} = 0 \tag{2}$$

The circuit equation is

$$\frac{d^2(LI)}{dt^2} + \frac{d(RI)}{dt} + \frac{\beta_{33}(h_s-h_b)}{2b_s(x_2-x_1)} I + \frac{h_{31}(h_s^2-h_b^2)}{2}\left(\frac{\partial^3 w}{\partial x^2 \partial t}\Big|_{x=x_p}\right) = 0 \tag{3}$$

Here, $w(x,t)$ is the transverse displacement of the beam, E_b is the beam elastic modulus, E_s is the PZT elastic modulus, L_b is the beam length, m_b is the beam mass per unit length, c_b is a uniform damping constant, $F(x,t)$ is the external disturbance force, and I_b and I_s are the beam and PZT layer moments of inertia, respectively. Also, (x_2-x_1) is the length of the PZT, $x_p = (x_1+x_2)/2$, R is the variable resistance, L is the variable inductance, I is the current, and δ and H are the Dirac Delta and Heavyside's functions, respectively. The other parameters are defined in the nomenclature.

Assume an approximate solution in the form of finite series,

$$w(x,t) = \sum_{i=1}^{N} \phi_i(x) q_i(t) \tag{4}$$

we can discretize equations (1)-(3) into the following form using the Galerkin's method.

$$M_b \frac{d^2 q}{dt^2} + C_b \frac{dq}{dt} + \Omega_0 q - [\Omega_1{}^T q + (\frac{d(LI)}{dt} + RI)\Omega_2] \frac{d(\phi(x_2)-\phi(x_1))}{dx} = \hat{F}(t) \tag{5}$$

$$\frac{d^2(LI)}{dt^2}+\frac{d(RI)}{dt}+\frac{\beta_{33}(h_s-h_b)}{2b_s(x_2-x_1)}I+\frac{h_{31}(h_s^2-h_b^2)}{2}\left(\frac{d^2\phi(x_p)}{dx^2}\right)^T\frac{dq}{dt}=0 \tag{6}$$

where q and ϕ are Nx1 vectors consisting of the generalized displacements $q_i(t)$ and the comparison functions $\phi_i(x)$, respectively. Here, we select $\phi_i(x)$ to be the eigenfunctions of a fixed-free beam. M_b is a diagonal matrix with constant elements m_b, C_b is a diagonal matrix with constant elements c_b, Ω_2 is a scalar:

$$\Omega_2=\frac{2b_s^2M_sh_{31}}{A_s\beta_{33}} \tag{7}$$

The elements of Ω_0 are:

$$\Omega_0(i,i)=E_bI_b\beta_i^4+2E_sI_s\int_{x_1}^{x_2}\left(\frac{d^2\phi_i(x)}{dx^2}\right)^2dx,\ i=1,2,...,N, \tag{8}$$

where β_i is the solution of the characteristic equation

$$\cos(\beta_iL_b)\cosh(\beta_iL_b)=-1, \tag{9}$$

and

$$\Omega_0(i,j)=2E_sI_s\int_{x_1}^{x_2}\frac{d^2\phi_i(x)}{dx^2}\frac{d^2\phi_j(x)}{dx^2}dx,\quad if\ i\neq j. \tag{10}$$

Here, j=1,2,3,....,N. The elements of vector Ω_1 are:

$$\Omega_1(j)=\frac{2b_s^2M_s^2h_{31}^2}{A_s(x_2-x_1)\beta_{33}}\left(\frac{d\phi_j(x_2)}{dx}-\frac{d\phi_j(x_1)}{dx}\right). \tag{11}$$

If consider only the ith mode and let L be a finite constant and c_b=R=0, the frequency response function of the system can be derived:

$$\frac{q_i(j\omega)}{\hat{F}(j\omega)} = \frac{(-C_1 L\omega^2+1)}{(-M\omega^2+K)(-C_1 L\omega^2+1)-C_3 C_2 L\omega^2} \tag{12}$$

where,

$$M=M_b(i,i),\quad K=\Omega_0(i,i)-\Omega_1(i)\frac{d(\phi_i(x_2)-\phi_i(x_1))}{dx},$$
$$C_1=\frac{2b_s(x_2-x_1)}{\beta_{33}(h_s-h_b)},\quad C_3=\Omega_2, \tag{13}$$
$$C_2=\frac{h_{31}b_s(x_2-x_1)(h_s+h_b)}{\beta_{33}}\frac{d^2\phi_i(x_p)}{dx^2}\frac{d(\phi_i(x_2)-\phi_i(x_1))}{dx}.$$

It is clear that one can make $q_i = 0$ by tuning L to be $1/(C_1\omega^2)$, which is a phenomenon typically observed in systems with vibration absorbers.

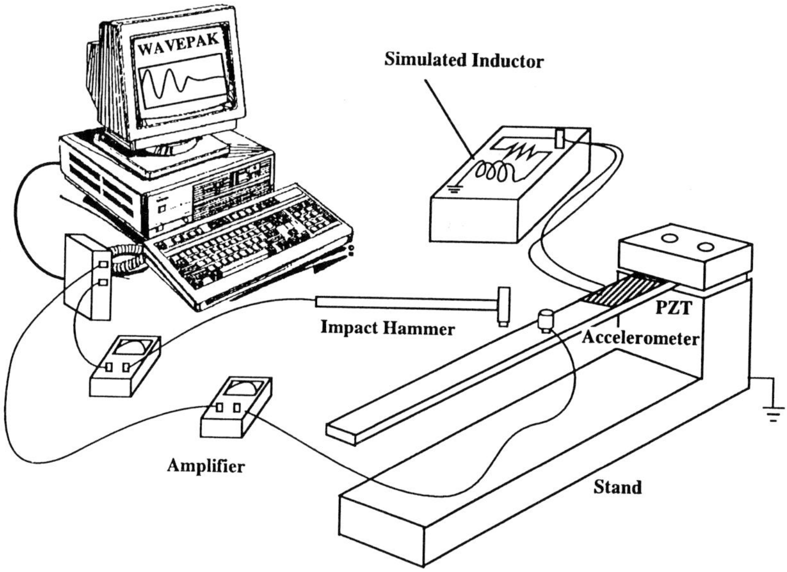

Figure 3. Schematic of experimental set up

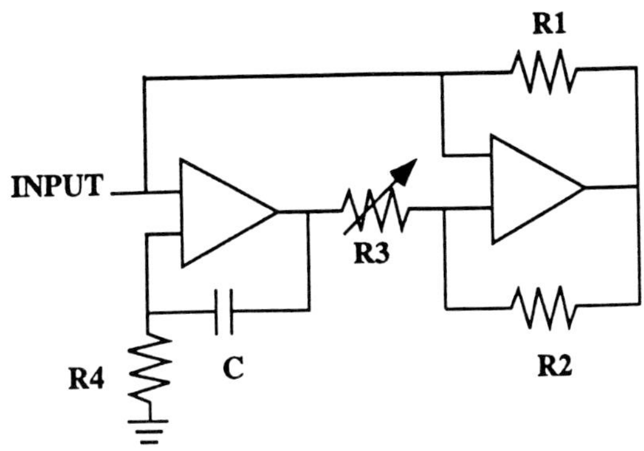

Figure 4. Circuit for simulating inductance

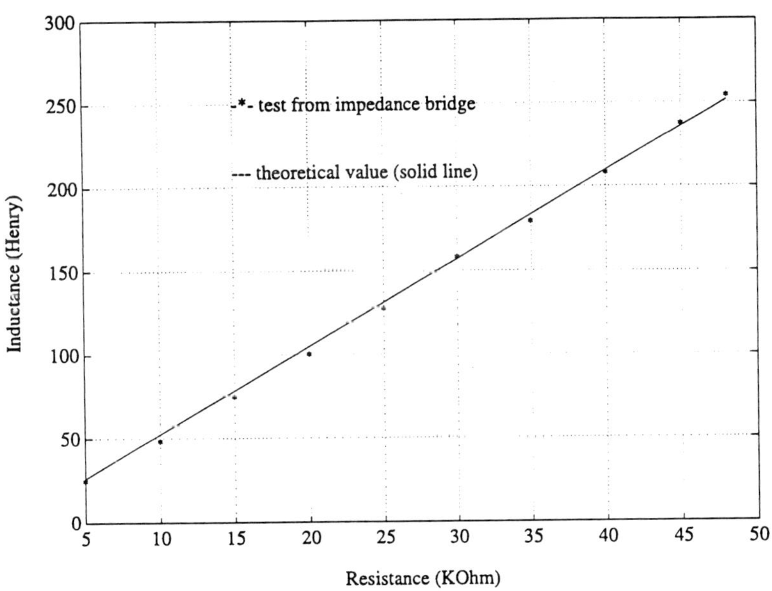

Figure 5. Simulated inductance L vs. variable resistance R_3

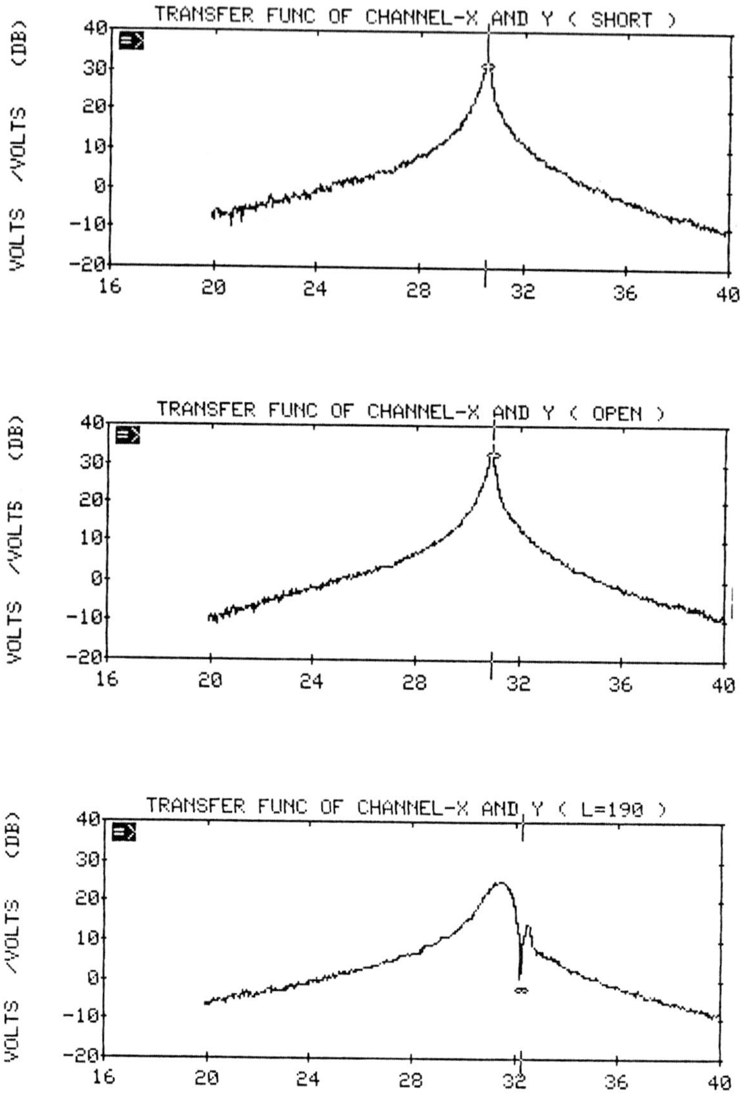

Figure 6. Frequency response function of beam (a) short circuit, (b) open circuit, and (c) L = 190 Henry

4. Experimental Observation

A lab fixture is set up for feasibility study of the concept (Figure 3). The experimental hardware consists of a cantilever beam (6061-T6 aluminum, 30 cm x 2.54 cm x 0.32 cm) with the a pair of piezoelectric ceramics (PZT) bonded near the fixed end. The PZTs were shunted with an external RL circuit. An OP Amp based device presented by Edberg and Bicos[7] was used to simulated the inductance in the shunted circuit (Figure 4). Figure 5 shows that the theoretically predicted inductance

$$L = \frac{R_1 R_3 R_4}{R_2} C \tag{14}$$

compared closely with the inductance measured using an impedance bridge.

Frequency response functions of the structure were identified through impact tests. The beam is excited by an instrumented hammer at the free end. The transverse vibration signals were picked up by an accelerometer. The signals were processed through a PC-based Wavepak Spectrum Analyzer. Figure 6 illustrates the frequency response function for a short circuit case, an open circuit case, and a shunted circuit case with L = 190 Henry. It is shown that the first resonant frequency shifted higher when the circuit is open (compared to short circuit). This is due to the electrical-mechanical coupling effect of the piezoelectric material[3]. When the circuit is shunted with a finite value of L, the absorber effect discussed in the previous section can be observed. As shown in Figure 6, a smaller frequency response amplitude can be clearly identified between two peaks.

5. Control Laws Development

Previous studies[3,6,7], as well as the discussions given in sections 3 and 4 of this paper, have demonstrated the feasibility of using the piezoelectric materials as "electrical vibration absorbers". Research efforts so far have been focused on passive or quasi-static circuits, which are only useful for structures under harmonic disturbances within a narrow and dominant frequency range. However, mechanical systems in general are excited by various types of non-harmonic and transient disturbances, such as forces with varying frequencies, impact loadings, and random excitations. To ensure that the controller is effective under different operating conditions, real-time algorithms to on-line adjust the adaptable circuit based on feedback is needed. The focus of this research is to synthesize such control systems to suppress vibration of structures under uncertain and transient disturbances.

In concept, vibration suppression could be achieved by minimizing the structure energy. Using the cantilever beam example, the following paragraphs discuss the relationship between the control actions and the energy levels of the main structure and the total system (beam, PZT, and the electrical circuit). Control algorithms are derived based on these discussions.

5.1. A Discussion on System and Structure Energies

One can derive the total system energy E_{nt} to be:

$$E_{nt}=\int_0^{L_b}\frac{m_b}{2}\left(\frac{\partial w}{\partial t}\right)^2+\frac{E_bI_b}{2}\left(\frac{\partial^2 w}{\partial x^2}\right)^2+\left[E_sI_s\left(\frac{\partial^2 w}{\partial x^2}\right)^2+2h_{31}b_sM_s\left(\frac{\partial^2 w}{\partial x^2}\right)D+ A_s\beta_{33}D^2\right](H(x-x_1)-H(x-x_2))dx+\frac{L}{2}I^2 \quad (15)$$

Here, the first two terms represent the beam kinetic and strain energies, respectively. The third term represents the strain energy of the PZT. The fourth and fifth terms in the equation denote the electrical-mechanical energies of the PZT, and the last term represents the energy stored in the inductance at a given instance. The free-response system energy rate can thus be derived:

$$\frac{dE_{nt}}{dt}=-\left(R+\frac{1}{2}\frac{dL}{dt}\right)I^2-\int_0^{L_b}c_b\left(\frac{\partial w}{\partial t}\right)^2dx \quad (16)$$

One can also define the mechanical energy of the main structure to be

$$E_{nb}=\frac{1}{2}\int_0^{L_b}\left[m_b\left(\frac{\partial w}{\partial t}\right)^2+E_bI_b\left(\frac{\partial^2 w}{\partial x^2}\right)^2+2E_sI_s\left(\frac{\partial^2 w}{\partial x^2}\right)^2(H(x-x_1)-H(x-x_2))\right]dx \quad (17)$$

For free response, we can further derive

$$\frac{dE_{nb}}{dt}=\frac{2h_{31}b_sM_s}{(h_s-h_b)\beta_{33}}\left[\frac{\partial^2 w(x_2,t)}{\partial x\partial t}-\frac{\partial^2 w(x_1,t)}{\partial x\partial t}\right] \left[I\left(\frac{dL}{dt}+R\right)+L\frac{dI}{dt}+\frac{h_{31}(h_s^2-h_b^2)}{2(x_2-x_1)}\frac{\partial w(x,t)}{\partial x}\Big|_{x_1}^{x_2}\right]-\int_0^{L_b}\left[c_b\left(\frac{\partial w}{\partial t}\right)^2\right]dx \quad (18)$$

Note that not only the circuit inductance and resistance, but also the *rate of change* of the adaptable inductance will affect the system and structural energies. From examining equation (16), we see that one can reduce E_{nt} by setting (R + (dL/dt)/2) to be greater than zero. While this will ensure local stability of the system, it might not be minimizing the energy flowing into the main structure (see equation (18)). In other words, we might not be suppressing the structural vibration most effectively, which is the main purpose of the controller.

5.2. Control Algorithms Synthesis

From the above discussions, it is recognized that an effective approach would be to reduce the total system energy while constraining the energy flowing into the main structure. Since there are two objectives to be accomplished and they could contradict each other, an algorithm is needed to balance between E_{nb} and E_{nt}. A fuzzy controller was developed previously for this purpose when only the variable inductance was used in the circuit[8]. With both the inductance and resistance adjustable, less complicated schemes can be developed. Tow cases are discussed in this paper as presented in the following sections.

5.2.1 General case study

For general system disturbances, the control strategy is designed and presented in the following paragraphs.

(a) Let

$$\Delta_1 = \frac{2h_{31}b_s M_s}{(h_s - h_b)\beta_{33}}\left(\frac{\partial^2 w(x_2,t)}{\partial x \partial t} - \frac{\partial^2 w(x_1,t)}{\partial x \partial t}\right) \tag{19}$$

If $\Delta_1 I < 0$, we want dL/dt and R to be as positive as possible to reduce E_{nb} under the given constraints (see equation (18)). Since positive dL/dt and R will also reduce the total system energy (equation (16)), the two criteria do not contradict each other. We thus let dL/dt = dL_{max} and R= R_{max}, where dL_{max} and R_{max} are the upper bounds of dL/dt and R, respectively. If L is already at its maximum L = L_{max}, dL/dt is set to zero.

(b) If $\Delta_1 I > 0$, we want dL/dt to be as negative as possible to reduce E_{nb}. We therefore set dL/dt to be -dL_{max}.

Examining equations (16) and (18), the two functions containing the control actions are identified to be:

$$f_{nb} = \Delta_1 I\left(R + \frac{dL}{dt}\right),$$
$$f_{nt} = -I^2\left(R + \frac{1}{2}\frac{dL}{dt}\right) \tag{20}$$

Substituting dL/dt with $-dL_{max}$ into equation (20):

$$f_{nb}=\Delta_1 I(R-dL_{max}),\qquad f_{nt}=-I^2\left(R-\frac{dL_{max}}{2}\right) \tag{21}$$

We want both f_{nb} and f_{nt} to be negative to reduce E_{nb} and E_{nt}. That is,

$$f_{nb}<0 \;\Rightarrow\; R<dL_{max},\qquad f_{nt}<0 \;\Rightarrow\; R>\frac{dL_{max}}{2}. \tag{22}$$

Therefore, we want

$$dL_{max}>R>\frac{dL_{max}}{2} \tag{23}$$

From examining equation (21), one can also see that the resistance which could cause the same amount of reduction in E_{nb} and E_{nt} will satisfy $f_{nb} = f_{nt}$. That is,

$$\Delta_1 I(R-dL_{max})=-I^2\left(R-\frac{dL_{max}}{2}\right) \tag{24}$$

$$\Rightarrow\quad R=\frac{\Delta_1+\frac{I}{2}}{\Delta_1+I}dL_{max} \tag{25}$$

If L is already at its minimum $L = L_{min}$, dL/dt is set to zero and let $R=R_{min}$ (the lower bound or minimum value of R).

5.2.2. Quasi-steady-state case study

For cases where external disturbances consist of a dominant, slow-varying excitation frequency $\omega_f(t)$, a feedforward action is specified for L :

$$L=\frac{1}{C_1\omega_f^2},\ or\qquad \frac{dL}{dt}=\frac{-2}{C_1\omega_f^3}\frac{d\omega_f}{dt} \tag{26}$$

In other words, L or dL/dt is used to reject the quasi-steady-state disturbance based on the discussions in sections 3 and 4. R is then designed to control the transient vibrations through examining the energy terms. Since (26) is derived based on the assumption that R=0, we want the smallest R possible that satisfies the energy criteria:

(a) If $\Delta_1 I < 0$ and $dL/dt > 0$, both f_{nb} and f_{nt} will be negative for positive R (see equation (20)). The resistance is therefore set to R_{min}.

(b) If $\Delta_1 I < 0$ and $dL/dt < 0$. To satisfy $f_{nt} < 0$, we want $R > |dL/dt|/2$. On the other hand, R needs to be greater than $|dL/dt|$ to have $f_{nb} < 0$. R is thus set to be equal to $(1+\eta)|dL/dt|$, where η is a small positive design variable.

(c) If $\Delta_1 I > 0$ and $dL/dt > 0$, $f_{nt} < 0$ for positive R. However, f_{nb} will be positive and increase with increasing R. R thus is set to R_{min} to minimize f_{nb}.

(d) If $\Delta_1 I > 0$ and $dL/dt < 0$, to ensure $f_{nb} < 0$ and $f_{nt} < 0$, one can derive

$$\left|\frac{dL}{dt}\right| > R > \frac{1}{2}\left|\frac{dL}{dt}\right| \tag{27}$$

or let

$$R = \left(\frac{1}{2}+\eta\right)\left|\frac{dL}{dt}\right| \tag{28}$$

6. Analysis and Results

To evaluate and analyze the control system performance, computer simulations are carried out on the beam structure. The system parameter values used are given in Table 1 unless stated otherwise.

6.1. Impulse Response

Given an impulse input (magnitude 0.02 N-s) at $x = x_f = 0.9\ L_b$, Figures 7 and 8 illustrate the structure transient response and mechanical energy. Comparing to an uncontrolled, short circuit case, the reductions of the vibration amplitude and energy level are significant when the feedback controller is used.

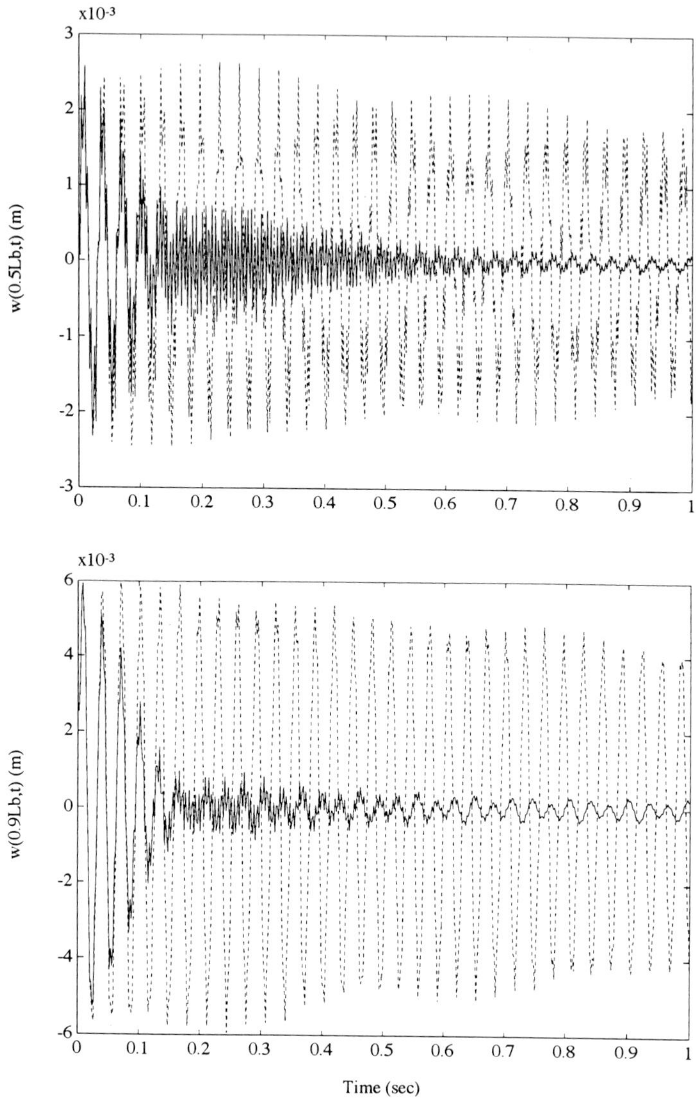

Figure 7. Impulse response of beam, uncontrolled (dashed line) and semi-active (solid line) cases.

Table 1. System Parameters

A_s	6.452×10^{-6} m^2	b_s	0.0254 m
c_b	0.148 N-s/m^2	dL_{max}	6000 Henry/sec
E_b	7×10^{10} Pa	E_s	6.9×10^{10} Pa
e_o	0.039 m	h_b	0.0016 m
h_s	0.0018 m	h_{31}	7.644×10^8 N/C
I_b	6.74×10^{-11} m^4	I_s	1.89×10^{-11} m^4
L_b	0.3 m	L_{max}	1000 Henry
L_{min}	1 Henry	m_b	0.21 kg/m
N	3	R_{max}	6 kΩ
R_{min}	0	t_a	1.0 second
x_f	0.9 L_b	x_p	0.03429 m
x_1	0.06604 m	x_2	0.00254 m
β_{33}	7.328×10^7 volt-m/coulomb	η	0.01
ω_0	80 rad/sec	ω_{ss}	345 rad/sec
ρ	0.00042 kg		

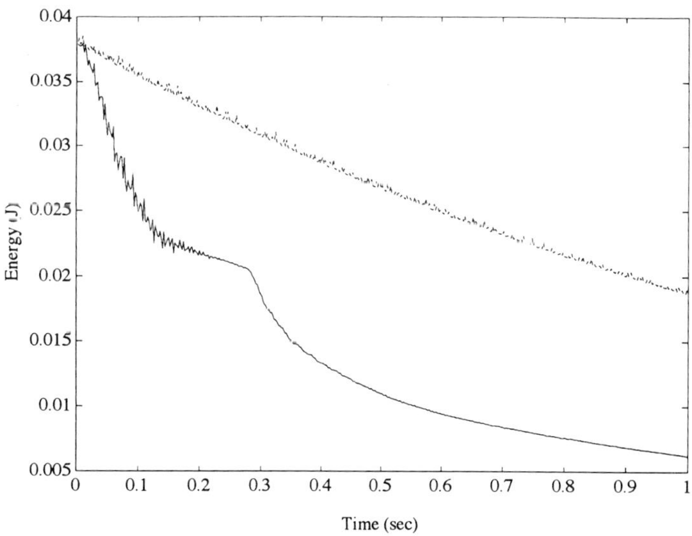

Figure 8. Structural mechanical energy. Impulse response of beam, uncontrolled (dashed line) and semi-active (solid line) cases.

6.2. Start-Up Response

In this case, the forcing function F(x,t) is defined to emulate a rotor start-up process with constant acceleration α:

$$F(x,t)=[\rho e_o(\omega_0+\alpha t)^2\sin(\omega_0 t+\frac{\alpha t^2}{2})-\rho e_o\alpha\cos(\omega_0 t+\frac{\alpha t^2}{2})]\delta(x-x_f),\quad for\ 0\le t<t_a \tag{29}$$

Here, ω_0 is the rotor speed at t = 0, and t_a is the time required to achieve the new operating speed ω_{ss}. ρ and e_o are the unbalance mass and eccentricity of the rotating element, respectively.

As discussed in section 5.2.2, we here vary the inductance to minimize the structure quasi-steady-state vibration amplitude by following the excitation frequency during the start-up process. That is, let $L = 1/(C_1(\omega_0+\alpha t)^2)$. The resistance is adjusted according to the control law presented in section 5.2.2.

Given a ramp time t_a = 1.0 second and x_f = 0.9 L_b, the structure response with feedback is compared to a passive case where the inductance is tuned for the steady-state frequency according to equation (26) ($L=1/C_1\omega_{ss}^2$). The large vibration amplitude, caused by the "transient resonance" while the rotor passes its first critical speed, is effectively suppressed when the adaptive circuit with feedback control is used (Figure 9). The structural mechanical energy is also reduced significantly (Figure 10).

7. Conclusions

A new structural vibration control method, using piezoelectric materials with real-time adaptable electrical networks, is investigated. A model of an adaptive beam structure is derived and used as an example in this study. With an adjustable circuit, a variable resistance and the changing rate of a variable inductance are used as control inputs. Based on energy analysis, parametric control laws are discussed for different case studies. The major idea is to reduce the total system energy while constraining the energy flowing into the main structure. The performance of the system is illustrated through analyzing the adaptive beam example.

8. Acknowledgement

This research is supported by the US Army Research Office. The Penn State graduate students who have made direct contributions to this study are J. S. Lai and W. K. Yu.

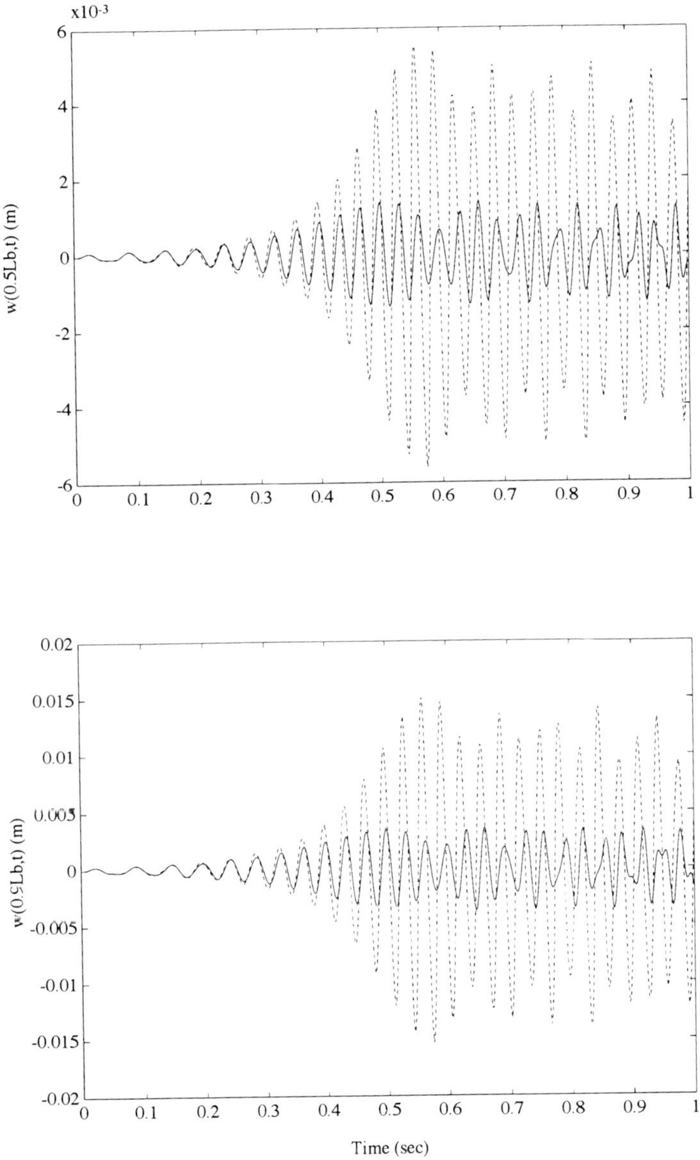

Figure 9. Beam time response of start-up scenario, passive (dashed line) and semi-active (solid line) cases.

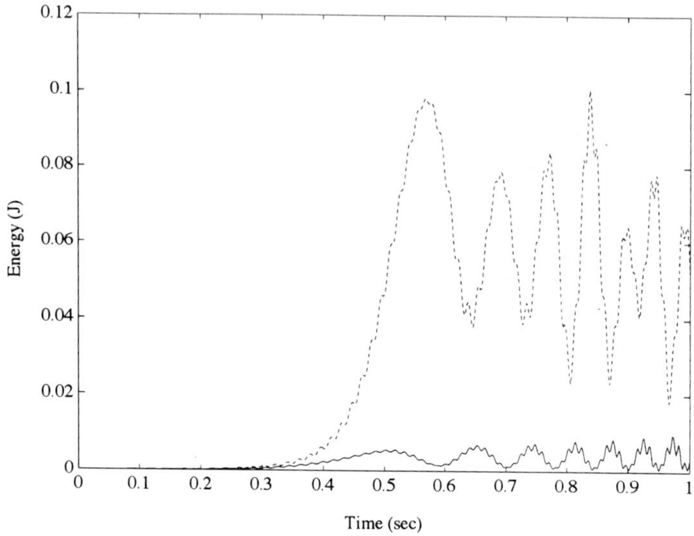

Figure 10. **Structural mechanical energy of start-up scenario, passive (dashed line) and semi-active (solid line) cases.**

9. Nomenclature

A_s	=	$b_s(h_s-h_b)$ (Cross sectional area of PZT)
b_s	=	Width of PZT and beam
C	=	Capacitance in the simulated inductance circuit
C_b	=	Damping matrix in equation (5)
c_b	=	Viscous damping constant of beam
D	=	Electrical displacement
E	=	Electrical field
E_b	=	Elastic modulus of beam
E_{nb}	=	Mechanical energy of structure
E_{nt}	=	Total system energy
E_s	=	Elastic modulus of PZT

e	=	Elastic stiffness of PZT
e_o	=	Eccentricity of rotor
F	=	Forcing function
h_b	=	Distance from neutral axis of beam to outside surface of beam
h_s	=	Distance from neutral axis of beam to outside surface of PZT
h_{31}	=	Electric displacement - stress coefficient of PZT
I	=	Current
I_b	=	Beam moment of inertia
I_s	=	PZT moment of inertia
L	=	Inductance
L_b	=	Beam length
M_b	=	Mass matrix in equation (5)
M_s	=	$(h_s^2-h_b^2)/2$
m_b	=	Beam mass per length
$q(t)$	=	Generalized displacement vector
$q_i(t)$	=	ith generalized displacement
R	=	Resistance of the PZT shunted circuit
R_1	=	Resistance in the simulated inductance circuit
R_2	=	Resistance in the simulated inductance circuit
R_3	=	Resistance in the simulated inductance circuit
R_4	=	Resistance in the simulated inductance circuit
t	=	Time
w	=	Beam transverse displacement
x_p	=	Mid-point of PZT along the x direction
x_1	=	Left end of PZT
x_2	=	Right end of PZT
α	=	Acceleration of rotor
β_{33}	=	Dielectric impermeability coefficient of PZT
η	=	Positive constant parameter
ϵ	=	Mechanical strain of PZT
ρ	=	Unbalance mass of rotor
τ	=	Material stress of PZT
$\phi_i(x)$	=	Comparison functions
Ω_0	=	Matrix in equation (5)
Ω_1	=	Matrix in equation (5)
Ω_2	=	Scalar in equation (5)
ω_{ss}	=	Steady-state operating frequency of rotor

10. References

1. M. J. Balas, "Trends in Large Space Structure Control Theory: Fondest Hopes, Wildest Dreams", *IEEE Trans of Automatic Control*, AC-27(3), 1982, pp.522-535.

2. L. Meirovitch, H. Baruh, H. Oz, "A Comparison of Control Techniques for Large Flexible Systems", *AIAA Journal of Guidance and Control*, **6(4)**, 1983, pp.302-310.

3. N.W. Hagood and A. von Flotow, "Damping of Structure Vibrations with Piezoelectric Materials and Passive Electrical Networks", *Journal of Sound and Vibration*, **146(2)**, 1991, pp.243-268.

4. G. Lesieutre and C.L. Davis, "Frequency-Shaped Passive Damping using Resistively-Shunted Piezoelectric Ceramics", *Proc. of ADPA/AIAA/ASME/SPIE Conf. of Active Materials and Adaptive Structures*, 1991, Alexandria, VA.

5. G. Lesieutre, S. Kurtz, S. Yoshikawa, "Resistively-Shunted Piezoceramics for Passively Damped Structural Composite Materials", *Proc. of the Intl. Symp. on Damping of Multiphase Inorganic Materials*, TMS Fall Meeting, 1992, Chicago, IL.

6. J.J. Hollkamp and T.F. Starchville, "A Self-Tuning Piezoelectric Vibration Absorber", *Proc. of the 35th SDM Conf.*, AIAA-94-1790, 1994, Hilton Head, SC.

7. D.L. Edberg and A.S. Bicos, "Design and Development of Passive and Active Damping Concepts for Adaptive Space Structures", *Proc. of ADPA/AIAA/ASME/SPIE Conf. of Active Materials and Adaptive Structures*, 1991, Alexandria, VA.

8. K.W. Wang and J.S. Lai, "Semi-Active Vibration Control of Flexible Structures via Adaptive Materials", *U.S. Army Research Office Smart Structure Workshop*, 1993, Arlington, TX.

11. Appendix

Using Hamilton's Principle, one can construct

$$\int_{t_1}^{t_2} [\delta T_b - \delta U_b - \delta U_s + \delta W_1 + \delta W_2 + \delta W_d] = 0 \tag{A1}$$

where, T_b is the beam kinetic energy, U_b is the beam strain energy, U_s is the mechanical and electrical energy of the piezoelectric material, δW_2 is the virtual work term from the external voltage V(t), and δW_1 and δW_d are the virtual works done by the external force F(x,t) and damping, respectively.

For one-dimensional structures with uni-axial loading, the constitutive equation of the piezoelectric materials can be written as

$$\begin{bmatrix}\tau\\E\end{bmatrix}=\begin{bmatrix}e & -h_{31}\\-h_{31} & \beta_{33}\end{bmatrix}\begin{bmatrix}\epsilon\\D\end{bmatrix}$$

where D is the electrical displacement (charge/area in the z-direction), E is the electric field (volts/length along the z-direction), ϵ is the mechanical strain in the x-direction, and τ is the material stress in the x-direction. e is the elastic stiffness, β_{33} is the dielectric constant, and h_{31} is the piezoelectric constant of the PZT. Based on the above constitutive equation, and assuming D is constant along the PZT thickness for thin materials, one can derive

$$U_s=\int_0^{L_b}\int_{A_s}(\tau\epsilon+ED)=\int_0^{L_b}[E_sI_s(\frac{\partial^2 w}{\partial x^2})^2+2h_{31}b_sM_s(\frac{\partial^2 w}{\partial x^2})D+ A_s\beta_{33}D^2](H(x-x_1)-H(x-x_2))dx$$

The other functions are presented as follows:

$$T_b=\frac{1}{2}\int_0^{L_b}m_b(\frac{\partial w}{\partial t})^2dx$$

$$U_b=\frac{1}{2}\int_0^{L_b}E_bI_b(\frac{\partial^2 w}{\partial x^2})^2dx$$

$$\delta W_d=-\int_0^{L_h}c_b\frac{\partial w(x,t)}{\partial t}\delta w(x,t)dx$$

$$\delta W_1=\int_0^{L_b}F(x,t)\delta w(x,t)dx$$

$$\delta W_2=2V(t)b_s(x_2-x_1)\delta D$$

Note that the voltage is related to the external circuit:

$$V=-\frac{d(LI)}{dt}-RI$$

Substituting the above equations into equation (A1), and further assume that the field within and electrical displacement on the surface are uniform for the piezoelectric material[3]:

$$V=E(h_s-h_b),$$
$$I=2b_s(x_2-x_1)\frac{dD}{dt},$$

one can derive the system model equations (1) - (3).

Wave Motion, Intelligent Structures and Nonlinear Mechanics, pp. 135–162
edited by A. Guran and D. J. Inman
Series on Stability, Vibration and Control of Structures: Vol. 1

A TRANSFER FUNCTION METHOD FOR MODELING AND CONTROL OF GYROSCOPIC DYNAMIC SYSTEMS

BINGEN YANG

Department of Mechanical Engineering, University of Southern California
Los Angeles, CA 90089, U.S.A.

1. Introduction

Gyroscopic dynamic systems are the basic element of such diverse mechanical systems as rotating machinery, high speed precision machines, robots, large space structures, manufacturing and material processing equipment, and sports and entertainment facilities. For years, vibration and dynamics of gyroscopic systems have been extensively studied by Ziegler[31], Mote[17], Meirovitch[13], Hagedorn[7], Huseyin[9], D'Eleuterio and Hughes[4], Païdoussis[18], Inman[10], Yang[25], Guran[5], and many others. In the recent two decades, the development of advanced technologies in all those engineering applications has aroused an ever-increasing interest in active vibration control of gyroscopic systems; for instance see Johnson[11], Balas[2], Juang and Balas[12], Hughes and Skelton[8], Meirovitch and Öz[15], Radcliffe and Mote[19], Ulsoy[21], and Yang and Mote[28].

A gyroscopic system is different from a conventional, self-adjoint mechanical system in its gyroscopic forces, which are described by skew-symmetric matrices or operators in equations of motion. These forces are associated with Coriolis acceleration or mass transport, and render vibration modes complex, which physically means a non-constant phase in the distribution of system response. Because of this, classical modal analysis for conventional systems are not applicable to gyroscopic systems. While state-space techniques are available, they are not convenient in dealing those problems which have to be formulated in the original physical coordinates, for instance the dynamic problems of constrained/combined gyroscopic systems.

This paper introduces a transfer function method for gyroscopic dynamic systems. Our purpose is two-fold. First, we use the method to solve certain dynamic problems which otherwise are difficult with state-space techniques. Second, we attempt to develop a new frequency domain technique for active control of gyroscopic systems. To this end, two problems have been selected in this study: (a) eigenvalue inclusion phenomena for constrained/modified gyroscopic systems; (b) vibration control of gyroscopic systems by direct velocity feedback. The problems themselves are interesting topics which are worthy of research efforts. The material to be presented in this paper is mainly based on the author's recent research results[23,24]. In Section 2, an integral form of the response of the gyroscopic system is derived, based on which, a transfer function formulation is established. A key feature of this formulation is that the system in the physical coordinates is expressed by a series of the complex non-orthogonal modes. With the transfer function formulation, the problems of eigenvalue inclusion and direct velocity feedback for gyroscopic systems are investigated in Sections 3 and 4, respectively.

2. System Description

2.1 Equations of Motion

Consider a linear, n-DOF gyroscopic system described by

$$M\ddot{x}(t)+G\dot{x}(t)+Kx(t)=f(t), \quad x(0)=a_0, \quad \dot{x}(0)=b_0 \tag{2.1}$$

where $x(t)$ is the displacement vector, $f(t)$ is the external force vector, the mass matrix M is symmetric and positive definite, the gyroscopic matrix G, normally associated with Coriolis acceleration or mass transport, is skew-symmetric, and the stiffness matrix K is symmetric and positive definite. Equation (2.1) can be a model of lumped parameter systems, or a discretized model of distributed gyroscopic systems from any discretization procedures, such as finite element analysis, Rayleigh-Ritz method, Galerkin's method, and assumed-modes method[14]. When $G=0$, (2.1) is a classical non-gyroscopic system.

One example is a flexible rotating shaft whose transverse vibration. By a single-lumped-mass model, the two mutually perpendicular shaft deflections q_1 and q_2, which are measured in a coordinate system that is fixed to the rotating shaft, are described by (2.1) with

$$M=m\begin{bmatrix}1 & 0\\ 0 & 1\end{bmatrix}, \quad G=2m\Omega\begin{bmatrix}0 & -1\\ 1 & 0\end{bmatrix}, \quad K=(k-m\Omega^2)\begin{bmatrix}1 & 0\\ 0 & 1\end{bmatrix}, \quad x=\begin{pmatrix}q_1\\ q_2\end{pmatrix} \tag{2.2}$$

where Ω is the shaft rotation speed, and m and k are the effective mass and stiffness of the shaft, respectively. Here the matrix G describe the Coriolis forces of the rotating shaft.

2.2 Eigenvalue Problems

The eigenvalue problem associated with (2.1) is described by

$$(\lambda^2 M+\lambda G+K)v=0. \tag{2.3}$$

The eigensolutions have the form

$$\lambda_{\pm k}=\pm i\omega_k, \quad v_{\pm k}=v_k^R \pm i v_k^I, \quad i=\sqrt{-1}, \quad k=1,2,\ldots,n \tag{2.4}$$

where $\omega_k>0$, and $v_k^R, v_k^I \in R^n$. The ω_k is the kth natural frequency and v_k is the corresponding mode shape distribution. The eigensolutions can be obtained through numerical solution of an equivalent eigenvalue problem in state-space form

$$A\phi_k = \lambda_k \phi_k \tag{2.5}$$

where

$$A = \begin{bmatrix} -M^{-1}G & -M^{-1}K \\ I & 0 \end{bmatrix} \in R^{2n\times 2n}, \qquad \phi_k = \begin{pmatrix} \lambda_k v_k \\ v_k \end{pmatrix} \in C^{2n} \tag{2.6}$$

with the v_k of the original problem (2.3) given by the lower half of the $2n$-vector ϕ_k.

The complex eigenvectors v_k indicate a non-constant phase behavior of the gyroscopic system in vibration; namely, the displacements at different points do not reach their maxima or minima simultaneously. This special feature, which is due to the skew-symmetric G, renders the eigenvectors non-orthogonal. As such, the conventional modal analysis can not be directly applied to represent the system response $z(t)$. In this paper, bi-orthogonality relations for the eigenvectors in the state-space are derived to predict the system response.

The adjoint eigenvalue problem to (2.5) is

$$A^T \psi_k = \overline{\lambda}_k \psi_k \tag{2.7}$$

where the over-bar represents complex conjugation, and A^T is the transpose of the matrix A. The bi-orthogonality relations among ϕ_k and ψ_k are given as follows

$$\langle \psi_j, \phi_k \rangle = 2\delta_{jk}, \quad \langle \psi_j, A\phi_k \rangle = 2\lambda_j \delta_{jk} \tag{2.8}$$

where the inner product is defined as

$$\langle a,b \rangle = \overline{a}^T b, \qquad \forall a,b \in C^{2n}. \tag{2.9}$$

While the bi-orthogonality relations (2.8) are well known, the previous analyses usually stop here without further exploring the relationship between ψ_k and v_k. Note that the associated eigenvectors ϕ_k in (2.6) are represented by the eigensolutions (ω_k, v_k) of the original eigenvalue problem (2.3). It is natural to ask if the adjoint eigenvectors can also be explicitly expressed in terms of (ω_k, v_k). To answer this question, write

$$\psi_k = \begin{pmatrix} \chi_1 \\ \chi_2 \end{pmatrix}, \quad \chi_1, \chi_2 \in C^n \,. \tag{2.10}$$

Plugging the above into $\langle \psi_k, A\phi_k \rangle = \langle A^T \psi_k, \phi_k \rangle$ and using (2.6) leads to

$$\chi_1 = \lambda_k M v_k, \quad \chi_2 = K v_k. \tag{2.11}$$

Thus, the adjoint eigenvectors can indeed be written in terms of (ω_k, v_k); i.e.,

$$\psi_k = \begin{pmatrix} \lambda_k M v_k \\ K v_k \end{pmatrix}, \quad \psi_{-k} = \overline{\psi}_k, \quad k = 1,2,\ldots,n \tag{2.12}$$

With the above result, the bi-orthogonality relations (2.8) become

$$\bar{v}_j^T\left(\omega_j\omega_k M + K\right)v_k = 2\delta_{jk}, \quad j,k = 1,2,\ldots,n. \tag{2.13}$$

The presence of the skew symmetric matrix G makes the system (2.1) non-self-adjoint; the system response $z(t)$ can not be evaluated by the classical modal analysis. The explicit representation of the bi-orthogonal eigenvectors ϕ_k and ψ_k, however, makes it possible to express $x(t)$ in terms of the non-orthogonal mode shapes v_k; see the next subsection.

2.3 System Response

An integral representation of the system response is derived using the relationships between the state-space eigenvectors and vibration modes obtained in the previous subsection. To this end, Eq. (2.1) is transformed into the equivalent state-space form

$$\dot{\eta}(t) = A\eta(t) + Q(t), \quad \eta(0) = \eta_0 \tag{2.14}$$

where

$$\eta(t) = \begin{pmatrix} \dot{x}(t) \\ x(t) \end{pmatrix}, \quad Q(t) = \begin{pmatrix} M^{-1}f(t) \\ 0 \end{pmatrix}, \quad \eta_0 = \begin{pmatrix} b_0 \\ a_0 \end{pmatrix}. \tag{2.15}$$

and the matrix A has been given in (2.6). The state vector $z(t)$ is represented as

$$\eta(t) = \sum_{k=\pm 1}^{\pm n} \alpha_k(t)\phi_k \tag{2.16}$$

where $\alpha_k(t)$ are complex modal coordinates. Substitution of the above series into Eq. (2.14) and use of the bi-orthogonality conditions (2.8) yields a set of independent first-order equations

$$\dot{\alpha}_k(t) = \lambda_k\alpha_k(t) + q_k(t), \quad k = \pm 1, \pm 2, \ldots, \pm n \tag{2.17}$$

where $\alpha_k(0) = \frac{1}{2}\langle \psi_k, z_0\rangle$, and $q_k(t) = \frac{1}{2}\langle \psi_k, Q(t)\rangle$. Solution of (2.17) and substitution of $\alpha_k(t)$ into (2.16) gives

$$\eta(t) = S(t)\eta_0 + \int_0^t S(t-\tau)Q(\tau)d\tau \tag{2.18}$$

where the $2n \times 2n$ matrix

$$S(t) \equiv \begin{bmatrix} s_{11}(t) & s_{12}(t) \\ s_{21}(t) & s_{22}(t) \end{bmatrix} = \frac{1}{2}\sum_{k=\pm 1}^{\pm n} e^{\lambda_k t}\begin{bmatrix} \omega_k^2 v_k \bar{v}_k^T M & \lambda_k v_k \bar{v}_k^T K \\ \lambda_k v_k \bar{v}_k^T M & v_k \bar{v}_k^T K \end{bmatrix}. \tag{2.19}$$

It follows that

$$x(t) = s_{21}(t)b_0 + s_{22}(t)a_0 + \int_0^t s_{21}(t-\tau)M^{-1}f(\tau)d\tau. \tag{2.20}$$

By (2.18), it is easy to show that

$$s_{22}(t) = \dot{s}_{21}(t) + s_{21}(t)M^{-1}G. \tag{2.21}$$

Thus, the response of the gyroscopic system is expressed by

$$x(t) = g(t)(Ga_0 + Mb_0) + \dot{g}(t)Ma_0 + \int_0^t g(t-\tau)f(\tau)d\tau \tag{2.22a}$$

where the Green's function $g(t)$, an n by n real matrix, is given by

$$g(t) = s_{21}(t)M^{-1} = \frac{1}{2}\sum_{k=\pm 1}^{\pm n} e^{\lambda_k t}\bar{\lambda}_k v_k \bar{v}_k^T. \tag{2.22b}$$

2.4 Transfer Function

Laplace transform of the integral formulation (2.22) with respect to t gives

$$X(s) = H_o(s)\{F(s) + F_I(s)\} \tag{2.23a}$$

where $X(s)$ and $F(s)$ are the Laplace transforms of $x(t)$ and $f(t)$, respectively, s is a complex parameter, and $F_I(s) = (G + sM)a_0 + Mb_0$ representing the initial conditions. The n by n complex matrix $H_o(s)$ is the Laplace transform of the system Green's function

$$H_o(s) = \mathcal{L}[g(t)] = \frac{1}{2}\sum_{k=\pm 1}^{\pm n} \frac{-\lambda_k}{s-\lambda_k} v_k \bar{v}_k^T \tag{2.23b}$$

where $\bar{\lambda}_k = -\lambda_k$ has been used. We shall call $H_o(s)$ the open-loop transfer function of the gyroscopic system.

With the modal expansion of the system response (2.22) and the transfer function formulation (2.23), various dynamics and control problems of gyroscopic systems can be conveniently and efficiently tackled, as shall be discussed in Secs. 3 and 4.

3. Eigenvalue Inclusion Principles

3.1 Problem Statement

In his famous treatise *The Theory of Sound*, Rayleigh[20] enunciated a theorem about the eigenvalues of modified non-gyroscopic systems, which may be stated as follows:

Theorem 3.1. Consider the linear, discrete non-gyroscopic system

$$M\ddot{x}(t) + Kx(t) = f(t), \quad M^T = M > 0, \quad K^T = K > 0 \tag{3.1}$$

whose natural frequencies are ω_k, arranged in ascending order of magnitude. If a spring is attached to the system or if one point of the system is fixed, the natural frequencies Ω_k of the modified system are such that

$$\omega_k \le \Omega_k \le \omega_{k+1}, \qquad k = 1, 2, \ldots \tag{3.2}$$

This theorem, now called the inclusion principle, has been used to develop algorithms for computing eigenvalues of modified vibrating systems, and to estimate the bounds of natural frequencies of structures.

Although modified gyroscopic systems encompass many mechanical systems and structures, few study has addressed the problem of eigenvalue inclusion for those systems, partially due to lack of effective ways to deal with non-orthogonality of the system eigenfunctions. The inclusion principle for non-gyroscopic systems was proven by the Rayleigh quotient method[20]. In the proof the method assumes and uses the orthogonal property of the system eigenfunctions or eigenvectors. Gyroscopic systems, however, are non-self-adjoint, and have non-orthogonal and complex eigenfunctions or eigenvectors[13]. The classical modal analysis is not valid for modal decoupling for gyroscopic systems. Therefore, Rayleigh's inclusion principle can not be directly extended to gyroscopic systems.

In this section, a transfer function method is applied to analyze the natural frequencies of the gyroscopic system (2.1) with various modifications/constraints. In stead of trying to formulate a Rayleigh quotient or to decouple the equations of motion, the method introduces the concepts of feedback control. The gyroscopic system under a modification is viewed as a feedback control system. The system without the modification is an open loop system. The system with the modification is a closed loop system; the modification is a feedback controller. Both the open loop and closed loop transfer functions are obtained by the modal expansion given in Sec. 2. The effect of constraint/modification on the system spectrum is studied through investigation of the close loop poles.

3.2 Transfer Function Formulation

Open Loop: the Unmodified System

The Laplace-transform of the unmodified system (2.1) is

$$(s^2M + sG + K)X(s) = F(s) + F_I(s) \tag{3.3}$$

where all the symbols have been defined in (2.23a). The solution to (3.3) is given by (2.23) in terms of the open-loop transfer function $H_o(s)$. The poles of $H_o(s)$ are the eigenvalues of the unmodified system (2.1), i.e., $\lambda_k = i\omega_k$.

Closed Loop: the Modified System

Assume that a modification is imposed on the unmodified system (2.1) at the jth point. The modification will generate a force, which is applied at the jth point and depends on the the jth component $x_j(t)$ of $x(t)$. This force can be considered as the control force from a feedback controller that has a sensor and an actuator both located at the jth point; i.e.,

$$F_c(s) = C(s)e_j e_j^T X(s) \tag{3.4}$$

Here $F_c(s)$ is the Laplace transform of the force due to the modification, e_j is an n-vector with 1 in the jth position and 0 elsewhere, and $C(s)$ is the transfer function of the modification (controller) in the form

$$C(s) = \mu \frac{N(s)}{D(s)} \tag{3.5}$$

where $N(s)$ and $D(s)$ are polynomials in s, and $\mu > 0$ is a gain parameter. Two examples of modifications are given as follows:

(a) a spring of coefficient μ attached to the system (2.1)

$$C(s) = -\mu, \quad N(s) = -1, \quad D(s) = 1; \tag{3.6}$$

(b) a mass μ added to the system (2.1)

$$C(s) = -\mu s^2, \quad N(s) = -s^2, \quad D(s) = 1. \tag{3.7}$$

With the above description the total external force applied to the gyroscopic system is

$$F(s) = F_c(s) + F_e(s) = \mu \frac{N(s)}{D(s)} e_j e_j^T X(s) + F_e(s) \tag{3.8}$$

where $F_e(s)$ is the Laplace transform of other external disturbance. Substituting (3.8) into (2.23a) gives

$$X(s) = H_{cl}(s)\{F_e(s) + F_I(s)\} \tag{3.9a}$$

where the closed-loop transfer function

$$H_{cl}(s) = [I - C(s)H_o(s)e_j e_j^T]^{-1} H_o(s) \in C^{n \times n} \tag{3.9b}$$

Through use of the matrix inversion formula

$$(I + ab^T)^{-1} = I - \frac{ab^T}{1 + b^T a}, \quad a, b \in C^n$$

the closed loop transfer function is derived as

$$H_{cl}(s) = N_{cl}(s) / \Delta(s) \tag{3.10a}$$

$$N_{cl}(s) = D(s)W_o(s) + \mu N(s)[H_o(s)e_j e_j^T H_o(s) - e_j^T H_o(s) e_j H_o(s)] \in C^{n \times n} \tag{3.10b}$$

$$\Delta(s) = D(s) - \mu N(s) \sum_{k=1}^{n} \frac{\omega_k^2}{s^2 + \omega_k^2} \left| v_{k,j} \right|^2 \in C \tag{3.10c}$$

where $v_{k,j}$ is the jth element of the eigenvector v_k. The eigenvalues of the modified gyroscopic system are the poles of the closed loop transfer function $H_{cl}(s)$.

Theorem 3.2. The eigenvalues of the modified system (23) are the roots of

$$\Delta(s) = 0$$

and λ_l for some l if $v_{l,j} = 0$ or $N(\lambda_l) = 0$.

Proof: Consider the limit of $H_{cl}(s)$ as $s \to \lambda_k$. Write $s - \lambda_k = \varepsilon e^{i\theta}$, $\varepsilon > 0$. Given a j (the location of the modification), if $v_{k,j} = 0$ for some k, by (2.23) and (3.10), as $s \to \lambda_k$

$$H_{cl}(s) \to \mathcal{B}_1 e^{-i\theta} / \varepsilon, \quad \mathcal{B}_1 \in C^{n \times n}, \; 0 < \eta_1 \le \|\mathcal{B}_1\| \tag{3.11}$$

where η_1 is a constant. Since $\varepsilon \to 0$, the limit in (3.11) is unbounded, implying that λ_k is an closed-loop pole. If $v_{k,j} \neq 0$, by (2.23) and (3.10), as $s \to \lambda_k$

$$H_{cl}(s) \to \mathcal{B}_2 / \left\{ \mu \lambda_k N(\lambda_k + \varepsilon e^{i\theta}) |v_{k,j}|^2 \right\} \tag{3.12}$$

where $\mathcal{B}_2 \in C^{n \times n}$, $0 < \eta_2 \le \|\mathcal{B}_2\| \le \eta_3 < +\infty$, and η_2 and η_3 are constants. Recall that $\mu >$ 0 and $\omega_k > 0$. The limit in (3.12) has two possibilities:

(i) $N(\lambda_k) = 0$, the limit in (3.12) is unbounded; λ_k is a pole of $H_{cl}(s)$.

(ii) $N(\lambda_k) \neq 0$, the limit in (3.12) is determinate and finite; λ_k is not a pole of $H_{cl}(s)$.

Note that $N_{cl}(s)$ in (3.10) has no singular points in the complex plane other than λ_k. Besides those λ_k which are the poles of $H_{cl}(s)$, all other poles of the modified system can only be the roots of $\Delta(s) = 0$. (Q.E.D.)

Theorem 3.2 relates to controllability of the gyroscopic system[27, 30]. If the modification is away from all nodes of the n modes of vibration, and has no pole-zero cancellation, the modified system is controllable for all modes; the modification will change all the system eigenvalues. If the modification is at a node of the kth mode or has a zero canceling the kth eigenvalue λ_k, the modified system is not controllable for the kth mode; the modification will not influence λ_k.

3.3 Eigenvalue Inclusion

With the transfer function formulation, eigenvalue inclusion principles are developed for six types of non-dissipative modifications. Without loss of generality, assume that the natural frequencies ω_k of the unmodified system are distinct. For the gyroscopic system (2.1) under a non-dissipative modification, the system eigenvalues (closed-loop poles) have the form $i\Omega_k$, where Ω_k are the natural frequencies of the modified system. In the following analyses the ω_k and Ω_k are arranged in ascending order of magnitude.

<u>Modification 1: A Spring Attached to the Gyroscopic System</u>

A spring of coefficient μ is attached to the system (2.1) at the jth point. The modification is described in (3.6). If the jth point is not a node of any mode of the unmodified system, i.e., $v_{k,j} \neq 0$ for k = 1,2,...,n, according to Theorem 3.2, the natural frequencies, Ω_k, are the roots of

$$h_1(\Omega) \equiv \Delta(i\Omega) = 1 + \mu \sum_{k=1}^{n} \frac{\omega_k^2}{-\Omega^2 + \omega_k^2} |v_{k,j}|^2 = 0 \tag{3.13}$$

where (3.6) and (3.10c) have been used. The function $h_1(\Omega)$ has the following properties:

$$h_1(0) > 0, \qquad \lim_{\Omega \to +\infty} h_1(\Omega) > 0$$

$$\lim_{\Omega \to \omega_k - 0} h_1(\Omega) = +\infty, \qquad \lim_{\Omega \to \omega_k + 0} h_1(\Omega) = -\infty \text{ for all } k$$

$$\frac{d}{d\Omega} h_1(\Omega) > 0 \text{ for } \Omega \in (0, +\infty) \text{ and } \Omega \neq \omega_k,\ k = 1, 2, \ldots, n.$$

Because $h_1(\Omega)$ is continuous and monotonically increasing on (ω_k, ω_{k+1}) and because $h_1(\omega_k+0)$ and $h_1(\omega_{k+1}-0)$ have different signs, there is at least and at most one root Ω_k of (25) in each interval (ω_k, ω_{k+1}), $k = 1, 2, \ldots, n-1$; see Figure 3.1. Also, there is a root $\Omega_n > \omega_n$. No root falls in the interval $(0, \omega_1)$. All these indicate that

$$0 < \omega_1 < \Omega_1 < \omega_2 < \Omega_2 < \cdots < \omega_n < \Omega_n.$$

Hence, addition of a spring increases all the system natural frequencies.

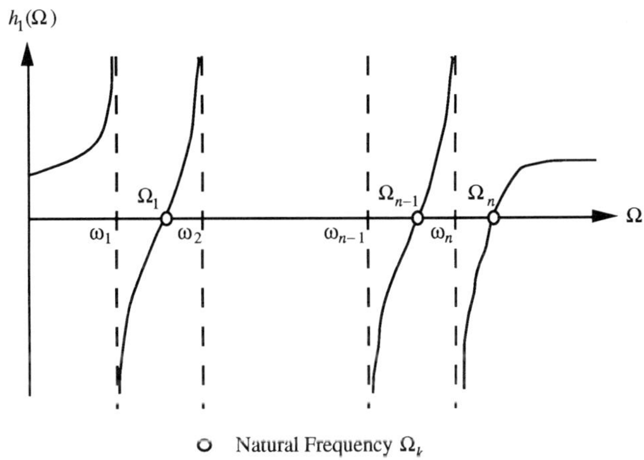

Fig. 3.1 Natural frequencies of the gyroscopic system under Modification 1, $v_{k,j} \neq 0$ for all k.

If the jth point is a node of the lth mode of the unmodified system, $v_{l,j} = 0$ and $h_1(\omega_l)$ is finite. According to Theorem 3.2, ω_l is a natural frequency of the modified system. Because $h_1(\Omega)$ is continuous and monotonically increasing on $(\omega_{l-1}, \omega_{l+1})$ and because $h_1(\omega_{l-1}+0)$ and $h_1(\omega_{l+1}-0)$ have different signs, $h_1(\Omega)$ has a root in $(\omega_{l-1}, \omega_{l+1})$. This means that the modified system has two natural frequencies, Ω_{l-1} and Ω_l, in $(\omega_{l-1}, \omega_{l+1})$. The locations of these two natural frequencies depend on the value of $h_1(\omega_l)$:

(a) $\omega_{l-1} < \Omega_{l-1} < \omega_l = \Omega_l < \omega_{l+1}$ for $h_1(\omega_l) > 0$;

(b) $\omega_{l-1} < \Omega_{l-1} = \omega_l = \Omega_l < \omega_{l+1}$ for $h_1(\omega_l) = 0$;

(c) $\omega_{l-1} < \Omega_{l-1} = \omega_l < \Omega_l < \omega_{l+1}$ for $h_1(\omega_l) < 0$.

Note that the modified system may have repeated eigenvalues even if the unmodified system has distinct eigenvalues. The above discussion is summarized as follows.

Theorem 3.3. If a spring is attached to the gyroscopic system (2.1) whose natural frequencies are ω_k, the natural frequencies Ω_k of the modified system are such that

$$0 < \omega_1 \le \Omega_1 \le \omega_2 \le \Omega_2 \le \cdots \le \omega_n \le \Omega_n. \tag{3.14}$$

Modification 2: A Mass Added to the Gyroscopic System

A mass μ is added to the system (2.1) at the jth point. The modification is described in (3.7). If the jth point is away from any node of all modes of the unmodified system, by Theorem 3.2, the natural frequencies Ω_k of the modified system are the roots of

$$h_2(\Omega) \equiv \Delta(i\Omega) = 1 - \mu\Omega^2 \sum_{k=1}^{n} \frac{\omega_k^2}{-\Omega^2 + \omega_k^2} \left|v_{k,j}\right|^2 = 0 \tag{3.15}$$

The function $h_2(\Omega)$ has the following properties:

$$h_2(0) > 0, \quad \lim_{\Omega \to +\infty} h_2(\Omega) > 0$$

$$\lim_{\Omega \to \omega_k - 0} h_2(\Omega) = -\infty, \quad \lim_{\Omega \to \omega_k + 0} h_2(\Omega) = +\infty \quad \text{for all } k$$

$$\frac{d}{d\Omega} h_2(\Omega) < 0 \text{ for } \Omega \in (0, +\infty) \text{ and } \Omega \ne \omega_k,\ k = 1, 2, \ldots, n.$$

With these properties it can be shown that

$$0 < \Omega_1 < \omega_1 < \Omega_2 < \omega_2 \cdots < \Omega_n < \omega_n.$$

Thus, addition of a lumped mass decreases the system natural frequencies.

If the jth point is a node of the lth mode of the unmodified system, based on a similar argument to that in Modification 1, the modified system has two natural frequencies, Ω_l and Ω_{l+1}, in $(\omega_{l-1}, \omega_{l+1})$:

(a) $\omega_{l-1} < \Omega_l = \omega_l < \Omega_{l+1} < \omega_{l+1}$ for $h_2(\omega_l) > 0$;

(b) $\omega_{l-1} < \Omega_l = \omega_l = \Omega_{l+1} < \omega_{l+1}$ for $h_2(\omega_l) = 0$;

(c) $\omega_{l-1} < \Omega_l < \omega_l = \Omega_{l+1} < \omega_{l+1}$ for $h_2(\omega_l) < 0$.

The results are summarized as follows.

Theorem 3.4. If a mass is added to the gyroscopic system (2.1) whose natural frequencies are ω_k, the natural frequencies Ω_k of the modified system are such that

$$0 < \Omega_1 \le \omega_1 \le \Omega_2 \le \omega_2 \le \cdots \le \Omega_n \le \omega_n. \tag{3.16}$$

Modification 3: One Point of the Gyroscopic System Fixed

Let the jth point of the gyroscopic system (1) be fixed, i.e.,

$$x_j(t) \equiv 0, \quad \forall t > 0 \tag{3.17}$$

where $x_j(t)$ is the jth component of $x(t)$. As a result one degree of freedom of the gyroscopic system is removed. The modification can be considered as the limiting case of Modification 1 as the spring coefficient μ approaches infinity. Thus, by (3.13) and Theorem 3.2, the natural frequencies Ω_k of the modified system are the roots of

$$h_3(\Omega) \equiv \sum_{k=1}^{n} \frac{\omega_k^2}{-\Omega^2 + \omega_k^2} \left| v_{k,j} \right|^2 = 0 \tag{3.18}$$

and ω_l for some l if $v_{l,j} = 0$. Similar to the proof of Theorem 3.3, the following conclusion can be reached.

Theorem 3.5. If the modification (3.17) is imposed on the gyroscopic system (2.1) whose natural frequencies are ω_k, the natural frequencies Ω_k of the modified system are such that

$$0 < \omega_1 \le \Omega_1 \le \omega_2 \le \Omega_2 \le \cdots \omega_{n-1} \le \Omega_{n-1} \le \omega_n \tag{3.19}$$

where one degree of freedom of the system (2.1) is removed.

Modification 4: A Mass-Spring System Connected to the Gyroscopic System

A mass m is connected to the system (2.1) at the jth point by a spring of coefficient μ. The displacement $\xi(t)$ of m is in the direction of the jth displacement $x_j(t)$. The motion of the modified system is described by

$$\begin{aligned} M\ddot{x}(t) + G\dot{x}(t) + Kx(t) &= \mu(\xi(t) - x_j(t))e_j + f_e(t) \\ m\ddot{\xi}(t) + \mu(\xi(t) - x_j(t)) &= 0 \end{aligned} \tag{3.20}$$

where e_j is defined in (3.4). Note that the one degree of freedom of the mass m is added to the modified system. Laplace transform of (3.20) and elimination of ξ lead to

$$(s^2 M + sG + K)X(s) = -\mu \frac{ms^2}{ms^2 + \mu} e_j e_j^T X(s) + F_e(s) + F_I(s) + F_{I\xi}(s)$$

where $F_{I\xi}(s)$ contains the initial conditions of $\xi(t)$. By (3.4) the transfer function of the modification is

$$C(s) = -\mu \frac{s^2}{s^2 + \alpha^2}, \quad \alpha^2 = \frac{\mu}{m}. \tag{3.21}$$

According to Theorem 3.2 the natural frequencies of the modified system are the roots of

$$\Delta(i\Omega) = -\Omega^2 + \alpha^2 - \mu\Omega^2 \sum_{k=1}^{n} \frac{\omega_k^2}{-\Omega^2 + \omega_k^2} \left| v_{k,j} \right|^2 = 0 \tag{3.22}$$

and ω_l for some l if $v_{l,j} = 0$. Rewrite (3.22) as

$$h_4(\Omega) = h_5(\Omega) \tag{3.23}$$

where
$$h_4(\Omega) \equiv -\Omega^2 + \alpha^2, \quad h_5(\Omega) \equiv \mu\Omega^2 \sum_{k=1}^{n} \frac{\omega_k^2}{-\Omega^2 + \omega_k^2} \left| v_{k,j} \right|^2 = 0. \tag{3.24}$$

The roots of (3.22) are the intersections of $h_4(\Omega)$ and $h_5(\Omega)$. With the properties

$$h_4(\Omega) > 0 \text{ for } \Omega \in [0, \alpha), \quad h_4(\alpha) = 0, \quad h_4(\Omega) < 0 \text{ for } \Omega \in (\alpha, +\infty)$$

$$h_5(0) = 0, \quad \lim_{\Omega \to +\infty} h_5(\Omega) < 0$$

$$\lim_{\Omega \to \omega_k - 0} h_5(\Omega) = +\infty, \quad \lim_{\Omega \to \omega_k + 0} h_5(\Omega) = -\infty \quad \text{for all } k$$

$$\frac{d}{d\Omega} h_5(\Omega) > 0 \text{ for } \Omega \in (0, +\infty) \text{ and } \Omega \neq \omega_k,\ k = 1, 2, \ldots, n$$

it can be shown that (see Figure 3.2)

(a) for $\alpha \in (0, \omega_1)$, $0 < \Omega_1 < \alpha < \omega_1 \le \Omega_2 \le \cdots \le \omega_{n-1} \le \Omega_n \le \omega_n \le \Omega_{n+1}$;

(b) for $\alpha > \omega_n$, $0 < \Omega_1 \le \omega_1 \le \cdots \le \Omega_n \le \omega_n < \alpha < \Omega_{n+1}$;

(c) for $\alpha \in (\omega_k, \omega_{k+1})$, $0 < \Omega_1 \le \omega_1 \le \cdots \le \Omega_k \le \omega_k$, $\omega_{k+1} \le \Omega_{k+2} \cdots \le \omega_n \le \Omega_{n+1}$

with
$$\begin{aligned} &\omega_k \le \Omega_{k+1} < \alpha \le \omega_{k+1}, && \text{if } h_5(\alpha) > 0 \\ &\omega_k < \alpha = \Omega_{k+1} \le \omega_{k+1}, && \text{if } h_5(\alpha) = 0 \\ &\omega_k < \alpha < \Omega_{k+1} \le \omega_{k+1}, && \text{if } h_5(\alpha) < 0; \end{aligned}$$

(d) for $\alpha = \omega_k$, $0 < \Omega_1 \le \omega_1 \le \cdots \le \Omega_k \le \alpha = \omega_k \le \Omega_{k+1} \cdots \le \omega_n \le \Omega_{n+1}$.

Thus, a statement is given as follows.

Theorem 3.6. If a mass-spring system is connected to the gyroscopic system (2.1) whose natural frequencies are ω_k, the natural frequencies Ω_k of the modified system are such that

$$0 < \Omega_1 \le \omega_1 \le \Omega_2 \le \omega_2 \le \cdots \le \Omega_n \le \omega_n \le \Omega_{n+1} \tag{3.25}$$

where one degree of freedom is gained for the modified system.

There are two limiting cases. First, if the spring coefficient μ goes to infinity, (3.21) becomes $C(s) = -ms^2$, Modification 2. This is expected: as μ goes to infinity, the mass m is rigidly connected to the system (2.1). Second, if the mass m approaches infinity, (3.21) reduces to $C(s) = -\mu$, Modification 1. This is also expected: as m approaches infinity, its displacement $\xi(t)$ vanishes rendering the spring attached to the fixed frame.

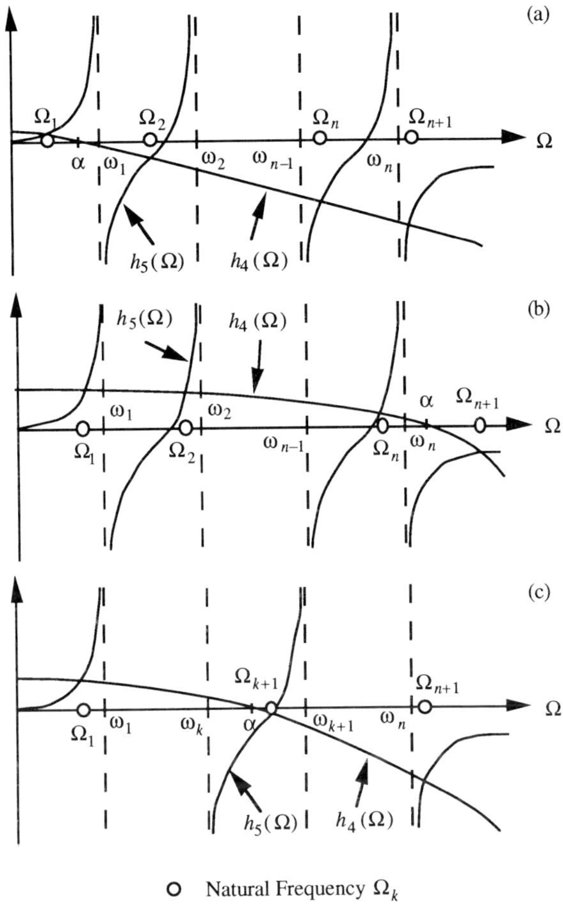

Fig. 3.2 Natural frequencies of the gyroscopic system under Modification 4. (a) $\alpha \in (0, \omega_1)$; (b) $\alpha > \omega_n$; (c) $\alpha \in (\omega_k, \omega_{k+1})$ and $h_5(\alpha) < 0$.

Modification 5: A Linear Constraint on the Displacements

The gyroscopic system (2.1) is constrained by

$$\sum_{k=1}^{n} d_k x_k(t) \equiv 0, \quad d_k \in R \quad \forall t > 0 \tag{3.26}$$

where $x_k(t)$ is the kth component of $x(t)$, and at least one of the coefficients d_k is nonzero. One example of (3.26) is that the jth and lth points of the gyroscopic system are rigidly connected, i.e., $x_j = x_l$.

Assume that d_l is nonzero for a specific l. Under the coordinate transformation

$$y(t) = Tx(t), \quad T = \begin{bmatrix} 1 & & & & & & \\ & \ddots & & & & & \\ & & 1 & & & & \\ d_1 & \cdots & d_{l-1} & d_l & d_{l+1} & \cdots & d_n \\ & & & & 1 & & \\ & & & & & \ddots & \\ & & & & & & 1 \end{bmatrix}, \quad \det T = d_l \neq 0 \tag{3.27}$$

the system (2.1) with the modification (3.26) is transformed into

$$\tilde{M}\ddot{y}(t) + \tilde{G}\dot{y}(t) + \tilde{K}y(t) = q(t) \tag{3.28a}$$

with the modification

$$y_l(t) \equiv 0 \quad \forall t > 0 \tag{3.28b}$$

where $y_l(t)$ is the lth component of the vector $y(t)$, $\tilde{M} = T^{-T}MT^{-1} = \tilde{M}^T$, $\tilde{G} = T^{-T}GT^{-1} = -\tilde{G}^T$, $\tilde{K} = T^{-T}KT^{-1} = \tilde{K}^T$, and $q = T^{-T}f$. In the new coordinates the gyroscopic system is under Modification 3; i.e, the lth point is fixed. Because the transformation T does not change the system eigenvalues, the following result is obtained by applying Theorem 3.5 to (3.28).

Theorem 3.7. If the modification (3.26) is imposed on the gyroscopic system whose natural frequencies are ω_k, the natural frequencies Ω_k of the modified system are such that

$$0 < \omega_1 \leq \Omega_1 \leq \omega_2 \leq \Omega_2 \leq \cdots \omega_{n-1} \leq \Omega_{n-1} \leq \omega_n \tag{3.29}$$

where due to the constraint one degree of freedom is removed.

Modification 6: A Spring Connecting Two Points of the Gyroscopic System

Without loss of generality, consider that a spring of coefficient μ is added to connect the first and second points of the gyroscopic system. The equation of motion for the modified system is

$$M\ddot{x}(t) + G\dot{x}(t) + Kx(t) = -\mu \begin{bmatrix} 1 & -1 & & & \\ -1 & 1 & & & \\ & & 0 & & \\ & & & \ddots & \\ & & & & 0 \end{bmatrix} x(t) + f(t). \tag{3.30}$$

Under the coordinate transformation

$$y(t) = Tx(t), \quad T = \begin{bmatrix} 1 & -1 & & \\ & 1 & & \\ & & \ddots & \\ & & & 1 \end{bmatrix}, \quad \det T = 1 \neq 0 \tag{3.31}$$

(3.30) is transformed into

$$\tilde{M}\ddot{y}(t) + \tilde{G}\dot{y}(t) + \tilde{K}y(t) = -\mu e_1 e_1^T y(t) + q(t) \tag{3.32}$$

where $\tilde{M} = T^{-T}MT^{-1} = \tilde{M}^T$, $\tilde{G} = T^{-T}GT^{-1} = -\tilde{G}^T$, $\tilde{K} = T^{-T}KT^{-1} = \tilde{K}^T$, $q = T^{-T}f$. In the new coordinates (3.32) is a gyroscopic system constrained by a spring (Modification 1) at the first point. Because the transformation T does not change the system eigenvalues, the following conclusion is drawn by applying Theorem 3.3.

Theorem 3.8. If a spring is added to connect any two points of the gyroscopic system whose natural frequencies are ω_k, the natural frequencies Ω_k of the modified system are such that

$$0 < \omega_1 \leq \Omega_1 \leq \omega_2 \leq \Omega_2 \leq \cdots \leq \omega_n \leq \Omega_n. \tag{3.33}$$

As an extreme case, when the spring coefficient μ goes to infinity, the two points of the gyroscopic system are rigidly connected; one degree of freedom is removed. Theorem 3.8 is still valid if the Ω_n is removed from (3.33). On the other hand, this limiting case is an example of Modification 5. Theorem 3.7 can be directly used here.

3.4 Remarks

Six types of non-dissipative modifications and their effects on the system natural frequencies have been studied through investigation of the close loop transfer function $H_{cl}(s)$. Many other modifications can be treated similarly. Also, the results presented herein can be easily extended to systems with repeated eigenvalues. Analyses follow the same steps as in this section. Results similar to Theorems 3.2 to 3.8 are expected.

The inclusion principle for the classical non-gyroscopic system (Theorem 3.1) has been proven using Rayleigh's quotient and the maximum-minimum characteristic of the system eigenvalues. Since the non-gyroscopic system is a special case of the gyroscopic system (2.1) with $G=0$, Theorems 3.3 and 3.5 in fact give a different proof of Theorem 3.1.

While this study is on the system natural frequencies, eigenvalue interlacing is also seen among the critical velocities of certain axially moving continua[6], where addition of a spring may decrease the critical velocities.

The transfer function method provides a systematic and convenient way to handle modification problems for gyroscopic systems. This method has been applied to develop eigenvalue inclusion principles for distributed gyroscopic systems, and combined gyroscopic systems composed of several component systems of many or infinite degrees of freedom[25, 26].

4. Vibration Control of Gyroscopic Systems via Direct Velocity Feedback

4.1 Background

Active vibration control of gyroscopic dynamic systems has received increased attention due to its wide industrial applications. Several control algorithms have been developed, among which are the independent modal space control[15], traveling wave control[19], and frequency domain control[28, 29]. In this section, an alternative control method for gyroscopic systems, namely direct velocity feedback, is developed. The purpose is to suppress the vibration of gyroscopic systems through energy dissipation.

Direct velocity feedback control for non-gyroscopic systems has been studied by Balas[1], who obtained a sufficient condition for the asymptotic stability of the control system. The condition requires that the number of sensors/actuators equal the total number of modes of the vibrating system, which is difficult to implement in practice. However, it is known that the asymptotic stability of a vibrating system can be achieved with a positive semi-definite damping matrix[9, 22]. This implies that all the modes can be stabilized by a smaller number of sensors and actuators. With this concept, the present work investigates stabilization of gyroscopic systems of many modes by a smaller number of sensors and actuators. A matrix decomposition and a coordinate transformation are proposed for stability analysis and controller design. Based on the modal information about the uncontrolled gyroscopic system, an asymptotic stability condition of the control system is derived. Also, a root locus sensitivity coefficient is defined as an index of damping introduced to each mode by the velocity feedback.

4.2 Problem Statement

The gyroscopic system (2.1) is controlled by velocity feedback with p pairs of colocated sensors and actuators; i.e.,

$$M\ddot{x}(t) + G\dot{x}(t) + Kx(t) = f_c(t) + f_e(t) \tag{4.1a}$$

$$f_c(t) = -BQB^T\dot{x}(t) \tag{4.1b}$$

where $B \in R^{n\times p}$ is a constant matrix dependent on the sensor/actuator locations, and $Q \in R^{p\times p}$ is the control gain matrix to be designed. Assume that the gain matrix Q is symmetric and positive semi-definite. Substituting (4.1b) into (4.1a) gives

$$M\ddot{x}(t) + (G + BQB^T)\dot{x}(t) + Kx(t) = f_e(t). \tag{4.2}$$

The control system (4.2) can be viewed as a damped gyroscopic system with BQB^T as the damping matrix. If rank $BQB^T = n$, following Balas[1], one can show that the system (4.2) is asymptotically stable. However, this requires that the number p of sensor/actuator pairs equal the total number n of modes of the gyroscopic system, which is difficult to achieve in practice, especially for large-scale systems. The objective of this study is to stabilize all n modes of the gyroscopic system via a smaller number of sensors and actuators.

4.3 Modal Stability Condition

Asymptotic stability of damped gyroscopic systems has been investigated by Hughes and Skelton[8]. They showed that the damped (controlled) gyroscopic system is asymptotically stable if and only if the undamped (uncontrolled) gyroscopic system is controllable with the damping matrix. This implies that a gyroscopic system of many degrees of freedom can be stabilized by a few colocated sensors and actuators. However, the stability condition of Hughes and Skelton does not tell which mode is or is not asymptotically stable, and how to design the damping matrix or the control gain matrix to attain asymptotic stability. In this section, the modal information of the uncontrolled gyroscopic system is utilized to extend the stability result of Hughes and Skelton, for the design of the control gain matrix Q.

Let the rank of the matrix BQB^T be r, $r \le p < n$. Because BQB^T is symmetric and positive semi-definite, it can always be written in a decomposed form[3]

$$BQB^T = \sum_{k=1}^{r} d_k \phi_k \phi_k^T \tag{4.3}$$

where d_k are real and positive scalars, and $\phi_k \in R^n$ are linearly independent vectors. Define a coordinate transformation

$$x(t) = \Phi^{-T} y(t), \quad \Phi = \begin{bmatrix} \phi_1 & \cdots & \phi_n \end{bmatrix} \in R^{n \times n} \tag{4.4}$$

where $\phi_1, \ldots, \phi_r$ are given in (4.3), and $\phi_{r+1}, \ldots, \phi_n$ are any linearly independent vectors that are orthogonal to $\text{Span}(\phi_1, \ldots, \phi_r)$. Accordingly, $\text{Span}(\phi_1, \ldots, \phi_n) = R^n$ and Φ^{-1} exists. With (4.4) Eq. (4.2) is transformed into an equivalent form

$$\tilde{M}\ddot{y}(t) + (\tilde{G} + \Lambda)\dot{y}(t) + \tilde{K}y(t) = q(t) \tag{4.5}$$

where
$$\Lambda = \text{diag}\begin{pmatrix} d_1 & \cdots & d_r & 0 & \cdots & 0 \end{pmatrix} \in R^{n \times n} \tag{4.6}$$

$\tilde{M} = \Phi^{-1} M \Phi^{-T} = \tilde{M}^T$, $\tilde{G} = \Phi^{-1} G \Phi^{-T} = -\tilde{G}^T$, $\tilde{K} = \Phi^{-1} K \Phi^{-T} = \tilde{K}^T$, $q(t) = \Phi^{-1} f_e(t)$, and $d_1, \ldots, d_r$ are from (4.3).

The eigenvalue problem and solutions (λ_k, v_k) of the uncontrolled gyroscopic system (i.e., (4.2) with $Q = 0$) have been studied in Sec. 2.2. In the new coordinates, the solutions of the eigenvalue problem associated with the uncontrolled system $\tilde{M}\ddot{y} + \tilde{G}\dot{y} + \tilde{K}y = 0$

$$\left(\gamma^2 \tilde{M} + \gamma \tilde{G} + \tilde{K}\right) w = 0 \tag{4.7}$$

have the following relationships with those of the original problem (2.3)

$$\gamma_k = \lambda_k, \quad w_k = \Phi^T v_k, \quad k = 1, \ldots, n. \tag{4.8}$$

The system (4.5) can be viewed as the uncontrolled system $\tilde{M}\ddot{y} + \tilde{G}\dot{y} + \tilde{K}y = 0$ connected to r dampers of coefficients d_k, k =1,..., r, that are located at the first r points, respectively. Here the word "damper" is borrowed for the equivalent system (4.5) although BQB^T in the original system (4.2) accounts for more general damping cases.

Theorem 4.1. The control system (3) is asymptotically stable if and only if the r-vector

$$\eta_l \equiv \begin{pmatrix} w_{l,1} & \cdots & w_{l,r} \end{pmatrix}^T \neq 0 \quad \text{for } l = 1,\dots,n \tag{4.9}$$

where $w_{l,k}$ is the kth element of the lth eigenvector w_l, i.e., $w_{l,k} = \phi_k^T v_l$ by (4.8).

Proof: Because of the non-singular matrix Φ, the system (4.2) is asymptotically stable if and only if the system (4.5) is asymptotically stable. According to Hughes and Skelton[8], the system (4.5) is asymptotically stable if and only if the uncontrolled system $\tilde{M}\ddot{y} + \tilde{G}\dot{y} + \tilde{K}y = 0$ is controllable with the r "dampers". By the controllability condition given by Yang and Mote[27, 30], the uncontrolled system is controllable if and only if the dampers are not all located at the nodes of every mode of the undamped system. Since $v_{l,k} = 0$ indicates that the kth damper is located at a node of the lth mode, the controllability condition is equivalent to (4.9). (Q.E.D.)

Using the modal information of the uncontrolled system, Theorem 4.1 conveniently checks the stability of each mode of the controlled gyroscopic system (4.2). If the r "dampers" are all located at the nodes of the lth mode, i.e., $\eta_l = 0$, the lth mode is not asymptotically stable and resonance occurs when a sinusoidal external force of frequency ω_l is applied to the gyroscopic system. If the qth "damper" is away from any node of the lth mode, i.e., $w_{l,q} = \phi_q^T v_l \neq 0$, the lth mode is asymptotically stable with the velocity feedback.

Theorem 4.1 also provides a simple way to design the control gain matrix Q. The idea is as follows. Assume that the matrix B is given and that the modal data $\{\omega_k, v_k\}$ of the uncontrolled gyroscopic system have been obtained. Write the gain matrix Q in a decomposed form

$$Q = \sum_{k=1}^{r} d_k \psi_k \psi_k^T \in R^{p\times p}, \quad \psi_k \in R^p, \quad d_k > 0 \tag{4.10}$$

such that

$$BQB^T = \sum_{k=1}^{r} d_k B\psi_k (B\psi_k)^T. \tag{4.11}$$

Note that $\phi_k = B\psi_k$, k =1,2,...,r. According to Theorem 4.1, the control system (4.2) is asymptotically stable if and only if

$$\begin{bmatrix} \psi_1 & \cdots & \psi_r \end{bmatrix}^T B^T v_l \neq 0, \quad l = 1,2,\dots,n. \tag{4.12}$$

Thus, the task is to find the linearly independent vectors $\psi_1,\dots,\psi_r$ that satisfy (4.12). Once these vectors are obtained, the control gain matrix Q will then be constructed by (4.10). Furthermore, if the number p of sensor/actuator pairs is identical to the rank r of the gain matrix Q, $\begin{bmatrix}\psi_1 & \cdots & \psi_r\end{bmatrix}$ in (4.12) is a non-singular, square matrix and the stability condition is simplified as

$$B^T v_l \neq 0, \quad l = 1,2,...,n. \tag{4.13}$$

4.4 Gyroscopic Systems with Repeated Eigenvalues

Theorem 4.1 is only valid for gyroscopic systems with distinctive eigenvalues. In this subsection the stability analysis in Sec. 4.3 is extended to systems with repeated eigenvalues. For the uncontrolled gyroscopic system (2.1) with repeated eigenvalues, the associated eigenvalue problem becomes

$$\left(\lambda_k^2 M + \lambda_k G + K\right) v_{kj} = 0, \quad k = 1,...,m, \quad j = 1,...,\beta_k, \quad \sum_{k=1}^{m} \beta_k = n \tag{4.14}$$

where $\lambda_k = i\omega_k$ is the kth eigenvalue of β_k multiplicity, and $v_{kj} \in C^n$, $j = 1,...,\beta_k$, are the corresponding linearly independent eigenvectors. Again, consider the gyroscopic system controlled by the velocity feedback (4.1b). Assume that $\text{rank} BQB^T = r$. Following the proof for Theorem 4.1, one obtains the modal stability condition for the system with repeated gyroscopic system.

Theorem 4.2. The control system (4.2) is asymptotically stable if and only if

(i) $r \geq \max\limits_{1 \leq k \leq m} \beta_k$, and

(ii) $\text{rank}\begin{bmatrix} w_{k1,1} & \cdots & w_{k\beta_k,1} \\ \vdots & & \vdots \\ w_{k1,r} & \cdots & w_{k\beta_k,r} \end{bmatrix} = \beta_k, \quad k = 1,2,...,m$

where $w_{kj,l} = \phi_l^T v_{kj}$, and ϕ_l are from (4.3).

Theorem 4.2 indicates that in order to stabilize all modes of vibration, the number of sensor/actuator pairs should be larger than the maximum eigenvalue multiplicity of the gyroscopic system.

4.5 Root Locus Sensitivity Coefficient

While Theorem 4.1 is applicable to the velocity feedback design for gyroscopic systems, it does not indicate how much damping introduced in each mode of vibration. In fact, as long as $\phi_1,...,\phi_r$ (or $\psi_1,...,\psi_r$) satisfy (4.9) (or (4.12)), $d_1,...,d_r$ can be any positive numbers. In this section the root locus sensitivity coefficient of the control system (4.2) is defined as an index of damping for the controller design.

Rewrite the matrix BQB^T as

$$BQB^T = \mu D = \mu \sum_{k=1}^{r} \tilde{d}_k \phi_k \phi_k^T, \quad \mu \tilde{d}_k = d_k, \quad \mu > 0. \tag{4.15}$$

In the Laplace transform domain, the control system (4.2) is described by

$$X(s) = H_{cl}(s) F_{el}(s) \tag{4.16}$$

with the closed-loop transfer function

$$H_{cl}(s) = \left(s^2 M + sG + K + \mu s D\right)^{-1} \in C^{n \times n} \tag{4.17}$$

where $F_{el}(s) = F_e(s) + (G + sM)a_0 + Mb_0$. The uncontrolled system ($\mu = 0$) is

$$X(s) = H_o(s) F_{el}(s) \tag{4.18}$$

with the open-loop transfer function

$$H_o(s) = \left(s^2 M + sG + K\right)^{-1} = \frac{1}{2} \sum_{k=\pm 1}^{\pm n} \frac{-\lambda_k}{s - \lambda_k} v_k \bar{v}_k^T \in C^{n \times n} \tag{4.19}$$

where (2.23b) has been used. With (4.17) and (4.19), it is easy to show that

$$H_{cl}(s) = \left(I + \mu s H_o(s) D\right)^{-1} H_o(s) \tag{4.20}$$

and that the characteristic equation of the closed-loop system is

$$\det\left(I + \mu s H_o(s) D\right) = 0. \tag{4.21}$$

Consider the root loci of the closed-loop system versus the gain parameter μ. Denote the root locus originating from the open-loop pole λ_k by $s_k(\mu)$, $s_k(0) = \lambda_k$. Define the sensitivity coefficient of the kth root locus as

$$\sigma_k \equiv -\mathrm{Re}\left(\frac{ds_k}{d\mu} |_{\mu=0} \right). \tag{4.22}$$

Claim: $$\sigma_k = \frac{1}{2} \omega_k^2 \sum_{j=1}^{r} \tilde{d}_j \left| w_{k,j} \right|^2, \quad w_{k,j} = \phi_j^T v_k, \quad k = 1, 2, \ldots, n. \tag{4.23}$$

Proof: At a point of the root locus s_k close to $\lambda_k = i\omega_k$, $s(\Delta\mu) = \lambda_k + \Delta s$, $|\Delta\mu| << 0$, $H_o(s) \approx -\frac{1}{2\Delta s} \lambda_k v_k \bar{v}_k^T$, and the characteristic equation (4.21) becomes

$$\det\left(I + \frac{\Delta\mu}{2\Delta s} \omega_k^2 v_k \bar{v}_k^T D \right) + O(\Delta\mu) = 0.$$

Note that for $a, b \in C^m$ and $A \in C^{m \times m}$, $\det\left(I + ab^T A\right) = 1 + b^T A a$. As $\Delta\mu \to 0$, $\Delta s / \Delta\mu \to ds_k / d\mu |_{\mu=0}$, and the above reduces to

$$1+\frac{1}{2}\omega_k^2\left(\frac{ds_k}{d\mu}|_{\mu=0}\right)^{-1}\bar{v}_k^T D u_k = 0$$

Substitution of (4.15) and (4.22) into the above equation yields (4.23). (Q.E.D.)

It is seen from (4.23) that all the root locus sensitivity coefficients are non-negative. According to Theorem 4.1 and Claim, $\sigma_k > 0$ for all k if and only if the gyroscopic system is asymptotically stable. Physically, $\sigma_k > 0$ indicates that damping has been introduced in the kth mode of vibration.

Now suppose that the stability condition (4.9) holds. Because $\sigma_k > 0$ for all k, as μ increases from zero, all the root loci $s_k(\mu)$ move into the left-half complex plane. Given a small gain $\mu = \Delta\mu$,

$$\operatorname{Re} s_k(\Delta\mu) = \operatorname{Re}\left\{\lambda_k + \left(\frac{ds_k}{d\mu}|_{\mu=0}\right)\Delta\mu + O(\Delta\mu^2)\right\} \approx -\sigma_k\Delta\mu. \tag{4.24}$$

With (4.23) and (4.24), one estimation of the damping ratio of the kth mode is

$$\xi_k \approx \frac{\sigma_k\Delta\mu}{2\omega_k} = \frac{1}{4}\omega_k\sum_{j=1}^{r} d_j\,|\,w_{k,j}\,|^2 = \frac{1}{4}\omega_k\sum_{j=1}^{r} d_j\,|\,\psi_j^T B^T v_k\,|^2 \tag{4.25}$$

where $d_j = \Delta\mu\tilde{d}_j$ has been used. It is seen that the larger the sensitivity coefficient σ_k, the larger the damping ratio ξ_k. Although $\Delta\mu$ is limited to small values, $d_j = \Delta\mu\tilde{d}_j$ in general are not infinitesimal. Therefore, the sensitivity coefficient σ_k can serve as an index of damping introduced to each mode of vibration by the velocity feedback (4.1b).

Based on (4.23) and (4.25), the controller design will choose proper d_j and ϕ_j (or ψ_j) to satisfy damping specification of some modes of interest, and at the same time to guarantee asymptotic stability of the whole control system by setting $\sigma_k > 0$, $k = 1,2,...,n$. With this scheme, pervasive damping for gyroscopic systems can be achieved by a smaller number of sensors and actuators.

4.6 Example

The theory is applied to a band saw that is controlled via two pairs of colocated sensors and actuators ($p = 2$); see Figure 4.1. The band saw is a gyroscopic dynamic system. The band transverse displacement $u(z,t)$ is modeled by the nondimensional equations[16]

$$\left\{\frac{\partial^2}{\partial t^2} + 2c\frac{\partial^2}{\partial z\partial t} + \frac{\partial^4}{\partial z^4} + (\kappa c^2 - T_0)\frac{\partial^2}{\partial z^2}\right\}u(z,t) = \sum_{j=1}^{2}\delta(z-a_j)h_j(t), \quad z\in(0,1)$$

$$w = 0 \quad \text{and} \quad \frac{\partial^2}{\partial z^2}w = 0 \quad \text{at } z = 0,1 \tag{4.26}$$

where c and T_0 are the dimensionless transport speed and tension of the band, respectively, $\kappa \in (0,1)$ is a constant dependent upon the pulley mounting system of the band saw, and $h_j(t)$ is the control input by the jth actuator located at $z = a_j \in (0,1)$. For direct velocity feedback with the colocated sensors and actuators, the control inputs are represented by

$$\begin{pmatrix} h_1(t) \\ h_2(t) \end{pmatrix} = -Q\frac{\partial}{\partial t}\begin{pmatrix} u(a_1,t) \\ u(a_2,t) \end{pmatrix}, \quad Q \in R^{2\times 2} \tag{4.27}$$

where Q is the gain matrix to be designed.

A six-term Galerkin's discretization ($n = 6$)

$$u(z,t) \approx \sum_{k=1}^{6} q_k(t) \sin k\pi z. \tag{4.28}$$

reduces (4.26) to the control system (4.2) with

$$x(t) = \{q_1(t) \quad \cdots \quad q_6(t)\}^T, \quad M = I_6$$

$$G = -G^T = [g_{jk}], \quad g_{jk} = \begin{cases} \dfrac{8jkc}{j^2 - k^2}, & \text{if } j + k = \text{odd number} \\ 0, & \text{otherwise} \end{cases}$$

$$K = \operatorname*{diag}_{1\le j\le 6}\left\{j^2\pi^2(j^2\pi^2 + T_0 - \kappa c^2)\right\}$$

$$B = 2\times\begin{bmatrix} \sin \pi a_1 & \cdots & \sin 6\pi a_1 \\ \sin \pi a_2 & \cdots & \sin 6\pi a_2 \end{bmatrix}^T \in R^{6\times 2}. \tag{4.29}$$

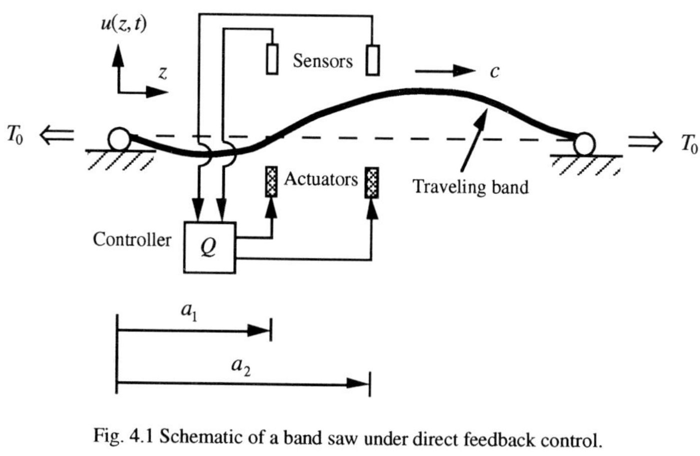

Fig. 4.1 Schematic of a band saw under direct feedback control.

In the numerical simulation, the band saw parameters and the locations of the sensors/actuators are chosen as follows

$$c = 10, \quad T_0 = 100, \quad \kappa = 0.5, \quad a_1 = 0.51, \quad a_2 = 0.29. \tag{4.30}$$

With the data, the natural frequencies of the uncontrolled band saw are calculated as follows:

$$\begin{aligned} &\omega_1 = 17.3678, \quad \omega_2 = 54.1024, \quad \omega_3 = 112.1830 \\ &\omega_4 = 189.6190, \quad \omega_5 = 308.2317, \quad \omega_6 = 484.5124. \end{aligned} \tag{4.31}$$

The corresponding eigenvectors v_k are given in Table 4.1.

The control gain matrix is of the form

$$Q = d_1\psi_1\psi_1^T + d_2\psi_2\psi_2^T \in R^{2\times 2}, \quad \psi_1, \psi_2 \in R^2, \quad d_1, d_2 > 0 \tag{4.32}$$

Selection of ψ_1, ψ_2, d_1 and d_2 is not unique. With no intention to find the optimal gain parameters in this paper, let us assume that

$$\psi_1 = \begin{pmatrix} 1 \\ -0.1 \end{pmatrix} \quad \psi_2 = \begin{pmatrix} -0.3 \\ 1 \end{pmatrix}. \tag{4.33}$$

Because ψ_1 and ψ_2 are linearly independent, Q is always of full rank. Thus, by (4.13) and the data given in Table 4.1, it is easy to show that all modes are stabilized for any positive d_1 and d_2. Suppose that some specific damping ratios ξ_1 and ξ_5 of the first and fifth modes are desired. To meet the damping specification, d_1 and d_2 can be obtained through use of (4.25):

$$\begin{pmatrix} d_1 \\ d_2 \end{pmatrix} = \begin{bmatrix} |\psi_1^T B^T u_1|^2 & |\psi_2^T B^T u_1|^2 \\ |\psi_1^T B^T u_5|^2 & |\psi_2^T B^T u_5|^2 \end{bmatrix}^{-1} \begin{pmatrix} 4\xi_1/\omega_1 \\ 4\xi_5/\omega_5 \end{pmatrix}. \tag{4.34}$$

Once d_1 and d_2 are evaluated, the control gain matrix Q can be constructed by (4.32). The computational effort in the above controller design is almost trivial.

Four sets of desired damping ratios are considered in the numerical simulation:

Case 1. $\xi_1 = 0.05$, $\xi_5 = 0.02$;
Case 2. $\xi_1 = 0.10$, $\xi_5 = 0.05$;
Case 3. $\xi_1 = 0.12$, $\xi_5 = 0.07$;
Case 4. $\xi_1 = 0.20$, $\xi_5 = 0.10$.

Table 4.2 gives the control gain matrix Q estimated based on (4.32-4.34), and the actual damping ratios of the first and fifth modes with the velocity feedback. The closed-loop eigenvalues in each of the above cases are listed in Table 4.3. It is seen that the proposed method provides a fairly good estimation of the control gain matrix for the desired damping ratios.

TABLE 4.1. Eigenvectors of the uncontrolled band saw (e-15 = 10^{-15}, i = $\sqrt{-1}$)

v_1		v_2		v_3	
1.6334e-15	+i3.9498e-02	1.5920e-16	+i1.0902e-02	2.0924e-17	-i3.5952e-03
1.2184e-02	-i5.0594e-16	-9.7582e-03	+i1.5480e-16	5.6659e-03	+i3.4716e-17
-6.4952e-17	-i1.6167e-03	1.1033e-16	+i7.1859e-03	-2.1374e-17	+i3.4802e-03
3.4311e-04	-i1.4290e-17	2.1972e-03	-i3.4203e-17	3.9940e-03	+i2.4479e-17
-4.8712e-18	-i1.2239e-04	1.7152e-20	+i2.0069e-05	1.1009e-17	-i1.8016e-03
5.1819e-05	-i2.5015e-18	2.0598e-04	-i2.6150e-18	-2.1918e-04	-i1.1649e-18

v_4		v_5		v_6	
1.1125e-18	-i6.5966e-04	-1.0780e-19	-i1.8376e-04	2.7418e-19	-i8.3331e-06
2.6845e-03	+i2.4306e-18	-3.1633e-04	+i4.3400e-19	-1.2857e-04	-i7.9579e-19
-3.3289e-18	+i3.4531e-03	8.0396e-19	-i1.6746e-03	-2.7587e-20	-i4.4909e-05
-1.7383e-03	-i1.8897e-18	2.4786e-03	+i1.1446e-18	-6.2855e-04	+i4.0902e-20
-2.2508e-18	+i2.4237e-03	-6.8985e-19	+i1.4985e-03	-5.3745e-20	-i1.5091e-03
1.2353e-03	+i1.5789e-18	1.4789e-03	+i2.4929e-19	1.7714e-03	+i4.5641e-19

TABLE 4.2. Control gain matrix and damping ratios

Case No.	Desired Damping Ratios	Q	Actual Damping Ratios
1	$\xi_1 = 0.05$ $\xi_5 = 0.02$	$\begin{bmatrix} 1.2002 & -1.0807 \\ -1.0807 & 3.3103 \end{bmatrix}$	$\xi_1 = 0.0503$ $\xi_5 = 0.0201$
2	$\xi_1 = 0.10$ $\xi_5 = 0.05$	$\begin{bmatrix} 1.8154 & -2.8062 \\ -2.8062 & 9.0295 \end{bmatrix}$	$\xi_1 = 0.1027$ $\xi_5 = 0.0521$
3	$\xi_1 = 0.12$ $\xi_5 = 0.07$	$\begin{bmatrix} 1.5934 & -4.0123 \\ -4.0123 & 13.2443 \end{bmatrix}$	$\xi_1 = 0.1230$ $\xi_5 = 0.0757$
4	$\xi_1 = 0.20$ $\xi_5 = 0.10$	$\begin{bmatrix} 3.6308 & -5.6124 \\ -5.6124 & 18.0590 \end{bmatrix}$	$\xi_1 = 0.2226$ $\xi_5 = 0.1149$

TABLE 4.3. Eigenvalues of the band saw under velocity feedback ($i = \sqrt{-1}$)

Mode No.	Case 1	Case 2	Case 3	Case 4
1	-1.7425+i17.315	-3.5786+i17.423	-4.4319+i18.020	-7.9827+i17.932
2	-4.4327+i53.994	-12.633+i52.922	-20.382+i50.495	-31.400+i45.559
3	-3.6581+i112.14	-7.1558+i111.44	-8.1473+i109.98	-14.682+i108.28
4	-2.2256+i189.59	-5.8697+i189.41	-8.6045+i189.10	-12.795+i188.54
5	-12.377+i307.51	-31.554+i302.96	-44.840+i296.05	-64.216+i279.39
6	-8.1620+i483.55	-18.996+i478.13	-23.929+i471.99	-28.499+i463.42

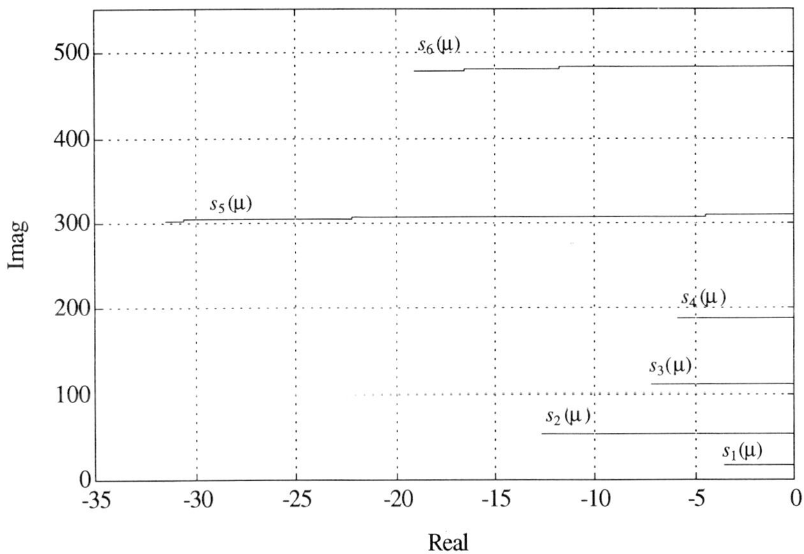

Figure 2. Root loci of the band saw under colocated velocity feedback:

$f_c(t) = -\mu BQB^T \dot{x}(t)$, $\mu \in [0,1]$, Q given in Case 2 of Table 4.2.

To see the whole picture of the closed-loop stability, the root loci of the gyroscopic system subject to the control force $f_c(t) = -\mu BQB^T \dot{x}(t)$, are plotted versus the parameter μ, $\mu \in [0,1]$; see Figure 4.2. Here, the gain matrix Q obtained in Case 2 of Table 4.2 is adopted. All the modes are stabilized by the direct velocity feedback. Each root locus starts at one open-loop pole. As $\mu \to 1$, the root loci approach the closed-loop poles of Case 2 listed in Table 4.2.

4.7 Remark

Colocated velocity feedback control of gyroscopic dynamic systems has been investigated. All modes of vibration can be stabilized by a smaller number of sensors and actuators. The control gain matrix can be designed to gain desired damping ratios for specific modes of interest, and at the same time to guarantee asymptotic stability of all other modes. Although this paper is focused on active control, the proposed method can also be applied to passive damping design and combined active and passive vibration control for gyroscopic systems.

5. Conclusions

A transfer function method has been applied to study eigenvalue inclusion principles and vibration control of gyroscopic dynamic systems. In this method, the state-space adjoint eigenvectors are given in terms of the eigensolutions from the original physical coordinate system; the system response can be expressed in an eigenfunction expansion even if the eigenvectors of the gyroscopic system are non-orthogonal. As a result constraints/modifications or control forces, which are described in the physical coordinates, can be conveniently treated in the transfer function formulation. The method has been shown to be effective in modeling and control of gyroscopic systems. Although this paper is limited to lumped gyroscopic systems, the transfer function method developed is readily extended to distributed gyroscopic systems whose motion is described by partial diffuremtial equations.

6. Acknowledgment

The work presented here was partially sponsored by the Powell Foundation, National Science Foundation, and USC Faculty Research and Innovation Fund.

7. References

1. Balas, M. J., 1979, "Direct Velocity Feedback Control of Large Space Structures," *AIAA Journal of Guidance, Control, and Dynamics*, Vol. 2, pp. 252-253.

2. Balas, M. J., 1982, "Trends in Large Space Structure Control Theory: Fondest Hopes, Wildest Dreams," *IEEE Transactions on Automatic Control*, Vol. AC-27, pp. 15-33.

3. Bellman, R., 1970, *Introduction to Matrix Analysis*, 2nd ed., McGraw-Hill, New York.

4. D'Eleuterio, G. M. T., and Hughes, P. C., 1984, "Dynamics of Gyroelastic Continua," *ASME Journal of Applied Mechanics*, Vol. 51, No. 2, pp. 415-422.

5. Guran, A., 1993, "Chaotic Motion of a Kelvin Type Gyrostat in a Circular Obit," *Acta Mechanica*, Vol. 98, pp. 51-61.

6. Guran, A., and Plaut, R. H., 1994, "Stability of a Fluid-Conveying Pipe with Flow-Dependent Support Stiffness," *ASME Journal of Applied Mechanics*, Vol. 61, No. 2, pp. 477-478.

7. Hagedorn, P., 1975, "Uber Die Instabilität Konservativer System Mit Gyroskopischen Kraften," *Arch. Rat. Mech. Anal.*, Vol. 58, pp. 1-9.

8. Hughes, P. C., and Skelton, R.E., 1979, "Stability, Controllability and Observability of Linear Matrix-Second-Order Systems," *JACC Proceedings*, Denver, pp. 56-62. Also in *ASME Journal of Applied Mechanics* [Vol. 47, June 1980, pp. 415-420].

9. Huseyin, K., 1978, *Vibration and Stability of Multiple Parameter Systems*, Sijthoff & Noordhoff International Publishing.

10. Inman, D. J., 1988, "A Sufficient Condition for the Stability of Conservative Gyroscopic Systems," *ASME Journal of Applied Mechanics*, Vol. 55, No. 4, pp. 895-898.

11. Johnson, T. L., 1982, "Progress in Modeling and Control of Flexible Spacecraft," *Journal of the Franklin Institute*, Vol. 315, pp. 495-520.

12. Juang, J. and Balas, M., 1978, "Dynamics and Control of Large Spinning Spacecraft," *AIAA/AAS Astrodynamics Conference*, Palo Alto, August. Also in *Journal of Astronautical Sciences* [Vol. 28, No. 1, 1980, pp. 31-48].

13. Meirovitch, L., 1974, "A New Method of Solution of the Eigenvalue Problem for Gyroscopic Systems," *AIAA Journal*, Vol. 12, No. 10, pp. 1337-1342.

14. Meirovitch, L., 1980, *Computational Methods in Structural Dynamics*, Sijthoff & Noordhoff.

15. Meirovitch, L., and Öz, H., 1980, "Modal-Space Control of Distributed Gyroscopic Systems," *Journal of Guidance and Control*, Vol. 3, No. 2, March-April, pp. 140-150.

16. Mote, C. D., Jr., 1965, "A Study of Bandsaw Vibrations," *J. Franklin Inst.*, Vol. 279, No. 6, pp. 329-346.

17. Mote, C. D., Jr., 1972, "Dynamic Stability of Axially Moving Materials," *Shock and Vibration Digest*, Vol. 4, No. 4, April, pp. 2-11.

18. Païdoussis, M. P., 1987, "Flow-Induced Instabilities of Cylindrical Structures," *Applied Mechanics Review*, Vol. 40, No. 2, pp. 163-175.

19. Radcliffe, C. J., and Mote, C. D., Jr., 1983, "Identification and Control of Rotating Disk Vibration," *ASME Journal of Dynamic Systems, Measurement, and Control*, Vol. 105, No. 1, pp. 39-45.

20. Rayleigh, J. W. S., 1945, *The Theory of Sound*, Dover, New York, Vol. 1, p. 119.

21. Ulsoy, A.G., 1984, "Vibration Control in Rotating or Translating Elastic Systems," *ASME Journal of Dynamic Systems, Measurement, and Control*, Vol. 106, No. 1, March, pp. 6-14.

22. Walker, J. A., and Schmitendorf, W. E., 1973, "A Simple Test for Asymptotic Stability in Partially Dissipative Symmetric Systems," ASME *Journal of Applied Mechanics*, Vol. 40, pp. 1120-1121.

23. Yang, B., 1992, "Eigenvalue Inclusion Principles for Discrete Gyroscopic Systems," *ASME Journal of Applied Mechanics*, Vol. 59, No. 2, Pt. 2, pp. S278-S283.

24. Yang, B., 1992, "Vibration Control of Gyroscopic Systems by Direct Velocity Feedback", ASME DSC-Vol. 38, pp. 331-335. Also to appear in *Jounal of Sound and Vibration*

25. Yang, B., 1992, "Eigenvalue Inclusion Principles for Distributed Gyroscopic Systems," *ASME Journal of Applied Mechanics*, Vol. 59, No. 3, pp. 650-656.

26. Yang, B., 1992, "Natural Frequencies of Combined Gyroscopic Systems," *Journal of Sound and Vibration*, Vol. 159, No. 1, pp. 23-37.

27. Yang, B., 1994, "Controllability and Observability of General Mechanical Systems," *ASME Journal of Vibration and Acoustics*, in press.

28. Yang, B., and Mote, C. D., Jr., 1991, "Frequency-Domain Vibration Control of Distributed Gyroscopic Systems," *ASME Journal of Dynamic Systems, Measurement, and Control*, Vol. 113, No. 1, pp. 18-25.

29. Yang, B., and Mote, C. D., Jr., 1991, "Active Vibration Control of the Axially Moving String in the s Domain," *ASME Journal of Applied Mechanics*, Vol. 58, No. 1, pp 189-196.

30. Yang, B., and Mote, C. D., Jr., 1991, "Controllability and Observability of Distributed Gyroscopic Systems," *ASME Journal of Dynamic Systems, Measurement, and Control*, Vol. 113, No. 1, pp. 11-17.14.

31. Ziegler, H., 1968, *Principles of Structural Stability*, Blaisdel Publishing Company, London, England.

Wave Motion, Intelligent Structures and Nonlinear Mechanics, pp. 163–193
edited by A. Guran and D. J. Inman
Series on Stability, Vibration and Control of Structures: Vol. 1

NONLINEAR CONTROL AND BOUNDARY TRANSITION OF CANTILEVERED DISTRIBUTED SYSTEMS

H. S. Tzou and D. D. Johnson
Department of Mechanical Engineering
Center for Robotics and Manufacturing Systems
University of Kentucky, Lexington, KY 40506

and

K. J. Liu
Active Noise and Vibration Tech., Phoenix, AZ

ABSTRACT

Distributed control of cantilevered continua (distributed systems) using fully distributed piezoelectric layers has been a classical problem for years. The equivalent control actuation is introduced at the free end of the cantilevered distributed systems, and the control action is equivalent to an (electric) counteracting moment determined by geometry, material properties, and control algorithms. In the classical negative proportional velocity feedback, the control effect is proportional to the feedback voltage, if the system is considered linear and the feedback voltage is reasonably low. In this study, nonlinear control effects and boundary transition of cantilever plates and beams with large feedback voltages are studied. Nonlinear control characteristics are investigated. Analytical solutions are compared with finite element simulations and experimental data.

1. Introduction

Distributed control of elastic continua, e.g., shells, plates, etc., using distributed piezoelectric sensor/actuator layers has attracted tremendous attention and interest in recent years. This is partially due to a new development of "smart" or "intelligent" structural systems possessing self sensing, diagnosis, actuation, and adaptation capabilities, and also partially due to the distributed control itself posting technical challenges for decades. The original activities of sensor/actuator integration concept was initialized from studies of distributed control of flexible beams using spatially distributed piezoelectric layers (Baily and Hubbard, 1985; Plumb, Hubbard, and Baily, 1987; Baz and Poh, 1988; Tzou, 1987; etc.) In the meantime, the intelligent structure concept was quickly emerging and expanded to other active electromechanical materials, e.g., shape memory alloys, electromagnetostrivtive materials, electrorheological materials, electrostrictive materials, etc. (Rogers, 1988; Crawley, Fanson, Wada, 1989; Gandhi and Thompson, 1992; Tzou and Anderson, 1992; Tzou and Fukuda, 1992; Tzou, 1993; Guran and Inman, 1994; Guran, 1994; Inman, 1990).

Distributed control of cantilever beams with surface laminated piezoelectric sensor and actuator layers has been a classical problem for a number of years (Baily and Hubbard, 1985; Crawley and de Luis, 1987; Plumb, Hubbard, and Baily, 1987; Baz and Poh, 1988; Tzou, 1987). Temperature effect to the dynamics and control of piezoelectric laminated beam was recently investigated (Tzou and Ye, 1993). Actuation and vibration control of plates with laminated piezoelectric layers was also investigated (Richetts, 1981 and 1989; Lee and Moon, 1988; Burke and Hubbard, 1990; Crawley and Lazarus, 1991; Dimitriadis, et al., 1991; Ha, et al., 1992; Hwang and Park, 1993; Liu, 1989; Tzou and Fu, 1994; Tzou and Tseng, 1990; etc.) Modeling and nonlinear control (via the negative velocity and Lyapunov controls) were investigated using analytical, experimental, and finite element techniques.

In this chapter, nonlinear vibration controls of piezoelectric laminated elastic plates and beams are studied and their boundary transition is investigated when large feedback voltages are imposed. The emphasis is placed on the control nonlinearity introduced by the large feedback voltage and boundary transitions. The *Laplace transform method* is employed for analytical solutions, the finite element method for simulation results, and laboratory experiments to validate fundamental control characteristics. Nonlinear control effects and boundary transition are studied.

2. System Modeling

Closed–loop system equations of plates and beams can be directly simplified from the generic shell electromechanical equations by using four system parameters: two Lamé parameters (A_1 and A_2) and two radii (R_1 and R_2) (Tzou, 1991, 1992, 1993). In general, plates are special cases of shells, and beams are special cases of plates. On the other hand, plates can be regarded as zero–curvature shells, i.e., infinite radii of curvature of two in–plane axes. Beams can be considered as one–dimensional (1D) rectangular plates in which the Poisson's ratio effects are neglected. In this study, only rectangular plates are considered. Accordingly, a straight beam can be considered as a slice of a rectangular plate. It is assumed that both plates and beams are surface laminated with fully distributed piezoelectric layers which are used as distributed sensors and actuators, respectively. The effective axes of piezoelectric layers are aligned with the in–plane coordinate axes. In the later derivations, mathematical modeling of plates is presented first, and the plate equation is reduced to a beam equation. Both plates and beams are assumed cantilevered, and their transverse vibrations are of interest in the later analysis. The electromechanical equations for a generic thin shell laminated with distributed piezoelectric sensor and actuator layers are briefly reviewed. Then, they are reduced to a transverse plate vibration equation and a transverse beam vibration equation when proper assumptions are employed.

2.1 Generic Shell Sensor/Actuator System

It is assumed that a generic thin shell is laminated with distributed piezoelectric sensor and actuator layers, Figure 1. The piezoelectric layers are perfectly bonded on the top and bottom surfaces of the elastic shell and the layers are much thinner than the elastic beam. Thus, physical properties of the bonding material and the piezoelectric layers are not considered in the analyses. The sensor layer gives a sensor signal proportional to the strain states, and then the signal is amplified and fed back to the distributed actuator layer from which distributed control actions are induced. Note that the elastic shell continuum follows the Kirchoff–Love thin shell assumptions.

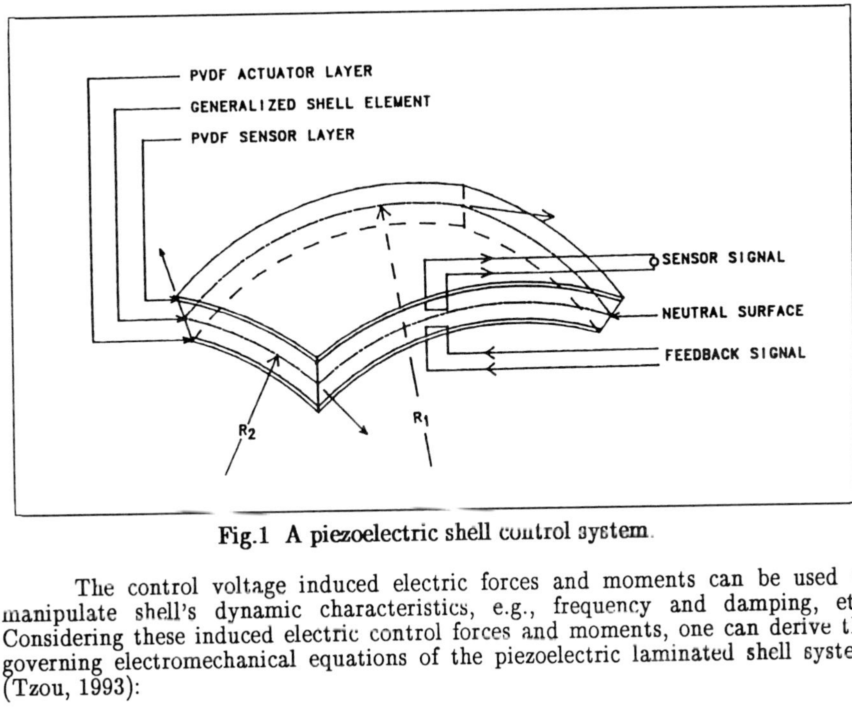

Fig.1 A piezoelectric shell control system.

The control voltage induced electric forces and moments can be used to manipulate shell's dynamic characteristics, e.g., frequency and damping, etc. Considering these induced electric control forces and moments, one can derive the governing electromechanical equations of the piezoelectric laminated shell system (Tzou, 1993):

$$
\begin{aligned}
&-(\tilde{N}_{11}A_2)_{,\alpha_1} - (\tilde{N}_{21}A_1)_{,\alpha_2} - \tilde{N}_{12}(A_{1,\alpha_2}) + \tilde{N}_{22}(A_{2,\alpha_1}) - \frac{1}{R_1}[(\tilde{M}_{11}A_2)_{,\alpha_1} \\
&\quad + (\tilde{M}_{21}A_1)_{,\alpha_2} + \tilde{M}_{12}(A_{1,\alpha_2}) - \tilde{M}_{22}(A_{2,\alpha_1})] + A_1A_2\,\rho\, h\, \ddot{u}_1 \\
&\quad = A_1A_2\, F_1\,, \qquad (1)
\end{aligned}
$$

$$- (\tilde{N}_{12}A_2)_{,\alpha_1} - (\tilde{N}_{22}A_1)_{,\alpha_2} - \tilde{N}_{21}(A_{2,\alpha_1}) + \tilde{N}_{11}(A_{1,\alpha_2}) - \frac{1}{R_2}[(\tilde{M}_{12}A_2)_{,\alpha_1} + (\tilde{M}_{22}A_1)_{,\alpha_2} + \tilde{M}_{21}(A_{2,\alpha_1}) - \tilde{M}_{11}(A_{1,\alpha_2})] + A_1A_2\,\rho\,h\,\ddot{u}_2 = A_1A_2\,F_2\,, \tag{2}$$

$$- \{[(\tilde{M}_{11}A_2)_{,\alpha_1} + (\tilde{M}_{21}A_1)_{,\alpha_2} + \tilde{M}_{12}(A_{1,\alpha_2}) - \tilde{M}_{22}(A_{2,\alpha_1})]/A_1\}_{,\alpha_1} - \{[(\tilde{M}_{12}A_2)_{,\alpha_1} + (\tilde{M}_{22}A_1)_{,\alpha_2} + \tilde{M}_{21}(A_{2,\alpha_1}) - \tilde{M}_{11}(A_{1,\alpha_2})]/A_2\}_{,\alpha_2} + A_1A_2\,[\tilde{N}_{11}/R_1 + \tilde{N}_{22}/R_2] + A_1A_2\,\rho\,h\,\ddot{u}_3 = A_1A_2\,F_3\,, \tag{3}$$

where A_1 and A_2 are the Lamé parameters determined by a fundamental form:

$$(ds)^2 = (A_1)^2(d\alpha_1)^2 + (A_2)^2(d\alpha_2)^2\,, \tag{4}$$

where ds is an arbitrary infinitesimal distance on the neutral surface; $d\alpha_1$ and $d\alpha_2$ are infinitesimal coordinate changes in the α_1 and α_2 directions, respectively. Note that $(\cdot)_{,\alpha_i} = \partial(\cdot)/\partial\alpha_i$; R_1 and R_2 are radii of two in–plane α_1 and α_2 axes; ρ is the mass density; h is the shell thickness; $\ddot{u}_i$ is the acceleration; F_i is the distributed mechanical excitation in the i–th direction; and the ~ terms include both elastic and feedback control effects, i.e., the resultant forces $\tilde{N}_{ij}$ and moments $\tilde{M}_{ij}$ include the original elastic components and the induced normal control forces and counteracting moments, i.e.,

$$\tilde{N}_{ij} = N_{ij} - d_{ij}h^aY_p\phi^a\,, \tag{5}$$

$$\tilde{M}_{ij} = M_{ij} - M^a_{ij}\,, \tag{6}$$

where N_{ij} and M_{ij} are the elastic membrane forces and bending moments; d_{ij} is the piezoelectric constant; h^a is the thickness of the piezoelectric actuator layer; Y_p is Young's modulus of the piezoelectric layer; and ϕ^a is the control voltage. Control moments M^a_{ij} are defined by

$$M^a_{ij} = r^a_i d_{ij} Y_p \phi^a\,, \tag{7}$$

where r^a_i is the moment arm and $r^a_i = (h+h^a)/2$; h^a is the piezoelectric actuator thickness; and the superscribe "a" denotes properties associated with the distributed actuator. Note that the in–plane twisting effect is usually assumed negligible, i.e.,

$\tilde{M}_{12} = M_{12}$ and $\tilde{N}_{12} = N_{12}$. In addition, material properties of the piezoelectric actuator layer are not considered. Note that depending on four system parameters, two Lamé parameters and two radii, the system equations can be simplified to many common geometries, e.g., spherical shells, cylindrical shells, conical shells, plates, rings, arches, beams, etc. (Tzou, 1993). For plates and beams discussed next, it is assumed that two thin layers of piezoelectric material are respectively bonded on the top and bottom surfaces of the elastic plate and beam. The effective axes of the piezoelectric layers are aligned with the coordinate axes to ensure its maximum piezoelectric effects in sensor and actuator applications. It is intended to use the distributed sensor signal as a feedback reference in a closed–loop velocity feedback control system. Note that the control action to the cantilevered plate and beam is introduced through an equivalent *boundary control.*

2.2 Plate Sensor/Actuator Systems

A rectangular plate is usually regarded as a zero–curvature shell, i.e., $1/R_1 = 1/R_2 = 0$. Figure 2 shows a rectangular plate defined in a Cartesian coordinate system. Since both X and Y axes are straight lines, their radii are $R_1 = R_2 = \infty$.

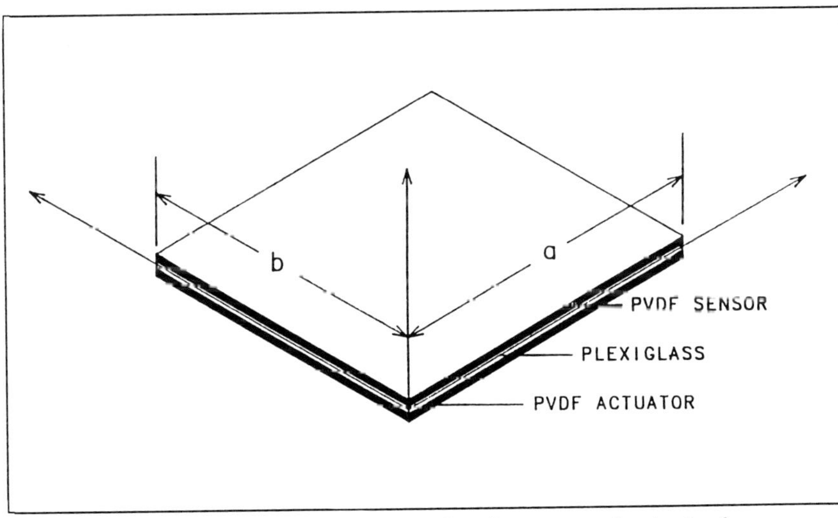

Fig.2 A cantilever plate with piezoelectric sensor and actuator layers.

Examining the rectangular plate in Figure 2, one can write a fundamental form for the rectangular plate:

$$(ds)^2 = (1)^2(dx)^2 + (1)^2(dy)^2 \,, \tag{8}$$

where dx and dy are infinitesimal coordinate changes. Comparing this with the standard fundamental form of the generic shell, one can define the Lamé parameters $A_1 = 1$ and $A_2 = 1$ for the rectangular plate. Substituting the four parameters into the generic shell electromechanical equations Eq.(1)–(3), one can derive the transverse governing equation of the plate:

$$\rho h \ddot{u}_3 + D\left[u_{3,xxxx} + 2\, u_{3,xxyy} + u_{3,yyyy} \right] - \left[M^a_{11,xx} + M^a_{22,yy} \right] = F_3 \,, \tag{9}$$

where h is the plate thickness; u_3 is the transverse displacement; D is the bending stiffness and $D = (Yh^3)/[12(1-\mu^2)]$; Y is Young's modulus of the plate; and μ is Poisson's ratio. M^a_{11} and M^a_{22} are control moments defined by

$$M^a_{11} = r^a_1\, d_{31}\, Y_p\, \phi^a \,, \tag{10a}$$

$$M^a_{22} = r^a_2\, d_{32}\, Y_p\, \phi^a \,. \tag{10b}$$

Since $h >> h^a$, material properties of the piezoelectric layers are not considered in the analysis. The control voltage ϕ^a can be either a reference voltage (open–loop) or the signal from the distributed piezoelectric sensor layer (closed–loop). The sensor signal ϕ^s is generated based on the direct piezoelectric effect proportional to a strain level experienced in the piezoelectric sensor layer, i.e., (Tzou, 1993)

$$\phi^s = -\frac{h^s}{A_s}\int_{A_s} \left[h_{31}S^s_{11} + h_{32}S^s_{22} \right] dA_s \,, \tag{11}$$

where h^s is the sensor layer thickness; the superscribe "s" denotes properties associated with the distributed sensor layer; A_s is the effective electrode area; h_{3i} are the piezoelectric constants (i=1,2); and S^s_{ii} are the normal strains (i=1,2). (Note that the in–plane shear strain effect is neglected.) In general, these strains can be divided into two components: 1) the membrane strains and 2) the bending strains. In the case of a transversely vibrating plate, the membrane strains are usually neglected. Accordingly, the sensor signal is primarily contributed by the bending strains induced by the transverse plate vibration. Thus, the sensor signal can be rewritten as

$$\phi^s = -\frac{h^s}{A_s}\int_{A_s} \left[h_{31}r^s_1 \left[u_{3,xx} \right] + h_{32}r^s_2 \left[u_{3,yy} \right] \right] dA_s \,, \tag{12}$$

where r^s_1 and r^s_2 are distances measured from the neutral surface to the sensor layer.

Usually, $r_1^s = r_2^s = (h+h^s)/2$ for surface laminated constant–thickness distributed sensor layers. Note that the sensor signal is an averaged signal over the effective electrode area. For a fully distributed constant–thickness piezoelectric sensor, the integration becomes

$$\phi^s = -\frac{h^s}{ab}\int_0^b\int_0^a\left[h_{31}r_1^s\left[u_{3,xx}\right] + h_{32}r_2^s\left[u_{3,yy}\right]\right]dxdy\,. \tag{13}$$

The velocity feedback control voltage can be defined by

$$\phi^a = \mathcal{G}\frac{\partial\phi^s}{\partial t} = -\mathcal{G}\frac{h^s}{A_s}\left\{\frac{\partial}{\partial t}\int_{A_s}\left[h_{31}r_1^s\left[u_{3,xx}\right] + h_{32}r_2^s\left[u_{3,yy}\right]\right]dA_s\right\}, \tag{14}$$

where $\mathcal{G}$ is the control gain – an amplification factor. This control voltage is used to induce control moments counteracting the plate vibration. Note that the in–plane membrane control forces do not appear in the governing equation because $R_1 = R_2 = \infty$.

2.3 Beam Sensor/Actuator Systems

Beams are special cases of plates, in which only 1–D behaviors are considered. A cantilever beam coupled with a distributed piezoelectric sensor and a distributed piezoelectric actuator is illustrated in Figure 3.

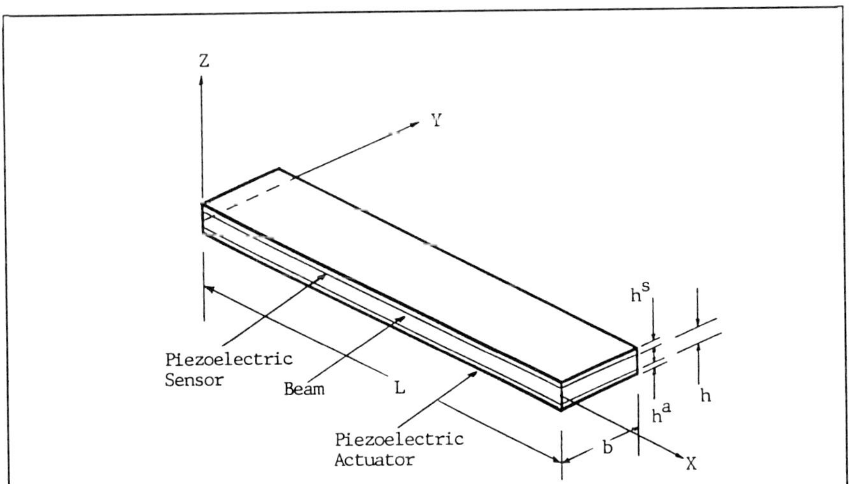

Fig.3 A cantilever beam with piezoelectric sensor and actuator layers.

Transverse vibration beam equation can be easily derived from a 1–D thin plate equation with an equivalent control moment (Lee and Moon, 1988; Liu, 1989; Tzou, 1993):

$$\rho h \ddot{u}_3 + D\, u_{3,xxxx} - M^a_{11,xx} = F_3 \,. \tag{15}$$

For a beam with a rectangular cross section, the equation is reduced to

$$\rho \tilde{A} \ddot{u}_3 + YI\, u_{3,xxxx} - b\,(M^a_{11,xx}) = bF_3 \,, \tag{16}$$

where $\tilde{A} = bh$; $I = bh^3/12$; b is the beam width and h is the beam thickness. Note that the Poisson's effect is neglected in the beam equation.

A sensor signal ϕ^s generated from the distributed sensor layer, covering from $x = x_1$ to $x = x_2$ and $x_2 > x_1$, and contributed by the bending effect can be estimated by the normal strain in the x direction:

$$\phi^s = -\frac{h^s}{x_2 - x_1} h_{31}\, r_1^s \left[u_{3,x} \right] \Big|_{x_1}^{x_2} \,, \tag{17}$$

where $(x_2 - x_1)$ is the sensor length. The sensor equation shows that the output signal is proportional to the slopes at both ends of the sensor layer. Note that the sensor signal is zero if the slopes at $x = x_1$ is equal to that at $x = x_2$, e.g., anti–symmetrical modes of a simply supported beam laminated with a symmetrically distributed sensor layer. For a fully distributed sensor, i.e., $x_1 = 0$ and $x_2 = L$, the sensor signal can also be derived easily. Next, a closed–loop velocity feedback control is discussed and control effectiveness is evaluated analytical and numerically. Laboratory experiments of a cantilever beam are also outlined.

3. Boundary Control and Solution Procesures

It is assumed that the negative proportional velocity feedback is considered in this study. Since the voltage is fully and uniformly distributed over the distributed actuator layer, counteracting control moments are not functions of spatial coordinates. Accordingly, control moment terms are dropped out from the governing equation and the distributed control is introduced through the boundaries. Thus, the distributed control becomes an equivalent boundary control in cantilevered beams and plates.

In this section, solution procedures, i.e., theoretical and finite element analyses, are discussed. Since beams are relatively easy to deal with, both theoretical and finite element solutions are presented. However, cantilever plates with boundary control moments are rather difficult to handle using the analytical technique. (Analytical solution procedures are not trivial even for the beams.) Thus, only the finite element method is discussed. Laboratory setup to verify the fundamental dynamics and control characteristics for the beam case is also discussed.

3.1 Beams

In this control algorithm, the sensor signal is differentiated, amplified, and then fed back into the distributed piezoelectric actuator. In this section, a closed–loop eigenvalue equation and its characteristic equation are derived. Note that the Laplace transform method is used to derive analytical solutions.

The velocity feedback signal ϕ^a can be defined as

$$\phi^a = \mathcal{G} \frac{\partial \phi^s}{\partial t} = - \frac{\mathcal{G} h^s h_{31} r_1^s}{L} \frac{\partial}{\partial t} \left[(u_{3,x}) \Big|_0^L \right] , \tag{18}$$

where $\mathcal{G}$ is the feedback gain. Using the velocity signal ϕ^a, calculating the distributed moment M_{11}^a, and substituting into the system equation, one can derive

$$\rho h \ddot{u}_3 + D\, u_{3,xxxx} - \left\{ \tilde{\xi} \frac{\partial}{\partial t} \left[(u_{3,x}) \Big|_0^L \right] \right\}_{,xx} = 0 , \tag{19}$$

where $\tilde{\xi} = -\,(\mathcal{G} h^s r_1^a d_{31} Y_p r_1^s h_{31})/L$. Note that the external mechanical excitation is set zero, i.e., $F_3 = 0$. For a uniformly distributed actuator layer, the distributed moment is not a spatial function, i.e., $\partial(\cdot)/\partial\alpha_i = 0$. In addition, the Poisson's ratio effect is usually neglected in a beam configuration. Thus, the beam equation is simplified to

$$\rho \tilde{A} \ddot{u}_3 + YI\, u_{3,xxxx} = 0 . \tag{20}$$

Since the distributed moment is not explicitly contained in the system equation, control actions are introduced through the electric boundary conditions – ***boundary controls*** – as discussed previously. Next, free vibration analysis of the beam with boundary control moments is discussed, and analytical solutions are derived.

Complex Eigenvalues

Assume the laminated beam system oscillates harmonically at its natural frequencies, i.e.,

$$u_3(x,t) = U_3(x)\, e^{j\hat{\omega}t} . \tag{21}$$

Then, the system equation becomes

$$U_3'''' - \lambda_{vk}^4\, U_3 = 0 , \tag{22}$$

where $U_3'''' = d^4U_3/dx^4$; λ_{vk} is the k–th eigenvalue of the system with the velocity feedback, denoted by the subscript v; and λ_{vk} is defined by $\lambda_{vk}^4 = (\rho\tilde{A}\,\hat{\omega}^2)/YI$. Boundary conditions at the free end are modified to reflect the velocity signal used in feedback controls. The fixed end boundary conditions are the same as before.

1) At clamped end (x = 0) :

a) Displacement: $U_3(0) = 0$; b) Slope: $U_3'(0) = 0$. (23a&b)

2) At free end (x = L) :

a) Moment: $M_{11}(L) = Lb\, M_{11}^a(L), \Longrightarrow -YI\, U_3''(L) = j\hat{\omega}\, Lb\tilde{\xi}\, U_3'(L)$;

b) Shear force: $Q(L) = 0, \Longrightarrow U_3'''(L) = 0$. (24a&b)

Note that the boundary control moment has a component of $j\hat{\omega}$ introduced by the velocity feedback and $j = \sqrt{-1}$. In order to satisfy the boundary condition, a complex frequency $\hat{\omega}$ and a complex transverse displacement U_3 need to be used in the later analysis. Assume both the displacement U_3 and the natural frequency $\hat{\omega}$ have a real component and an imaginary component:

$$U_3(x) = U_r(x) + jU_i(x) , \tag{25}$$

where the subscript *r* denotes the real component and *i* the imaginary component, and

$$\hat{\omega} = \omega_1 + j\omega_2 . \tag{26}$$

Substituting the complex frequency into Eq.(21) yields

$$u_3(x,t) = U_3(x)\, e^{j(\omega_1 + j\omega_2)t} . \tag{27}$$

Since $\hat{\omega}$ is complex, the real part of the exponential function, i.e., $j(j\omega_2)$, represents the decay rate (a convergence envelop or oscillation envelop); the imaginary part, i.e., $j\omega_1$, represents the damped natural frequency (Ewins, 1986; Zaveri, 1985). It is assumed that the beam system is lightly damped. Concerning the complex transverse displacement $U_3(x)$, the magnitude of U_3 represents the displacement; the phase angle of U_3 represents the phase differences among various parts of a distributed structure (Zaveri, 1985).

Substituting complex U_3 and $\hat{\omega}$ into Eq.(22) and separating the real and imaginary parts of the equation, one can derive

$$[d^4U_r(x)/dx^4] - pU_r(x) + qU_i(x) = 0 , \tag{28a}$$

$$[d^4U_i(x)/dx^4] - pU_i(x) - qU_r(x) = 0 , \tag{28b}$$

where p and q are the real and imaginary parts of the k–th eigenvalue λ_{vk}^4; they are defined as:

$$\lambda_{vk}^4 = \frac{\rho h \hat{\omega}^2}{YI} = p + jq , \tag{29}$$

where $p = [\rho h(\omega_1^2 - \omega_2^2)]/YI$ and $q = 2\rho h\omega_1\omega_2/YI$. Taking the Laplace transform of Eqs.(28a&b) and applying the fixed end boundary conditions in Eq.(23), one can derive

$$(s^4 - p)\, U_r(s) + q\, U_i(s) = U_r'''(0) + s\, U_r''(0) , \tag{30a}$$

$$(s^4 - p)\, U_i(s) - q\, U_r(s) = U_i'''(0) + s\, U_i''(0) , \tag{30b}$$

where $U_r'''(0) = d^3U_r(0)/dx^3$, $U_r''(0) = d^2U_r(0)/dx^2$, $U_i'''(0) = d^3U_i(0)/dx^3$, and $U_i''(0) = d^2U_i(0)/dx^2$ are constants determined by the boundary conditions. Solving Eqs.(30a&b) simultaneously gives

$$U_r(s) = \frac{1}{(s^4-p)^2+q^2}\Big[(s^4-p)U_r'''(0) + (s^5-ps)U_r''(0) - qU_i'''(0) - qsU_i''(0)\Big], \tag{31a}$$

$$U_i(s) = \frac{1}{(s^4-p)^2+q^2}\Big[qU_r'''(0) + qsU_r''(0) + (s^4-p)U_i'''(0) + (s^5-ps)U_i''(0)\Big]. \tag{31b}$$

Taking the inverse Laplace transformation and using the partial fractions, one can derive the solutions of $U_r(x)$ and $U_i(x)$:

$$U_r(x) = e_1(x)U_r'''(0) + e_2(x)U_r''(0) + e_3(x)U_i'''(0) + e_4(x)U_i''(0)\,, \tag{32a}$$

$$U_i(x) = f_1(x)U_r'''(0) + f_2(x)U_r''(0) + f_3(x)U_i'''(0) + f_4(x)U_i''(0)\,, \tag{32b}$$

where $e_n(x)$ and $f_n(x)$ are defined by exponential functions:

$$e_1 = \Big[m_1e^{z_1x} + m_2e^{z_2x} + m_3e^{z_3x} + m_4e^{z_4x} + m_5e^{z_5x} + m_6e^{z_6x} + m_7e^{z_7x} + m_8e^{z_8x}\Big]\,, \tag{33a}$$

$$e_2 = \Big[m_9e^{z_1x} + m_{10}e^{z_2x} + m_{11}e^{z_3x} + m_{12}e^{z_4x} + m_{13}e^{z_5x} + m_{14}e^{z_6x} + m_{15}e^{z_7x} + m_{16}e^{z_8x}\Big]\,, \tag{33b}$$

$$e_3 = \Big[m_{17}e^{z_1x} + m_{18}e^{z_2x} + m_{19}e^{z_3x} + m_{20}e^{z_4x} + m_{21}e^{z_5x} + m_{22}e^{z_6x} + m_{23}e^{z_7x} + m_{24}e^{z_8x}\Big]\,, \tag{33c}$$

$$e_4 = \Big[m_{25}e^{z_1x} + m_{26}e^{z_2x} + m_{27}e^{z_3x} + m_{28}e^{z_4x} + m_{29}e^{z_5x} + m_{30}e^{z_6x} + m_{31}e^{z_7x} + m_{\ 2}e^{z_8x}\Big]\,; \tag{33d}$$

$$f_1 = \Big[n_1e^{z_1x} + n_2e^{z_2x} + n_3e^{z_3x} + n_4e^{z_4x} + n_5e^{z_5x} + n_6e^{z_6x} + n_7e^{z_7x} + n_8e^{z_8x}\Big]\,, \tag{34a}$$

$$f_2 = \Big[n_9 e^{z_1 x} + n_{10} e^{z_2 x} + n_{11} e^{z_3 x} + n_{12} e^{z_4 x} + n_{13} e^{z_5 x} + n_{14} e^{z_6 x} + n_{15} e^{z_7 x} + n_{16} e^{z_8 x}\Big] , \tag{34b}$$

$$f_3 = \Big[n_{17} e^{z_1 x} + n_{18} e^{z_2 x} + n_{19} e^{z_3 x} + n_{20} e^{z_4 x} + n_{21} e^{z_5 x} + n_{22} e^{z_6 x} + n_{23} e^{z_7 x} + n_{24} e^{z_8 x}\Big] , \tag{34c}$$

$$f_4 = \Big[n_{25} e^{z_1 x} + n_{26} e^{z_2 x} + n_{27} e^{z_3 x} + n_{28} e^{z_4 x} + n_{29} e^{z_5 x} + n_{30} e^{z_6 x} + n_{31} e^{z_7 x} + n_{32} e^{z_8 x}\Big] , \tag{34d}$$

where z_1, ... z_8, m_1, ... m_{32}, and n_1, ... n_{32} can be determined by the partial fraction expansion for a given λ_{vk} (Liu, 1989). Substituting the complex displacement, Eq.(25), and the complex frequency into the boundary conditions and separating the real and imaginary parts of these equations, one can derive the following four equations.

$$\text{a)} \quad U_3'''(L) = 0 \Longrightarrow U_r'''(L) = 0 \text{ and } U_i'''(L) = 0 ; \tag{35}$$

$$\begin{aligned} \text{b)} \quad & -YI\, U_3''(L) = j\hat{\omega}\, Lb\tilde{\xi}\, U_3'(L) \\ \Longrightarrow \quad & -YI\, U_r''(L) = \tilde{\gamma}\, U_r'(L) - \tilde{\eta}\, U_i'(L) \\ & \text{and } -YI\, U_i''(L) = \tilde{\eta}\, U_r'(L) + \tilde{\gamma}\, U_i'(L) , \end{aligned} \tag{36}$$

where $j\hat{\omega}\ Lb\tilde{\xi} = \tilde{\gamma} + j\tilde{\eta}$. Substituting Eqs.(32a&b) and Eqs.(33)–(34) into Eqs.(35)–(36) and simplifying yields an eigenvaluc (matrix) expression.

$$\begin{bmatrix} A(1,1) & A(1,2) & A(1,3) & A(1,4) \\ A(2,1) & A(2,2) & A(2,3) & A(2,4) \\ A(3,1) & A(3,2) & A(3,3) & A(3,4) \\ A(4,1) & A(4,2) & A(4,3) & A(4,4) \end{bmatrix} \begin{Bmatrix} U_r'''(0) \\ U_r''(0) \\ U_i'''(0) \\ U_i''(0) \end{Bmatrix} = \{0\} , \tag{37}$$

where $A(1,1),\cdot\cdot, A(4,4)$ are the coefficients. Since $\{U_r'''(0),\ U_r''(0),\ U_i'''(0),\ U_i''(0)\}^t \neq 0$, the determinant of the coefficient matrix equals zero, i.e.,

$$\det \begin{bmatrix} A(1,1) & A(1,2) & A(1,3) & A(1,4) \\ A(2,1) & A(2,2) & A(2,3) & A(2,4) \\ A(3,1) & A(3,2) & A(3,3) & A(3,4) \\ A(4,1) & A(4,2) & A(4,3) & A(4,4) \end{bmatrix} = 0 \,. \tag{38}$$

Expansion of the above determinant gives a *characteristic equation* of the laminated beam system with the velocity feedback, and roots λ_{vk}'s of the characteristic equation are the *eigenvalues* of the beam system. Note that λ_{vk}'s are complex, from which natural frequencies $\hat{\omega}_k$'s ($\hat{\omega}_k = \omega_1 + j\omega_2$) can be derived. The complex frequency $\hat{\omega}$ provides the information about the damped natural frequency and damping ratio of the beam system.

Finite Element Modeling

A finite beam element with three degree–of–freedom (DOF) at each node, i.e., two in–plane translation DOF's and one rotation DOF (with respect to the z–axis), is used to model the beam. From the mathematical modeling, it is known that the equivalent control action comes from a counteracting moment at the free end. In addition, since the proportional velocity feedback is used, an equivalent rotational damper is lumped at the free end to simulate the control action. The equivalent damping constant c can be derived from the boundary conditions as

$$c = -\mathcal{G}\,(b\, h^s r_1^a\, d_{31} Y_p r_1^s\, h_{31}) \,. \tag{39}$$

Note that the equivalent damping constant is determined by a number of geometry and material parameters and the gain factor. As long as geometry and material are determined, the only control variable is the gain factor. Figure 4a shows the finite beam element and Figure 4b illustrates the finite element model.

Laboratory Experiments

A piezoelectric polymeric polyvinylidene fluoride (PVDF) laminated plexiglas beam is fabricated and tested in laboratory to investigate its distributed control effectiveness. The control voltage is taken from the distributed sensor output, phase shifted, and then fed back into the distributed actuator layer. Control gains are kept increasing until near the breakdown voltage above which the piezoelectric polarization is destroyed. Thus, the control effectiveness induced by small and large control voltages can be investigated. Figure 5 illustrates the experimental setup.

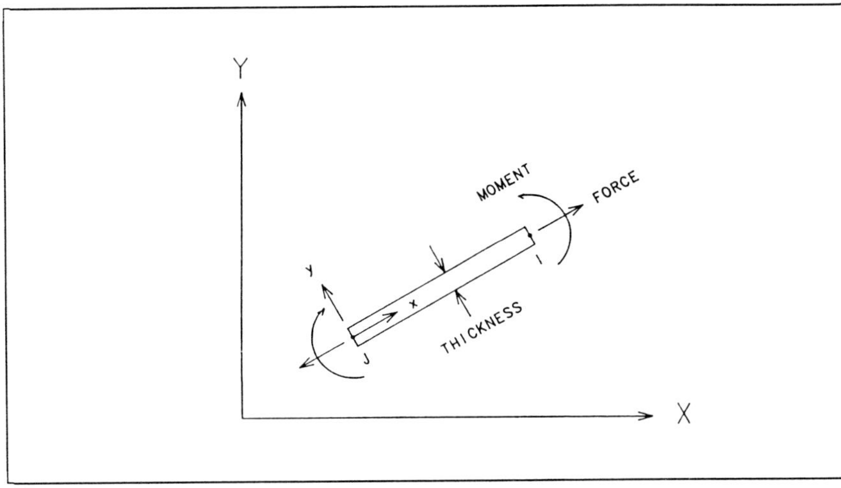

Fig.4a Finite beam element.

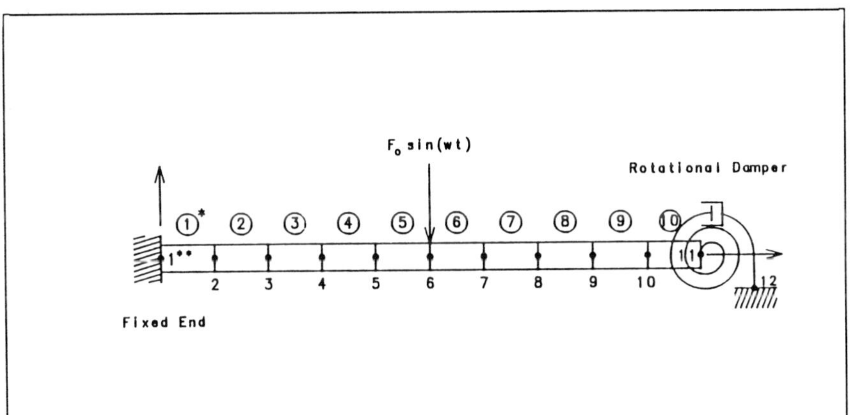

Fig.4b Beam finite element model.

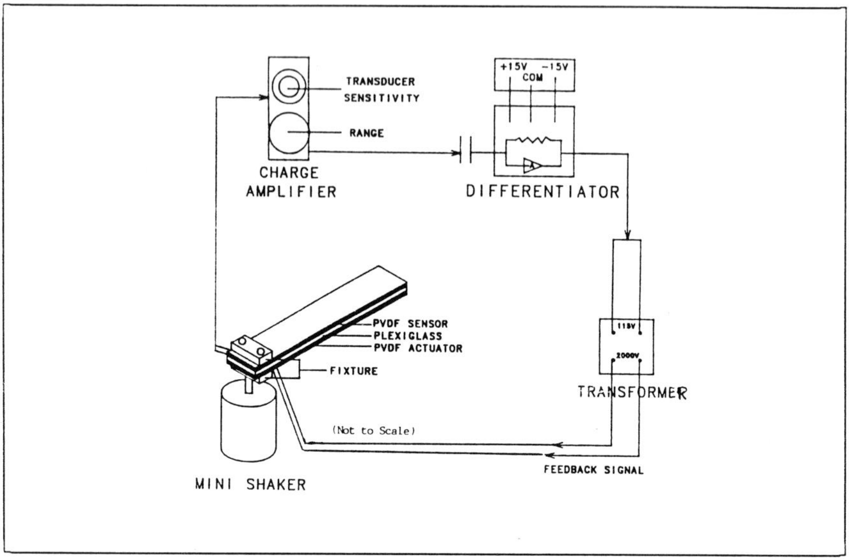

Fig.5 Experimental setup.

3.2 Plates

Cantilever plates have received extensive studies because of their practical importance (Classen and Thorne, 1963; Leissa, 1969). Although exact solutions are difficult to obtain, many numerical methods can be employed to examine the vibration of cantilever plates (Classen and Thorne, 1963; Gorman, 1976; Leissa, 1969). In order to approach the analytical solution, the governing differential equations with the piezoelectric boundary conditions are formulated. It is assumed that the fixed edge is at $x = 0$, and the other three edges are free. The major control effect is introduced at the far free edge, i.e., $x = a$. The plate length is assumed longer than the width, i.e., $a>b$, thus, the first plate mode is a bending mode.

Mathematical Modeling

Again, since the control voltage is uniformly distributed on the piezoelectric layer, control actions do not appear in the system equation.

$$\rho h \ddot{u}_3 + D\left[u_{3,xxxx} + 2\, u_{3,xxyy} + u_{3,yyyy} \right] = F_3 \,. \tag{40}$$

In addition, the sensor signal can be simplified to

$$\phi^s = -\frac{h^s}{a} h_{31} r_1^s \left[u_3(a,y)\right]_{,x} . \tag{41}$$

Note that the output signal is proportional to the slope at x = a; this sensor signal is processed and fed back to the actuator to suppress the plate vibration. Since the voltage is uniformly distributed, the resulting control action goes to the boundary conditions as discussed previously.

Again assuming zero external excitation and harmonic oscillation at natural frequencies, i.e.,

$$u_3(\alpha_1,\alpha_2,t) = U_3(x,y)\, e^{j\omega t} , \tag{42}$$

one can rewrite the governing equation as

$$D\left[U_{3,xxxx} + 2\, U_{3,xxyy} + U_{3,yyyy} \right] - \rho h \omega^2\, U_3 = 0 . \tag{43}$$

Boundary conditions are defined by

1) At fixed edge (x=0,y) :

a) Displacement: $U_3(0,y) = 0$, (44a)

b) Slope: $U_3(0,y)_{,x} = 0$. (44b)

2) At free cdgc (x–a,y) :

a) Moment: $\overset{*}{M}_{11}(a,y) = a\, M_{11}^a(a,y)$

$$\Longrightarrow -D\left[U_3(a,y)_{,xx} + \mu\, U_3(a,y)_{,yy}\right] = a\, \tilde{\xi}_1\, j\omega\, U_3(a,y)_{,x} , \tag{45a}$$

b) Sheai Furce: $\overset{*}{Q}(a,y) = 0$

$$\Longrightarrow U_3(a,y)_{,xxx} + (2-\mu) U_3(a,y)_{,xyy} = 0 . \tag{45b}$$

3) At free edges (x,y=0 & b) :

a) Moment: $\overset{*}{M}_{22}(x,0) = b\, M_{22}^a(x,0)$

$$\Longrightarrow -D\left[U_3(x,0)_{,yy} + \mu\, U_3(x,0)_{,xx}\right] = b\, \tilde{\xi}_2\, j\omega\, U_3(a,0)_{,x} , \tag{46a}$$

b) Shear Force: $\overset{*}{Q}(x,0) = 0$

$$\Longrightarrow U_3(x,0)_{,yyy} + (2-\mu)\, U_3(x,0)_{,yxx} = 0 ; \tag{46b}$$

a) Moment: $\overset{*}{M}_{22}(x,b) = b\, M^{a}_{22}(x,b)$

$$\Longrightarrow\ -D\,[U_3(x,b)_{,yy} + \mu\, U_3(x,b)_{,xx}] = b\,\tilde{\xi}_2\, j\omega\, U_3(a,b)_{,x}\,, \qquad (46c)$$

b) Shear Force: $\overset{*}{Q}(x,b) = 0$

$$\Longrightarrow\ U_3(x,b)_{,yyy} + (2-\mu)\, U_3(x,b)_{,yxx} = 0\,. \qquad (46d)$$

Note that $\tilde{\xi}_1 = (-\mathcal{G} h^s r^a_1 d_{31} Y_p r^s_1 h_{31})/a$ and $\tilde{\xi}_2 = (-\mathcal{G} h^s r^a_2 d_{32} Y_p r^s_2 h_{32})/b$ which are related to counteracting control moments. In practice, $r_1 = r_2$ for uniform–thickness piezoelectric layers. In addition, $d_{31} = d_{32}$ and $h_{31} = h_{32}$ for piezoelectric layers made of hexagonal symmetrical piezoelectrics.

As it can be observed from analytical solution procedures of beams, it seems rather difficult to solve the plate equations with associated boundary conditions analytically. Of course, one may follow the same procedures to attack the plate problem. However, the level of complexity is beyond the purpose of this chapter. Thus, finite element analysis is conducted to investigate the dynamics and controls.

Finite Element Modeling

A quadrilateral shell element having four nodes with six DOF's at each node is used to model the plate. The six DOF's are three translation DOF's and three rotation DOF's. This shell element allows both membrane and bending deformations, and it permits both in–plane and normal loads. Figure 6a shows the quadrilateral shell element, and Figure 6b illustrates the plate finite element model.

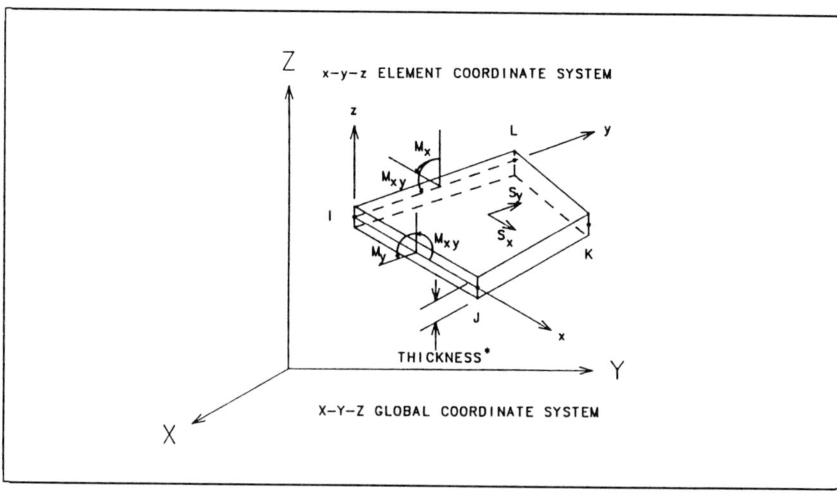

Fig.6a Finite quadrilateral shell element.

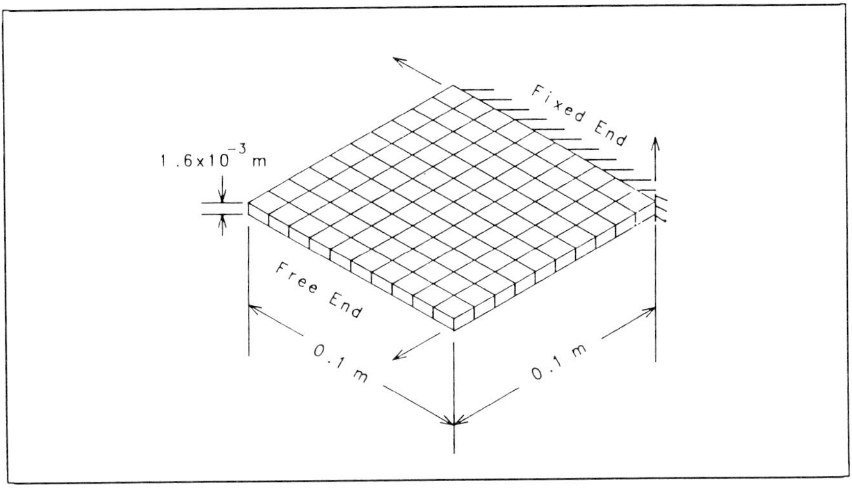

Fig.6b Plate finite element model.

4. Results and Discussion

In this section, nonlinear distributed controls of a beam and a plate using the velocity feedback are studied. Both beam and plate are surface laminated with fully distributed sensor and actuator layers; they are also cantilevered. The control voltage keeps increasing up to the breakdown voltage. Complex natural frequencies are calculated and controlled damping variations evaluated. Boundary transition of the cantilever beam subjected to high feedback voltages is investigated. The beam case is presented first, and followed by the plate case. All material and physical properties are summarized in Table 1.

4.1 Beams

Complex frequencies and damping ratios of the piezoelectric laminated beam are studied. Experimental results are presented, and analytical and finite element solutions are compared. The original first three natural frequencies, in the complex plane, are presented in Figures 7–9. (Note that these uncontrolled natural frequencies are on the real axis and the imaginary part is zero.) Controlled complex frequencies for the corresponding modes are illustrated in Figures 10–12. Shifts of natural frequencies due to the velocity feedback can be observed in these figures. Variations of the first three complex frequencies due to the velocity feedback are summarized in Table 2.

Table 1 Properties of the piezoelectric laminated models.

1a. Plexiglas	
Y (Young's modulus)	3.1028×10^9 (N/m^2)
ρ (Mass density)	1190.0 (Kg/m^3)
h (Thickness)	1.6×10^{-3} (m)
a (Length–Plate)	0.1 (m)
b (Width–Plate)	0.1 (m)
b (Width–Beam)	0.01 (m)
L (Length–Beam)	0.1 (m)
μ (Poisson's ratio)	0.3
D (Bending stiffness)	$(Yh^3/[12(1-\mu^2)] = 1.1638$ (N–m)
I (Area moment of inertia)	$bh^3/12 = 3.4133 \times 10^{-12}$ (m^4)

1b. Polyvinylidene Fluoride (PVDF) Sensor/Actuator Layer	
Y_p (Young's modulus)	2.00×10^9 (N/m^2)
ρ_p (Mass density)	1800.0 (Kg/m^3)
h^s (Thickness of sensor)	40 (μm)
h^a (Thickness of actuator)	40 (μm)
r_1^s (Sensor distance)	$(h^s+h)/2 = 8.2 \times 10^{-4}$ (m)
r_1^a (Actuator distance)	$(h^a+h)/2 = 8.2 \times 10^{-4}$(m)
d_{31} (Piezoelectric constant)	2.3×10^{-11} $(\frac{m/m}{V/m})$
h_{31} (Piezoelectric constant)	4.32×10^8 $(\frac{V/m}{m/m})$

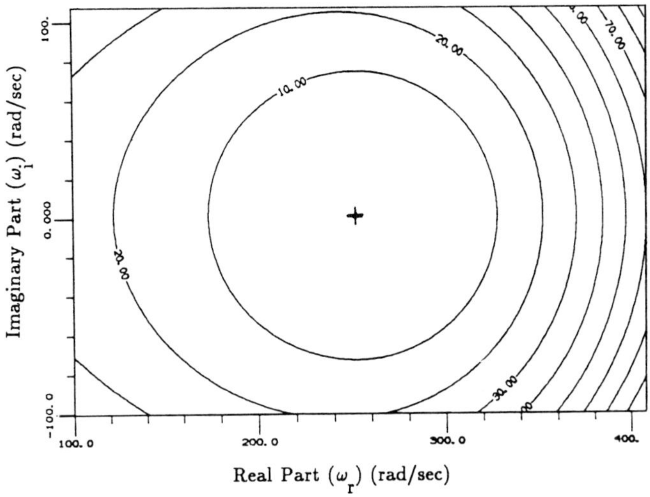

Fig.7 Uncontrolled complex natural frequency of the first beam mode.

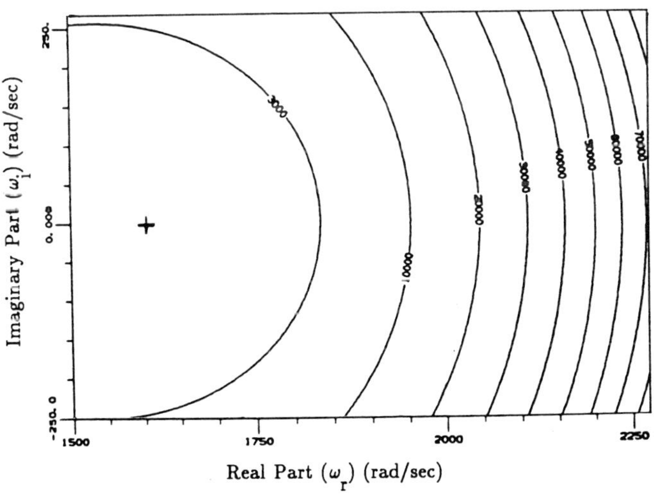

Fig.8 Uncontrolled complex natural frequency of the second beam mode.

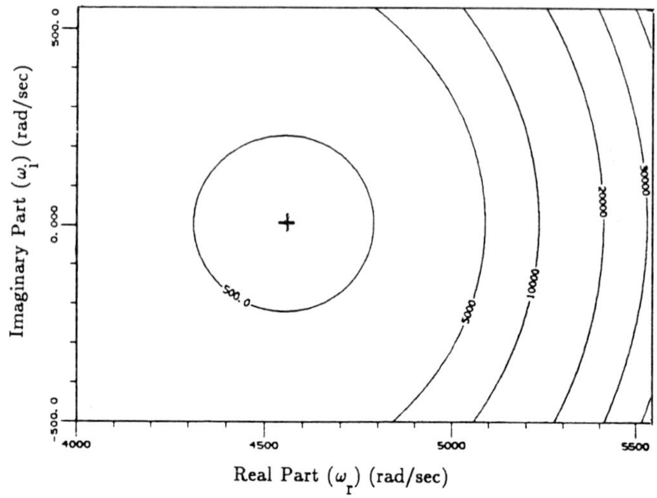

Fig.9 Uncontrolled complex natural frequency of the third beam mode.

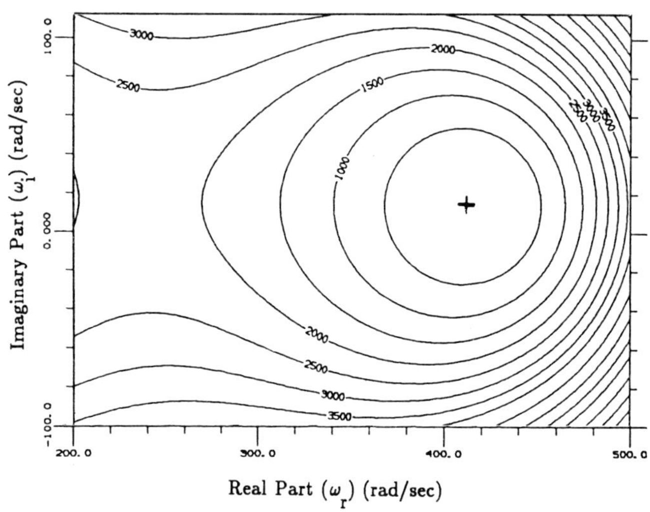

Fig.10 Controlled complex frequency of the first beam mode.

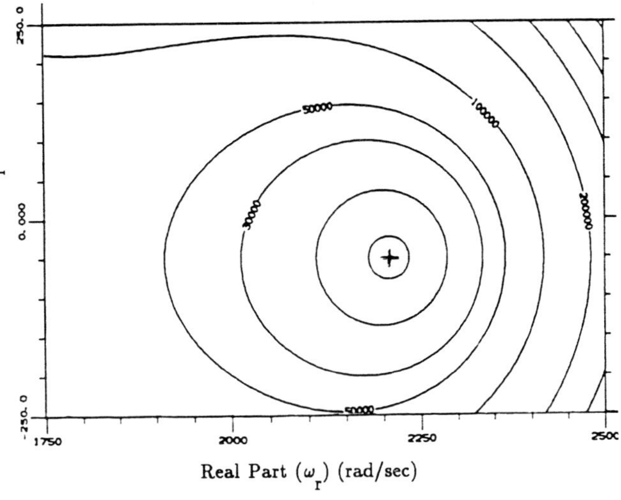

Fig.11 Controlled complex frequency of the second beam mode.

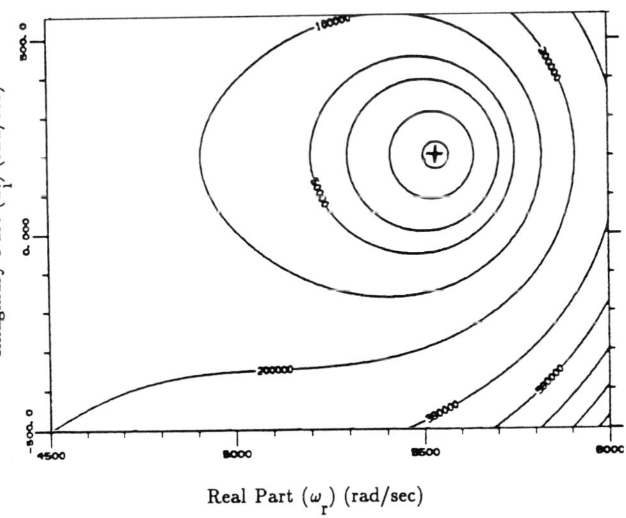

Fig.12 Controlled complex frequency of the third beam mode.

Table 2 Relationship between complex frequency and velocity feedback gain, (f_1,f_2) (Hz)

Velocity Feedback Gain	Natural Frequency f (Hz)		
	1st mode	2nd mode	3rd mode
0.00	(41.7, 0.0)	(261.6, 0.0)	(732.4, 0.0)
100.00	(51.7, 16.2)	(351.7, 31.2)	(881.2, 31.2)
200.00	(62.7, 11.1)	(357.0, 16.0)	(884.9, 15.8)
300.00	(64.8, 7.7)	(357.9, 10.7)	(885.5, 10.5)
400.00	(65.6, 5.7)	(358.4, 8.0)	(885.7, 7.8)
500.00	(65.9, 4.6)	(358.6, 6.4)	(885.9, 6.4)
600.00	(66.1, 3.8)	(358.6, 5.3)	(885.9, 5.3)
700.00	(66.1, 3.3)	(358.6, 4.6)	(885.9, 4.5)
800.00	(66.2, 2.9)	(358.7, 3.5)	(885.9, 4.0)
900.00	(66.2, 2.6)	(358.7, 3.2)	(885.9, 3.5)
1000.00	(66.2, 2.4)	(358.7, 2.9)	(885.9, 3.2)

As discussed previously, a complex frequency of the k–th mode is defined as

$$\hat{\omega}_k = \omega_1 + j\omega_2 , \tag{47}$$

where ω_2 defines the envelope of damped oscillation ($\omega_2 = \zeta_k\omega_n$); ω_1 is the damped natural frequency [$\omega_1 = \omega_n(1-\zeta_k^2)^{1/2}$] (Ewins, 1986); and ω_n denotes the undamped natural frequency of the laminated beam without any feedback control. Thus, the damping ratio ζ_k for the k–th mode can be estimated by (Ewins, 1986; Zaveri, 1985; Harris and Crede, 1976)

$$\zeta_k = [\omega_2^2/(\omega_2^2 + \omega_1^2)]^{1/2} \cong \frac{\omega_2}{\omega_1} . \tag{48}$$

Damping enhancement of the beam system due to the velocity feedback can be calculated in this way and their results are plotted. In addition, analytical solutions are compared with finite element results in Figures 13–15. It can be observed that controlled damping ratio increases to an optimal point and then drops down as feedback gain increases. Analytical solutions are favorably compared with finite element results. However, the analytical predictions are higher than the finite element results. This is due to non–ideal modeling deficiencies in the finite element technique. The maximum differences are 9.8% for the first mode, 11.2% for the second mode, and 17.3% for the third mode, respectively at their peaks. Beside its peak value, the rest of curve is closely compared for all three modes.

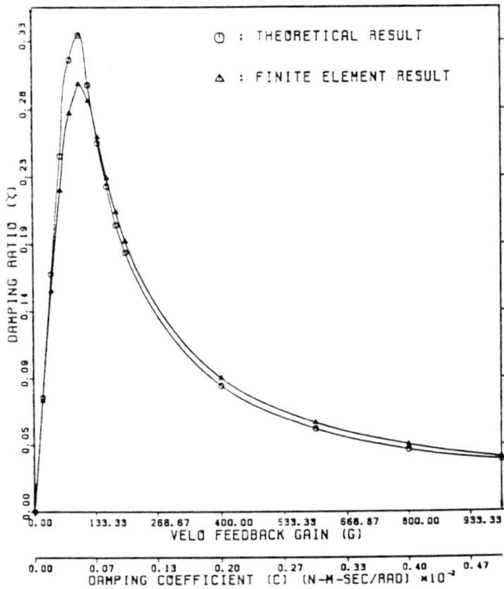

Fig.13 Controlled damping ratio versus feedback gain (first beam mode).

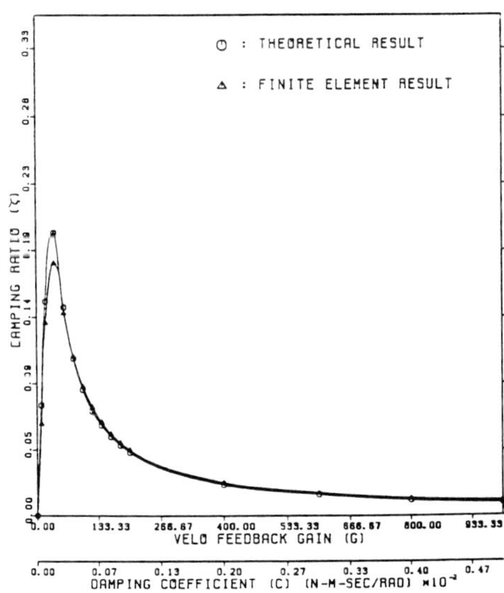

Fig.14 Controlled damping ratio versus feedback gain (second beam mode).

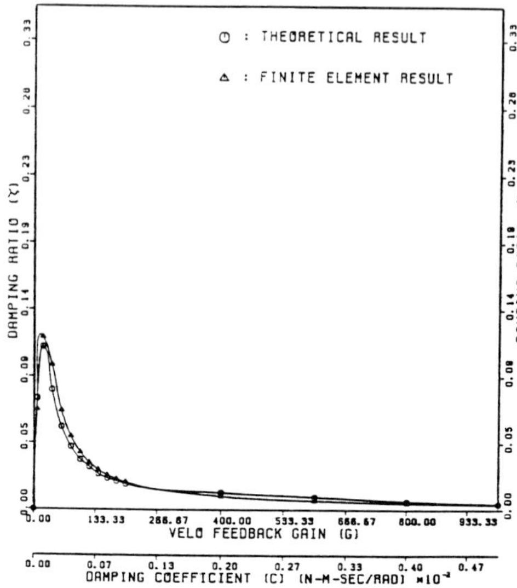

Fig.15 Controlled damping ratio versus feedback gain (third beam mode).

Laboratory experiments of a plexiglas beam laminated with PVDF layers are also conducted with a number of feedback gains (Johnson, 1990). Figure 16 shows the experimental results giving the same characteristics as those of analytical and finite element analyses.

Boundary Transition

It is observed that the damping ratio increases to an ultimate value and drops down afterwards. (It is assumed that the piezoelectric di–pole molecular structure still sustains the high feedback voltages.) This phenomenon is caused by the high boundary feedback control effects at the free end. As discussed previously, the distributed piezoelectric actuator basically applies a counteracting moment at the free end of the cantilever beam when a high feedback voltage is injected into the actuator. As the feedback voltage becomes higher and higher, the free end is further constrained and the boundary condition is gradually changing to a sliding–roller boundary condition (Liu, 1989; Tzou, 1993). In order to verify this proposition, the first three natural frequencies of a <u>fixed/sliding–roller beam</u> are calculated and compared with the controlled natural frequencies (at Gain = 1000) of the cantilever beam, Table 3. In addition, the first three (undamped) natural frequencies of a <u>double fixed–end beam</u> are calculated using the finite element method and they are also listed in Table 3. Basically, these data support the physical explanations mentioned above. Experimental results of the physical model also suggested the same control phenomenon (Johnson, 1990).

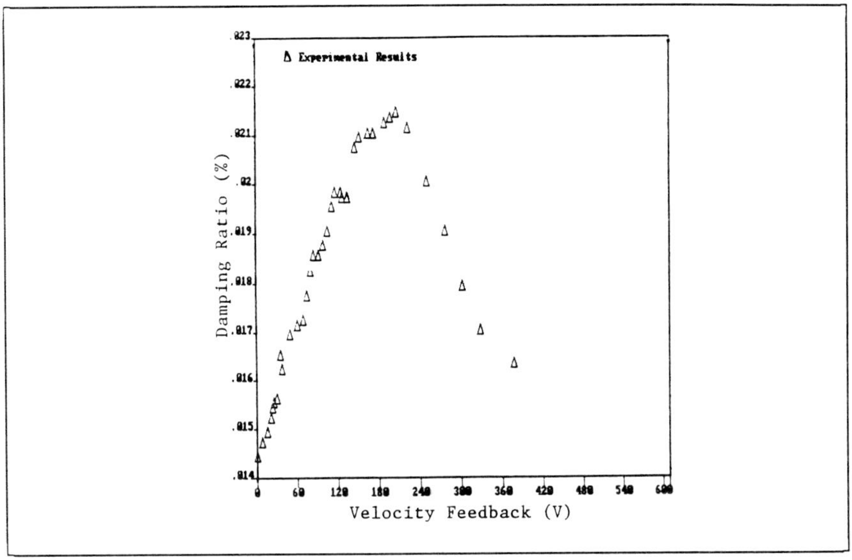

Fig.16 Experimental results.

Table 3 Comparison of natural frequencies for different boundary conditions.

Beams	Natural Frequency f (Hz)			Solution Technique
	1st	2nd	3rd	
1) Cantilever Beam/Actuator	65.8 65.9	355.9 357.0	878.8 884.3	(Theory) (Finite Element)
2) Fixed–Roller Beam	65.8	355.8	878.6	(Finite Element)
3) Double Fixed –end Beam	66.2	358.7	885.9	(Finite Element)

In conclusion, analytical solutions are well compared with finite element results. In addition, experimental results also suggest the same control characteristics.

4.2 Plates

As discussed previously, due to the complexity of analytical procedures, only finite element results are presented in this section. (Based on the beam case, it is confident that finite element results basically provide similar control characteristics.) Figure 17 shows the damping ratio and control gain relations for the first three modes.

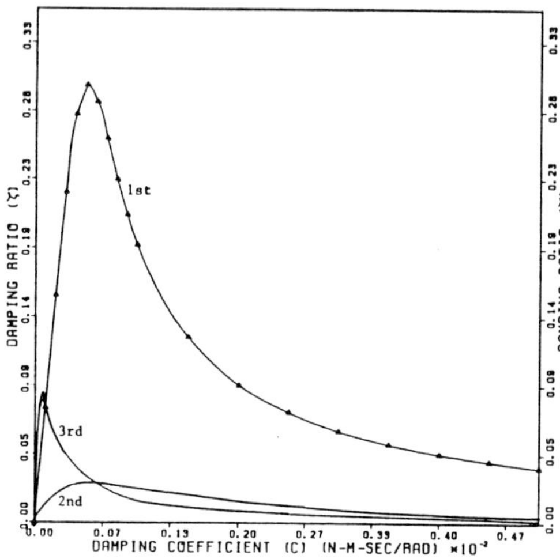

Fig.17 Controlled damping ratio versus feedback gain (three plate modes).

Figure 17 suggests that the control effectiveness is significant for plate bending modes, e.g., 1st and 3rd, and it is relatively insignificant for twisting modes, e.g., 2nd and 4th (not shown) because the main control action is along the x–axis. Note that the overall convergence (i.e., envelop of time history responses) of a modal oscillation is determined by its natural frequency and damping ratio. Usually, higher natural modes converge much faster than the lower natural modes because their natural frequencies.

5. Conclusions

Distributed nonlinear controls of PVDF laminated cantilever beams and plates were investigated in this study. The laminated cantilever beam/plate had a distributed piezoelectric sensor and a distributed actuator, and both layers were surface bonded. Closed–loop feedback controls of the beam/plate using the velocity signals were evaluated.

In the beam case, analytical solutions were compared with finite element results, and experimental data were also presented. Results suggested that the velocity feedback control was effective. In the velocity feedback control, the controlled damping ratio increased to an optimal value and then gradually dropped down as the feedback gain continuously increased. This was caused by the additional constraint imposed by the boundary control moment at the free–end. The free–end boundary condition was gradually changing to a sliding–roller boundary condition as proved by the finite element analyses and laboratory experiments.

In the plate case, finite element results suggested the same conclusion. Controlled damping ratios increased to an optimal value and then dropped down. Distributed control was relatively significant for bending modes (e.g., 1st, 3rd, 5th, etc.) and relatively insignificant for twisting modes (e.g., 2nd, 4th, etc.) because the primary control action was a counteracting control moment resisting the bending oscillation and the charge distribution on the piezoelectric actuator layer.

6. Acknowledgement

This research was partially supported by a grant from the Army Research Office (DAAL03–91–G–0065), Technical Monitor: Dr. G. L. Anderson. This research was also partially supported by an NSF–REU grant. Contents of the information do not necessarily reflect the position or the policy of the government, and nor should official endorsement be inferred.

7. References

Baily, T. and Hubbard, J.E., 1985, Distributed Piezoelectric Polymer Active Vibration control of a Cantilever Beam," *J. of Guidance, Control, and Dynamics*, Vol.8, No.5, pp.605–611.

Baz, A. and Poh, S., 1988, "Performance of an Active Control System with Piezoelectric Actuators," *Journal of Sound & Vibration*, Vol.126, No.8, pp.327–343.

Burke, S. and Hubbard, J.E., 1990, "Distributed Transducer Control Design for Thin Plates," *Electro-Optical Materials for Switches, Coatings, Sensor Optics, and Detectors* (1990), SPIE Vol.1307, pp.222–231.

Classen, R.W. and Thorne, C.J., 1963, "Vibrations of Rectangular Cantilever Plates, *J. of Aerospace Sciences*, Vol.29, pp.1300–1963.

Crawley, E.F. and de Luis, J., 1987, "Use of Piezoelectric Actuators as Elements of Intelligent Structures," *AIAA Journal*, Vol.25, No.10, pp.1373–1385.

Crawley, E.F. and Lazarus, K.S., 1991, "Induced Strain Actuation of Anisotropic Plates," *AIAA Journal*, Vol.29, No.6, pp.944–951.

Dimitriadis, E.K., Fuller, C.R., and Rogers, C.A., 1991, Piezoelectric Actuators for Distributed Vibration Excitation of Thin Plates, ASME *Journal of Vibration and Acoustics*, Vol.113, No.1, pp.100–107.

Ewins, D.J., 1986, *Modal Testing: Theory and Practice*, Research Studies Press, England.

Gandhi, M.V. and Thompson, B.S., 1992, *Smart Materials and Structures*, Chapman and Hall, New York.

Gorman, D.J., 1976, "Free Vibration Analysis of Cantilever Plates by the Method of Superposition," *Journal of Sound & Vibration*, Vol.49, No.4, pp.453–467.

Guran A. and Inman, D.J., 1994, *Smart Structures, Nonlinear Dynamics, and Control*, Prentice Hall, Englewood Cliffs, N.J.

Guran, A., 1994, Control of Buckling Load in Elastic Structures by Means of Shape Memory Alloys, *Archives of Applied Mechanics*. (To appear)

Ha, S.K., Keilers, C., and Chang, F.–K., 1992, "Finite Element Analysis of Composite Structures Containing Distributed Piezoceramics Sensors and Actuators," *AIAA Journal*, Vol.30, No.3, pp.772–780.

Harris, C.M. and Crede, C.E., 1976, *Shock and Vibration Handbook*, 2nd Ed., McGraw–Hill, 1976.

Hwang, W.–S. and Park, H.C., 1993, "Finite Element Modeling of Piezoelectric Sensors and Actuators," *AIAA Journal*, Vol.31, No.5, pp.930–937.

Inman, D.J., 1990, "Control/Structures Interaction: Effect of Actuator Dynamics," *Mechanics and Control of Large Flexible Structures*, Editor: Junkins, J.L., AIAA Press, pp.507–535.

Johnson, D., 1990, *An Experimental Study on Feedback Control of a Piezoelectric Laminated Cantilever Beam*, NSF–REU Report, Department of Mechanical Engr., University of Kentucky, August 1990.

Lee, C. K. and Moon, F.C., 1988, "Modal Sensors/Actuators," IBM Research Report, RJ 6306 (61975), June.

Leissa, A.W., 1969, *Vibration of Plates*, NASA Technical Report.

Liu, K.J, 1989, *A Theoretical Study on Distributed Active Vibration Suppression of Mechanical Systems with Integrated distributed Piezoelectric Sensor and Actuator*, MSME Thesis, Dept. of Mech. Engr., Univ. of Kentucky, 08/1989.

Plump, J.M., Hubbard, J. E., and Baily, T., 1987, "Nonlinear Control of a Distributed System: Simulation and Experimental Results," ASME *J. Dynamic Systems, Measurement, and Control*, pp.133–139.

Ricketts, D., 1981, Model for a Piezoelectric Polymer Flexural Plate Hydrophone, *Journal of Acoustic Society of America*, Vol.70, No.4, pp.929–935.

Ricketts, D., 1989, The Frequency of Flexible Vibration of Completely Free Composite Piezoelectric Polymer Plates, *Journal of Acoustic Society of America*, Vol.80, No.3, pp.2432–2439.

Rogers, C.A., Ed., 1988, *Smart Materials, Structures, and Mathematical Issues*, Technomic Pub.

Tzou, H. S., 1987, "Active Vibration Control of Flexible Structures Via Converse Piezoelectricity", *Developments in Mechanics*, Vol.14–C, pp.1201– 1206.

Tzou, H.S., 1991, Distributed Modal Identification and Vibration Control of Continua: Theory and Applications, *ASME Journal of Dynamic Systems, Measurements, and Control*, Vol.113, No.3, pp.494–499.

Tzou, H.S., 1992, A New Distributed Sensation and Control Theory for "Intelligent" Shells, *Journal of Sound and Vibration*, Vol.152, No.2, pp.335–350.

Tzou, H.S., 1993, *Piezoelectric Shells (Distributed Sensing and Control of Continua)*, Kluwer Academic Pub., Dordrecht/Boston/London.

Tzou, H.S. and Anderson, G.L. (Editors), 1992, *Intelligent Structural Systems*, 472 pages, (ISBN No.0–7923–1920–6), Kluwer Academic Publishers, Dordrecht/Boston, August 1992.

Tzou, H.S. and Fukuda T. (Editors), 1992, *Precision Sensors, Actuators, and Systems*, 478 pages, (ISBN No.0–7923–2015–8), Kluwer Academic Publishers, Dordrecht/Boston/London, December 1992.

Tzou, H.S. and Fu, H.Q., 1994, "A Study of Segmentation of Distributed Piezoelectric Sensors and Actuators, Part 1 and 2," *Journal of Sound & Vibration*, Vol.172, No.2, pp.247–276.

Tzou, H.S. and Tseng, C.I., 1990, "Distributed Piezoelectric Sensor/Actuator Design for Dynamic Measurement/Control of Distributed Parameter Systems: A Piezoelectric Finite Element Approach," *Journal of Sound & Vibration*, Vol.138, No.1, pp.17–34, April.

Tzou, H.S. and Ye, R., 1993, "Piezothermoelasticity and Control of Piezoelectric Systems," *Intelligent Materials, Structures, and Vibration*, Shanhipoor and Tzou (Ed.), Vol.58, pp.27–34, 1993 ASME Design Technical Conference, September 22–25, 1993. ASME Transactions *Journal of Vibration and Acoustics* (To appear)

Wada, B.K., Fanson, J.L. and Crawley, E.F., 1989, "Adaptive Structures," *Adaptive Structures*, Wada (Ed.), AD–Vol.15, pp.1–8, 1989 ASME WAM.

Zaveri, K., 1985, *Modal Analysis of Large Structures – Multi–Exciter Systems*, Bruel & Kjaer Technical Review. (PzBmBdryCntl.Shl493.Spie94.DistC194)

Wave Motion, Intelligent Structures and Nonlinear Mechanics, pp. 194–225
edited by A. Guran and D. J. Inman
Series on Stability, Vibration and Control of Structures: Vol. 1

ON NEURAL NETWORKS IN IDENTIFICATION AND CONTROL OF DYNAMIC SYSTEMS

MINH Q. PHAN
Department of Mechanical and Aerospace Engineering
Princeton University, Princeton, NJ 08544

JER-NAN JUANG
Structural Dynamics Branch
NASA Langley Research Center, Hampton, VA 23681

and

DAVID C. HYLAND
Government Aerospace Systems Division
Harris Corporation, Melbourne, FL 32902

ABSTRACT

The paper presents a tutorial discussion on the applicability of the neural networks in the identification and control of dynamic systems. Emphasis is placed on the understanding of how the neural networks handle linear systems and how the new approach is related to conventional system identification and control methods. Extensions of the approach to non-linear systems are then made. The paper provides a simple explanation of the basic concepts of the neural networks from a modern control perspective, including an interpretation of the linear networks in terms of new results in observer and system identification. Among the topics discussed are feedforward and recurrent networks in relation to the standard state-space and observer models, linear and non-linear auto-regressive models, linear predictors, one-step ahead control, and model reference adaptive control for linear and non-linear systems. Numerical examples are presented to illustrate the application of these important concepts.

1. Introduction

System identification and control are two related fields that have received considerable development in the last few decades. System identification deals with the problem of finding a mathematical description of a physical system from experimental data. Control theory devises ways to influence the system in a desirable and predictable manner. Typical control objectives are pointing control, vibration suppression, and tracking control. System identification provides the necessary mathematical model of a system for a particular control scheme to be designed. In turn, information gathered during the control process can be used to evaluate the validity of the assumed model. Existing system identification and control methods are based on mathematical systems theory, which first deals with deterministic then stochastic systems. For the most part, the systems under study are idealized. They are linear, time-invariant, and often assumed to be noise-free. When noises are present, they are assumed to be white, zero-mean, and with known characteristics. These assumptions are often justified because less idealized assumptions tend to render the analysis mathematically intractable.

In practice, however, all systems are affected by noises and non-linearities which may lead to instabilities for control laws that are based on idealized models. This motivates the development of a variety of control approaches which, if classified according to the amount of information required to design a controller, can be broadly divided in three classes. They are model-independent control, robust control, and adaptive control. A model-independent controller seeks to guarantee stability of the closed-loop system independently of the system model. As implied, such a design does not require that the system be known in advance. Robust controllers can tolerate certain specific variations about some nominal model, which is required to be known with somewhat accuracy, and variations about the nominal model need to be quantified. While a nominal model may be obtained analytically or experimentally, meaningful characterization of the variations about the nominal model is often difficult to obtain. In both model-independent control and robust control, there is a trade-off between stability robustness and performance. This is because performance requires that the system is known with certainty. If for some reason such knowledge is in error, then the designed optimal performance will not be achieved and instabilities may occur. Striking somewhat of a balance between the two control approaches is adaptive control which involves some level of on-line parameter estimation, where knowledge of the system being controlled is gained during the control process. The estimated parameters can be either the system parameters or the controller gains. In the former case, known as indirect adaptive control, the parameters representing a mathematical model of the system are identified on-line, and the control input is then computed. In the latter case, known as direct adaptive control, the system identification step is bypassed and the controller gains are directly updated at each time step. Adaptive control identifies the appropriate parameters of the system only for the purpose of control, thus offers a meaningful way to integrate system identification and control in one package. Adaptive control also offers the potential ability to handle systems with changing dynamics by constantly identifying and adjusting the control action accordingly.

Recently, there has been a substantial amount of interest in the field of neural networks. As a collection of interconnected neurons, a multi-layer neural network with appropriate weights has been shown to be able to approximate any input-output function. Consequently, the neural network is a natural candidate in the area of identification and control of both linear and non-linear systems. See, for example, the works of Narendra and Parthasarathy (1990), Hornik, Stinchcomb, and White (1989), Billings and Leontaritis (1985), Chen, Billings, and Grant (1990), and Hyland (1991). The neural networks are typically implemented in the adaptive form, and thus possess similar attributes of adaptive control. The main objective of this paper is to examine the implication of the neural networks approach for linear systems, and to see how this approach is similar to and different from conventional methods. Only after a firm grasp of how the neural networks treat the linear problem, extensions to the non-linear problem can then be made, and the potential benefits of the non-linear approach can be better understood and appreciated. The extent to which linear approaches can handle non-linear systems can also be revealed. To this end, basic concepts in neural network will be presented and whenever possible, direct connection to existing system identification and control methods are made. Linear system identification techniques and adaptive control theory will be heavily drawn upon to make this connection, Goodwin and Sin (1984), Ljung and Södertröm (1983). In this paper, we focus on the role of the neural networks as applicable to structural system identification and control as opposed to other fields such as pattern recognition, image processing, and computer science. The recently developed Observer/Kalman filter identification (OKID) algorithm will also be discussed in the context of the neural networks, Chen at el. (1992), Juang at el. (1993), and Phan at el. (1992, 1993). Potential applicability of the existing techniques to non-linear problems will be examined. Several numerical examples will be used to illustrate the basic concepts discussed in this paper. Since accelerometers are often used in structural system identification and control, a direct transmission term is included in input-output models that are discussed in this paper. Minor adjustments can be easily made when this term is not present which is the case treated in Narendra and Parthasarathy (1990).

2. The Neural Networks

A neural network is simply a set of interconnected individual units called neurons. Depending on the connection between the neurons, there are two basic types of networks known as the multi-layer feedforward networks and the recurrent networks, which will be described in this section.

2.1. A Neuron

As a basic building block for a neural network, an individual neuron has a finite number of scalar inputs and one scalar output. Associated with each input is a scalar weighting value. The input signals are weighted by these values and added at the summation junction. The combined signal is then passed through an activation function producing the output signal. The activation function $\gamma(x)$ can take a variety of forms, the most common one is a sigmoid function denoted by *sigm*(x),

$$sigm(x) = \frac{1-e^{-x}}{1+e^{-x}} \tag{1}$$

A plot of the sigmoid function is shown in Fig. 1 below. Generally, the activation function can be any non-decreasing differentiable function which has finite limits at both ends as shown in Fig. 1 below.

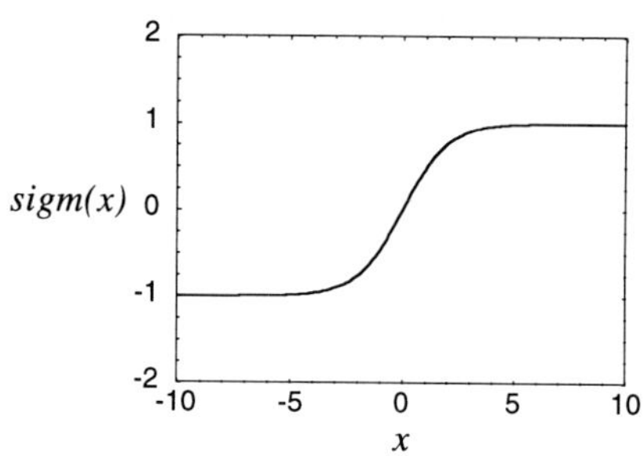

Figure 1: The sigmoid function.

Let r inputs of a neuron be denoted by $u_1, u_2, \ldots, u_r$ and the output denoted by y. Let the r weights be denoted by $w_1, w_2, \ldots, w_r$. The output of the neuron can be expressed mathematically as

$$y = \gamma(x)\ , \quad x = \sum_{i=1}^{r} w_i u_i \tag{2}$$

The neuron is shown schematically in Fig. 2 below with the sigmoid function as the activation function. For simplicity of notations, the network weights for the i-th layer may sometimes be presented collectively as $W_i = \{w_1, w_2, w_3, \ldots\}$. Also, for simplicity the arrows pointing to the right are omitted from the drawing.

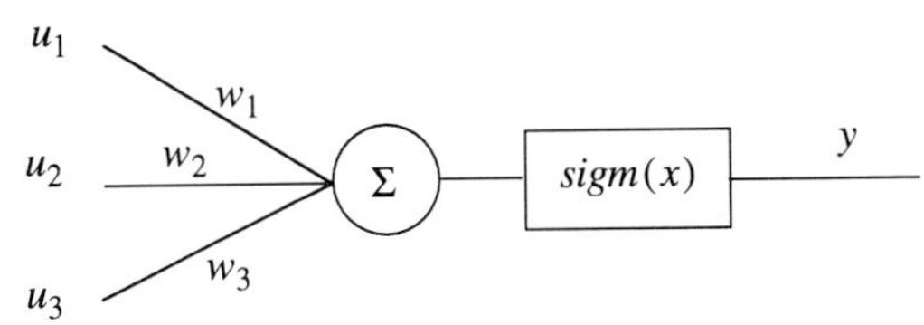

Figure 2: Schematic diagram of a single neuron.

Remark 2.1.1. The activation function is a limiter that bounds the incoming signal, which serves as a non-linear element in a neuron. The activation function given in Eq. (1) has a linear range about the origin, and is bounded between -1 and 1. If output of a neuron is bounded between $-\alpha$ and $+\alpha$ by taking $\bar{\gamma}(x) = \alpha\, sigm(x)$ as its activation function then Eq. (2) becomes

$$\bar{y} = \bar{\gamma}\left(\sum_{i=1}^{r} w_i u_i\right) = \alpha\, sigm\left(\sum_{i=1}^{r} w_i u_i\right) = \alpha\, y \tag{3}$$

which is the same as the output of a neuron with the original activation function multiplied by a constant gain α. Therefore, the activation function can be taken to be between -1 and 1 provided an additional factor is inserted after the neuron. In a network, this factor is absorbed into the weights of the following neurons that directly receive the output of this neuron as their inputs.

Remark 2.1.2. If the activation function is a linear function, $\gamma(x) = x$, then the neuron is a linear neuron. The input-output relationship of a linear neuron is

$$y = \sum_{i=1}^{r} w_i u_i \tag{4}$$

Equation (4) simply says that the output signal is a weighted (linear) combination of the input signals with the weighting coefficients being the weights of the neuron.

2.2. *Multi-Layer Feedforward Neural Network*

A multi-layer feedforward neural network consists of an input layer, a number of hidden layers, and an output layer. In a fully connected feedforward network, every neuron in each layer accepts as its inputs all signals coming from all neurons in the layer immediately preceding it (see Fig. 3). In a partially connected network, some of these connections are missing. This is equivalent to setting the corresponding network weights to zero. Figure 3 show a typical three-layer three-input three-output feedforward network with two hidden layers.

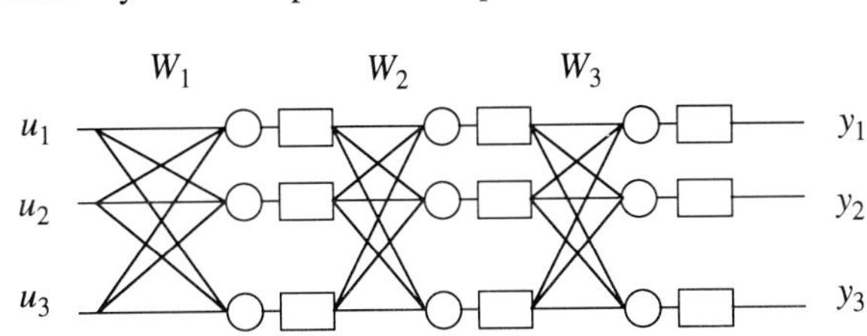

Figure 3: A three-layer three-input three-output feedforward neural network.

2.3. Recurrent Network.

A feedforward network with time delay feedback elements is called a recurrent network. The delay elements take the outputs of certain neurons in the network, delay them for a certain number of time steps, and feed back as input to the neurons. In other words, in a recurrent network, time delayed outputs of a certain number of neurons are the inputs to other neurons. A special one layer network where the delayed outputs of the neurons are fed back as inputs to themselves is called a Hopfield network (see Fig. 4).

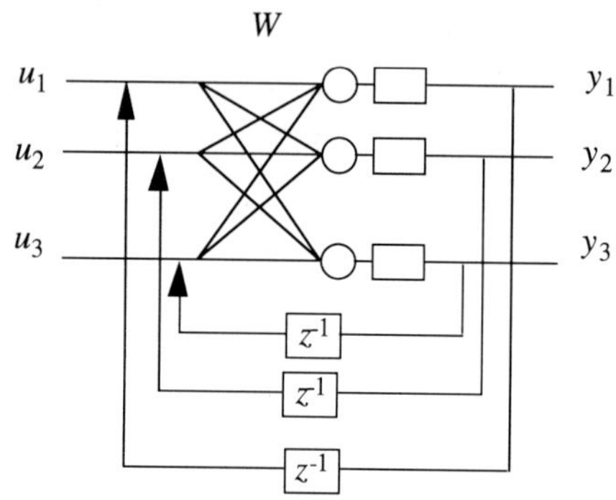

Figure 4: A three-input three-output Hopfield network.

Remark 2.3.1. Unlike a feedforward network, a recurrent network contains self-propagation dynamics. Due to the feedback mechanism, starting with some non-zero initial conditions, the output values of a recurrent network will evolve over time, thus simulating a dynamic system.

Remark 2.3.2. When compared to a recurrent network, a feedforward network is a static in the sense that it simply accepts a set of input values (or input pattern) and produces a set of output values (or output pattern) without any self-propagation mechanism. Thus it may seem that a recurrent network is preferred for modelling dynamic systems. In fact, the two types of networks simply represent two different ways of modelling a dynamic system. In identification, if a set of input-output data is already available, then the weights of a recurrent network can be identified by a feedforward network where the time-delayed outputs are treated as inputs. This seems a bit confusing, but should later be obvious when connections between these types of neural networks to standard ways of representing linear systems are made in the next section.

3. Neural Network Representation of Linear Systems

For each neuron, the only element that is non-linear is the activation function. If the activation is taken to be a linear function, $\gamma(x) = x$, then the network becomes linear. This is true no matter how complicated the neural network is. In this section, attention will be focused on linear networks. The relationship between these networks and conventional ways to represent linear systems will be discussed.

3.1. A Multi-Layer Feedforward Linear Network

Consider the feedforward network shown in Figure 3, with the activation function being a linear function. Furthermore, it is instructive to examine a simple case of a two-layer three-input one-output network shown in Figure 4 below

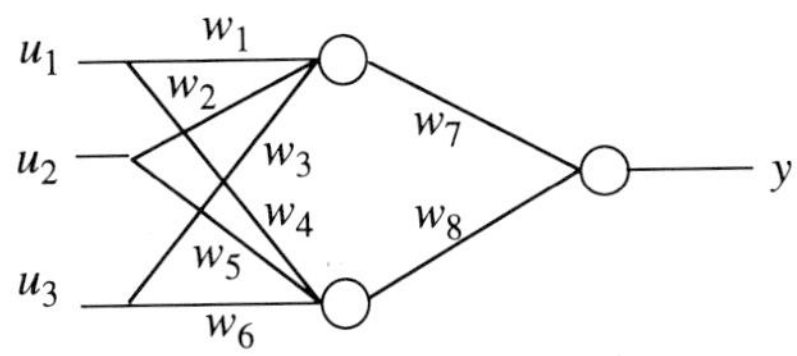

Figure 5: A two-layer three-input one-output feedforward network of linear neurons

The network weights between the individual connections are shown in the figure. Since the neurons are linear, each neuron is represented by a summation junction and the activation being a linear function is omitted. The output of the network in Fig. 5 is simply,

$$\begin{aligned} y &= w_7(w_1u_1 + w_2u_2 + w_3u_3) + w_8(w_4u_1 + w_5u_2 + w_6u_3) \\ &= (w_7w_1 + w_8w_4)u_1 + (w_7w_2 + w_8w_5)u_2 + (w_7w_3 + w_8w_6)u_3 \\ &= \overline{w}_1u_1 + \overline{w}_2u_2 + \overline{w}_3u_3 \end{aligned} \tag{5}$$

which is the output of a single neuron with the following weights:

$$\begin{aligned} \overline{w}_1 &\equiv w_7w_1 + w_8w_4 \\ \overline{w}_2 &\equiv w_7w_2 + w_8w_5 \\ \overline{w}_3 &\equiv w_7w_3 + w_8w_6 \end{aligned} \tag{6}$$

The above example can be immediately generalized to show that a single-output multi-layer neural network with linear neurons is equivalent to a single linear neuron with appropriate weights. The following remarks can be immediately made.

Remark 3.1.1. A multi-layer feedforward network of linear neurons is simply an over-parameterized set of linear equations where the over-parameterization takes the form of the type shown in Eqs. (6). This form is non-linear in the parameters. Thus the problem of determining these parameters from known input-output data is a non-linear parameter estimation problem even if the network is linear.

Remark 3.1.2. Since any single-output feedforward network of linear neurons is equivalent to a particular single linear neuron, there is no benefit in using an over-parameterized multi-layer linear network for linear system identification. In fact, in such an over-parameterized model, the network weights cannot be uniquely determined from input-output data. This is obvious, for example, in the case of the network shown in Fig. 5. The set of three equations in (6) contains eight unknowns. Thus the use of a complicated multi-layer linear network for linear system identification is therefore neither advantageous nor necessary. The same is true for the case of using a non-linear network to identify a linear system.

Remark 3.1.3. For a multi-output system, a multi-layer feedforward network of linear neurons is equivalent to a collection of single neurons arranged in parallel, each of which sharing the same number of inputs. In other words, any multi-output multi-layer feedforward network of linear neurons has an equivalent multi-output single-layer network representation.

3.2. Feedforward Linear Network and the State Space Model

This section describes the relationship between the feedforward linear network and the state space model which is a common form of representing linear systems. The discrete-time state space model of an n-th order, m-input, q-output system is a set of n simultaneous first order difference equations of the form

$$\begin{aligned} x(k+1) &= Ax(k) + Bu(k) \\ y(k) &= Cx(k) + Du(k) \end{aligned} \tag{7}$$

where the dimensions of A, B, C, and D are $n \times n$, $n \times m$, $q \times n$, and $q \times m$, respectively. Solving for the output $y(k)$ in terms of the previous inputs yields

$$y(k) = \sum_{i=0}^{k} h_i u(k-i) \tag{8}$$

where the parameters,

$$h_0 = D \ , \quad h_k = CA^{k-1}B \quad , \quad k = 1,\ 2,\ 3,\ \ldots \tag{9}$$

are the *Markov parameters* of the system described by Eqs. (7), which are also the system pulse response samples. The Markov parameters are expressed in terms of the system discrete state space matrices A, B, C, D. Since the state vector is coordinate-dependent, the state space matrices are not unique for a given system but the Markov parameters are unique. Let the state vector be transformed by a coordinate transformation T, $z(k) = Tx(k)$, then the relationship between $u(k)$ and $y(k)$ via a new state vector $z(k)$ can be described by a new state space representation TAT^{-1}, TB, CT^{-1}, D. The system Markov parameters computed using the new state space matrices are the same as before, i.e.,

$$h_k = CT^{-1}\left(TAT^{-1}\right)^{k-1} TB = CA^{k-1}B \tag{10}$$

For an asymptotically stable systems, the pulse response can be neglected after a finite number of time steps, say p_s. The input-output description in Eq. (8) can be approximated by a finite number of Markov parameters

$$y(k) \approx h_0 u(k) + h_1 u(k-1) + h_2 u(k-2) + \cdots + h_{p_s} u(k-p_s) \tag{11}$$

where p_s is sufficiently large such that $CA^k B \approx 0$, $k \geq p_s$. Comparing Eq. (11) with the structure of the linear neurons immediately leads to the following remarks.

Remark 3.2.1. The elements of the Markov parameters are simply the weights of a single-layer linear network where inputs to the network include both current and past values of the input signal. Note that the time delayed inputs do not affect the neuron configuration because they are feedforward signals and thus can be treated as separate input channels. This case is shown in Fig. 6.

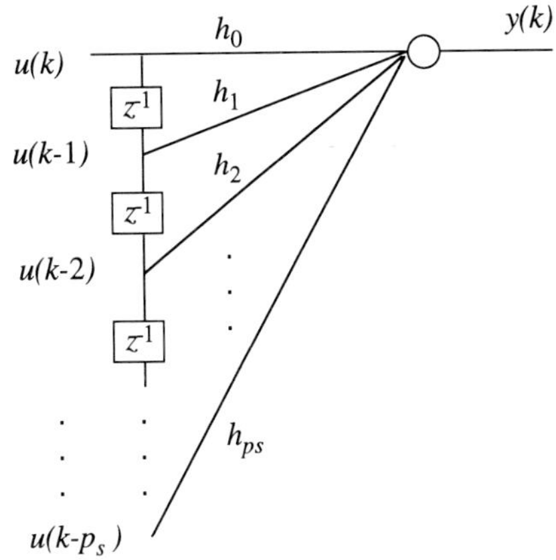

Figure 6: Representation of linear systems by a feedforward network with the system Markov parameters as network weights.

Remark 3.2.2. Since a general multi-layer feedforward network of linear neurons is equivalent to a single-layer network, the relationship between the weights of the multi-layer network which represents a linear system and the Markov parameters of its equivalent state space model is also immediately obvious. For example, if the network shown in Fig. 5 represents some linear system with three non-zero Markov parameters, then

$$\begin{aligned} w_1 &= w_{11}^{(2)}w_{11}^{(1)} + w_{21}^{(2)}w_{12}^{(1)} = h_0 \\ w_2 &= w_{11}^{(2)}w_{21}^{(1)} + w_{21}^{(2)}w_{22}^{(1)} = h_1 \\ w_3 &= w_{11}^{(2)}w_{31}^{(1)} + w_{21}^{(2)}w_{32}^{(1)} = h_2 \end{aligned} \tag{12}$$

provided that $u_1 = u(k),\ u_2 = u(k-1),\ u_3 = u(k-2)$.

Remark 3.2.3. In practice, if the system is lightly damped, a large number of system Markov parameters is needed to maintain Eq. (11) a valid approximation. This implies that the equivalent network representing the same system has a large number of input channels containing distant past input values, not a large number of hidden layers. In other words, it is not possible to represent such a system by simply adding extra neurons or extra hidden layers in the feedforward network. The fact that a large number of system Markov parameters is required to represent a lightly damped system of the form in Eq. (11) is a major weakness of the representation. The same can be said for the equivalent neural network representation.

3.3. Recurrent Linear Network and the Observer Model

This section shows the connection between the recurrent network and an observer of the system. Adding and subtracting the term *My(i)* to the right hand side of the state equation in Eq. (7) yields

$$\begin{aligned} x(k+1) &= Ax(k) + Bu(k) + My(k) - My(k) \\ &= (A+MC)x(k) + (B+MD)u(k) - My(k) \end{aligned} \tag{13}$$

If M is a matrix such that $A+MC$ is deadbeat of order p, i.e.,

$$(A+MC)^k \equiv 0 \ , \quad k \geq p \tag{14}$$

then for $k \geq p$, the output $y(k)$ can be expressed as

$$y(k) = \alpha_1 y(k-1) + \cdots + \alpha_p y(k-p) + \beta_0 u(k) + \beta_1 u(k-1) + \cdots + \beta_p u(k-p) \tag{15}$$

where

$$\begin{aligned} \alpha_k &= -C(A+MC)^{k-1}M \\ \beta_k &= C(A+MC)^{k-1}(B+MD) \ , \quad \beta_0 = h_0 = D \end{aligned} \tag{16}$$

The matrix M in the above development can be interpreted as an observer gain. The system considered in Eqs. (7) has an observer of the form

$$\begin{aligned} \hat{x}(k+1) &= A\hat{x}(k) + Bu(k) - M[y(k) - \hat{y}(k)] \\ \hat{y}(k) &= C\hat{x}(k) + Du(k) \end{aligned} \tag{17}$$

Besides the effect of noises, $\hat{y}(k)$ may differ from $y(k)$ if the actual initial condition $x(0)$ is not known and some different initial condition is assumed for $\hat{x}(0)$. Defining the state estimation error $e(k) = x(k) - \hat{x}(k)$, the equation that governs $e(k)$ is

$$e(k+1) = (A+MC)e(k) \tag{18}$$

For an observable system, the matrix M exists such that the eigenvalues of $A+MC$ may be placed in any desired (symmetric) configuration. If the matrix M is such that $A+MC$ is asymptotically stable, then the estimated state $\hat{x}(k)$ tends to the true state $x(k)$ as k tends to infinity for any initial difference between the assumed observer state and the actual system state. The matrix M can therefore be interpreted as an observer gain. The parameters defined as

$$\begin{aligned} \overline{Y}(k) &= C(A+MC)^{k-1}[B+MD \ , \ -M] \\ &= [\beta_k \ , \ \alpha_k] \end{aligned} \tag{19}$$

are the Markov parameters of an observer system, hence they are referred to as *observer Markov parameters*. Like the system Markov parameters, the observer Markov parameters are also invariant with respect to a coordinate transformation of the state vector. To see this, again let the state vector be transformed by a coordinate transformation T, $z(k) = Tx(k)$, then the observer is described by a new state space representation TAT^{-1}, TB, CT^{-1}, D and a new observer gain TM. The observer Markov parameters computed using these new state space matrices and the new observer gain are the same as before,

$$\begin{aligned} \overline{Y}(k) &= CT^{-1}(TAT^{-1} + TMCT^{-1})^{k-1}[TB + TMD \ , \ -TM] \\ &= C(A+MC)^{k-1}[B+MD \ , \ -M] \ , \quad k = 1, \ 2, \ 3, \ \ldots \end{aligned} \tag{20}$$

Notice that in Eq. (15), the output $y(k)$ is the open-loop response of the system, yet the coefficients α_k, β_k are related to an observer gain. Consider the special case where M is a deadbeat observer gain where all eigenvalues of $A+MC$ are zero, the observer Markov parameters will become identically zero after a finite number of terms. For lightly damped structures, this means that the system can be described by a reduced number of observer Markov parameters $\overline{Y}(k)$ instead of an otherwise large number of the usual system Markov parameters $Y(k)$. For this reason, the observer Markov parameters are important in linear system identification. By examining of the structure of Eq. (15), the following remarks can be made.

Remark 3.3.1. The input-output equation given in Eq. (15) can be represented by a recurrent network with a single layer of linear neurons. The number of neurons is equal to the number of outputs of the system. The inputs to the neurons consists of both the feedforward time-delayed input signals and the feedback time-delayed output signals. Figure 7 shows the configuration of such a network for a single-output system.

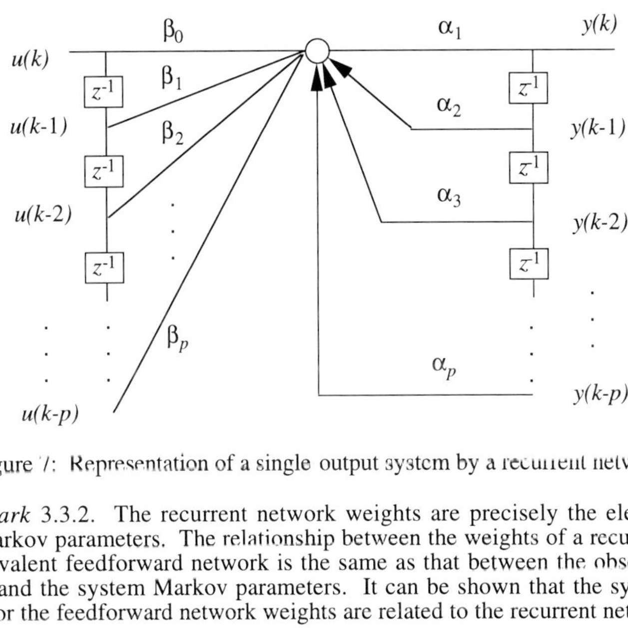

Figure 7: Representation of a single output system by a recurrent network

Remark 3.3.2. The recurrent network weights are precisely the elements of the observer Markov parameters. The relationship between the weights of a recurrent network and an equivalent feedforward network is the same as that between the observer Markov parameters and the system Markov parameters. It can be shown that the system Markov parameters or the feedforward network weights are related to the recurrent network weights by

$$h_k = \beta_k + \sum_{i=1}^{k} \alpha_i h_{k-i} \tag{21}$$

where $\alpha_k \equiv 0$, $\beta_k \equiv 0$ for $k > p$. To describe a system of order n, the number of observer Markov parameters p must be such that $qp \geq n$ where q is the number of outputs. Furthermore, the maximum order of a system that can be described with p observer Markov parameters is qp.[9] The implication of this result to the network configuration is that a recurrent network generally requires fewer number of parameters (or weights) than that required by an equivalent feedforward network. The two equivalent networks, however, have the same number of neurons. The minimum number of recurrent network weight matrices that can describe the system is $p_{\min}$, which is the smallest value of p such that $qp_{\min} \geq n$.

Remark 3.3.3. As mentioned previously, to represent lightly damped structures, the feedforward representation requires a large number of weights. Furthermore, it is not possible to represent a marginally stable or unstable system by a feedforward network. However, it is possible to represent such a system by a recurrent network. The implication of this fact for the system identification problem will be discussed further in later sections.

4. Identification of Linear Systems using Neural Networks

It has been shown that a general network of linear neurons is equivalent to a single neuron with appropriate weights. The problem of linear system identification using neural network is therefore reduced to finding these network weights from input-output data. The computation may be done off-line or on-line. In off-line computation, the input-output data is already available and a network representing the system is to be determined. On-line computation refers to the case where the network weights are continually updated as data is made available.

4.1. Parallel vs. Series-Parallel Identification Models

In previous consideration, it appears that the recurrent network is more advantageous in representing certain systems than the feedforward network. To identify the recurrent network weights one can simply use the feedforward network configuration with *actual* delayed system outputs appeared as inputs to the feedforward network. Consider two identification models shown in Figs. 8 and 9 below, which are known as parallel and series-parallel identification model. The block denoted by D represents the time delay elements.

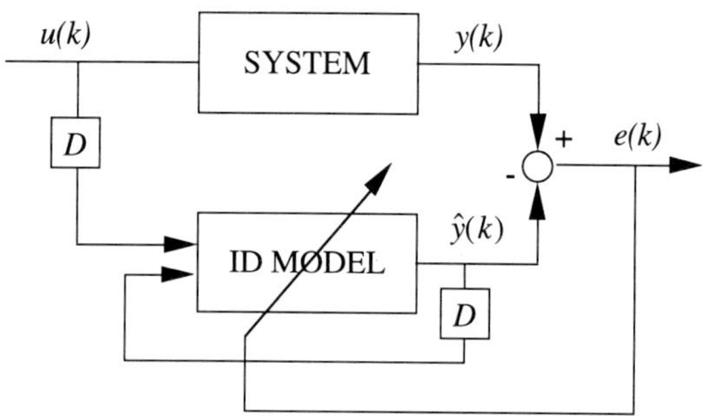

Figure 8: Identification using parallel model.

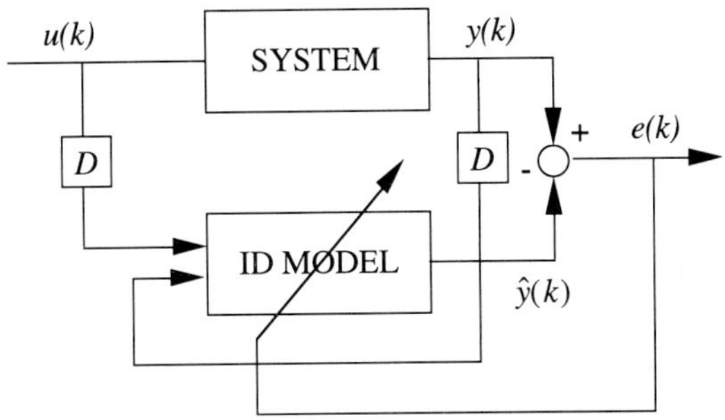

Figure 9: Identification using series-parallel model.

The basic difference between the two schemes is that in the parallel identification model, the estimated output $\hat{y}(k)$ is computed based on the model own previous (estimated) values whereas in the series-parallel model, it is based on actual output values. Mathematically, in the parallel model, the purpose of the identification is to obtain the estimates $\hat{\alpha}_k$, $\hat{\beta}_k$ of the coefficients α_k, β_k that minimize the estimation error, $e(k) = y(k) - \hat{y}(k)$, where the estimated output $\hat{y}(k)$ is computed from

$$\hat{y}(k) = \hat{\alpha}_1 \hat{y}(k-1) + \cdots + \hat{\alpha}_p \hat{y}(k-p) + \hat{\beta}_0 u(k) + \hat{\beta}_1 u(k-1) + \cdots + \hat{\beta}_p u(k-p) \quad (22)$$

In the series-parallel model, however, the estimated output is computed from

$$\hat{y}(k) = \hat{\alpha}_1 y(k-1) + \cdots + \hat{\alpha}_p y(k-p) + \hat{\beta}_0 u(k) + \hat{\beta}_1 u(k-1) + \cdots + \hat{\beta}_p u(k-p) \quad (23)$$

The difference between the two above equations is a subtle but important one. As discussed in the previous section, the estimated output of the model in Eq. (22) is the estimated open-loop prediction even though the coefficients of the model are related to an observer. On the other hand, the estimated output of the model in Eq. (23) is that of an observer. To see this, substitute the expression for $\hat{y}(k)$ to the estimated state equation in (17) produces

$$\hat{x}(k+1) = (A + MC)\hat{x}(k) + (B + MD)u(k) - My(k) \quad (24)$$

Since $\hat{y}(k) = C\hat{x}(k) + Du(k)$, one can obtain Eq. (23) assuming zero initial conditions for the observer. Therefore, $\hat{y}(k)$ in Eq. (23) represents the estimated output provided by the observer. The estimation error $\hat{e}(k)$ is the difference between the actual output and the estimated output provided by the observer. On the other hand, if the actual response $y(k)$ is replaced by the estimated value $\hat{y}(k)$ in Eq. (23) then the terms involving the observer gain M cancel each other identically for any arbitrary initial condition $\hat{x}(0)$,

$$\begin{aligned} \hat{x}(k+1) &= (A + MC)\hat{x}(k) + (B + MD)u(k) - M\hat{y}(k) \\ &= A\hat{x}(k) + Bu(k) \end{aligned} \quad (25)$$

Therefore, there is no longer any observer involved in the equation; $\hat{x}(k)$ now plays the role of the state vector $x(k)$ as in Eq. (7) and the estimated output $\hat{y}(k) = C\hat{x}(k) + Du(k)$ is the same as that produced by the open-loop model provided that the initial conditions for $\hat{x}(k)$ and $x(k)$ are identical. The quantity $\hat{y}(k)$ now represents the predicted output provided by the open-loop model alone, which is referred in this paper as open-loop prediction. In this case, the error $\hat{e}(k)$ is the difference between the actual output and the predicted output provided by the identified open-loop model.

Remark 4.1.1. First recall that the model structure in Eq. (15) subsumes an observer. If the parallel identification model is used in conjunction with the model structure of Eq. (15) then the prediction error that drives the parameter estimation scheme is simply the open-loop prediction error not the observer (output) estimation error. Consequently, the observer portion of the model cannot be identified. This fact accounts for the difficulties encountered in parallel model identification, namely, the conditions for which the scheme will converge are presently not known.

Remark 4.1.2. In the series-parallel identification model, since the actual instead of (open-loop) predicted output enters the model, a feedforward network with delayed input and actual output measurements can be used to identify the system. This consideration eliminates the use of a recurrent network which would introduce additional but unnecessary difficulties to the system identification problem, (see Fig. 10). Each output of the system is represented by a single linear neuron. A multiple-output system is represented by a single layer of neurons. The identified network can be used either as a feedforward or a recurrent network. In the former case, the network provides estimation of the response by an observer. In the latter case, it is an open-loop predictor. Again, this depends on whether actual or predicted output is used in computing the response.

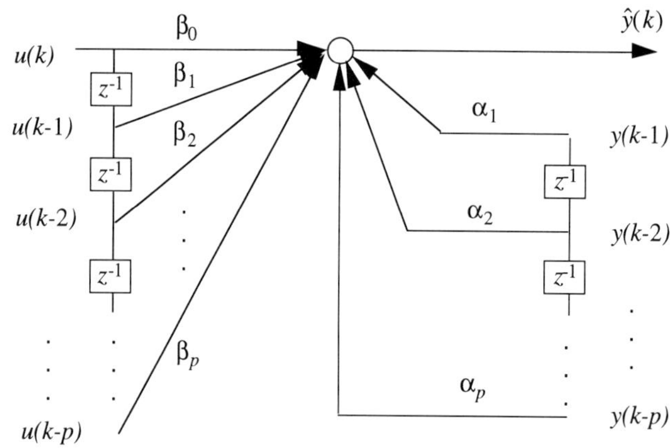

Figure 10: Feedforward representation of the series-parallel identification model by a single neuron for each output.

4.2. Identification of the Network Weights

This section shows how the weights of the network represented by Eq. (15) can be computed using a feedforward model. For linear systems, it is sufficient to use a one layer network having as many neurons as the number of outputs. This is a simple linear parameter estimation problem. The off-line computation is shown first, followed by an equivalent on-line computation. For simplicity, consider the case where the system starts from zero initial conditions. Equation (15) can be written as

$$y(k) = \sum_{i=1}^{p} [\beta_i \ , \ \alpha_i] \begin{bmatrix} u(k-i) \\ y(k-i) \end{bmatrix} + \beta_0 u(k) \tag{26}$$

where network weight matrices β_i , α_i are defined in Eq. (16). Writing Eq. (26) in matrix form for a set of input-output data N+1 samples long yields

$$\mathbf{y} = \mathbf{YV} \tag{27}$$

where

$$\mathbf{y} = [y(0) \quad y(1) \quad \cdots \quad y(p) \quad y(p+1) \quad \cdots \quad y(N)] \tag{28}$$

$$\mathbf{Y} = [\beta_0, \ \beta_1, \ \alpha_1, \ \beta_2, \ \alpha_2, \ \ldots, \ \beta_p, \ \alpha_p] \tag{29}$$

$$\mathbf{V} = \begin{bmatrix} u(0) & u(1) & \cdots & u(p) & u(p+1) & \cdots & u(N) \\ & \begin{bmatrix} u(0) \\ y(0) \end{bmatrix} & \cdots & \begin{bmatrix} u(p-1) \\ y(p-1) \end{bmatrix} & \begin{bmatrix} u(p) \\ y(p) \end{bmatrix} & \cdots & \begin{bmatrix} u(N-1) \\ y(N-1) \end{bmatrix} \\ & & \ddots & \vdots & \vdots & \vdots & \vdots \\ & & & \begin{bmatrix} u(0) \\ y(0) \end{bmatrix} & \begin{bmatrix} u(1) \\ y(1) \end{bmatrix} & \cdots & \begin{bmatrix} u(N-p) \\ y(N-p) \end{bmatrix} \end{bmatrix} \tag{30}$$

The network weight matrices are estimated using the equation

$$\hat{\mathbf{Y}} = \mathbf{yV}^{+} \tag{31}$$

where $(.)^{+}$ denotes the pseudo-inverse of the quantity in the parentheses. If the initial conditions are not zero then a slightly different equation must be used to solve for the network weights, that is

$$\hat{\mathbf{Y}} = \mathbf{y}_t\mathbf{V}_t^{+} \tag{32}$$

where $\mathbf{y}_t$ and $\mathbf{V}_t$ are obtained by deleting the first p columns in $\mathbf{y}$ and $\mathbf{V}$, respectively.

Remark 4.2.1. The least-squares solution in Eq. (31) or (32) minimizes the error between the actual output and the estimated output computed using the actual input and output data, i.e., the least-squares solution minimizes the residual $\hat{\mathbf{e}} = \mathbf{y} - \hat{\mathbf{y}}$ where $\hat{\mathbf{y}}$ is computed from $\hat{\mathbf{y}} = \hat{\mathbf{Y}}\mathbf{V}$ and $\hat{\mathbf{Y}}$ is given in Eq. (31). If Eq. (32) is used instead, then the least-squares solution minimizes $\hat{\mathbf{e}}_t = \mathbf{y} - \hat{\mathbf{y}}_t$ where $\hat{\mathbf{y}}_t = \hat{\mathbf{Y}}\mathbf{V}_t$. This computation, therefore, corresponds to the series-parallel identification scheme that minimizes the observer estimation error.

Remark 4.2.2. Ideally, the error between the actual output and the predicted output provided by the identified open-loop model is the proper error to be minimized for the identification of the system open-loop model. The above computation minimizes the observer estimation error instead. For a linear system, it turns out that in the absence of noises the open-loop system can be identified exactly by minimizing the observer estimation error. In the presence of noises, however, minimizing the observer estimation error does not necessarily implies that the open-loop prediction error is minimized. Therefore, it is possible that the observer model fits the data well but the open-loop model does not. Fortunately, if the order of the regression equation is chosen to be sufficiently large then simultaneous observer and system identification will still be achieved in the limit as the data record tends to infinity and the noises are white, Gaussian, and zero-mean (see Ref. 9).

Remark 4.2.3. The least-squares solution in Eq. (31) can be obtained by an on-line parameter estimation scheme. First, write each column in $\mathbf{V}$ as

$$\mathbf{V} = [\Gamma(0),\ \Gamma(1),\ \Gamma(2),\ \dots\ ] \tag{33}$$

so that at each time step k, Eq. (27) can be written as

$$y(k) = \mathbf{Y}\Gamma(k) \tag{34}$$

The recursive least-squares equation for the network weights is simply,

$$\hat{\mathbf{Y}}(k) = \hat{\mathbf{Y}}(k-1) + \left[y(k) - \hat{\mathbf{Y}}(k-1)\Gamma(k)\right]\left\{\frac{\Gamma(k)^T R(k-1)}{1+\Gamma(k)^T R(k-1)\Gamma(k)}\right\} \tag{35}$$

$$R(k) = R(k-1) - \frac{R(k-1)\Gamma(k)\Gamma(k)^T R(k-1)}{1+\Gamma(k)^T R(k-1)\Gamma(k)} \tag{36}$$

where $\hat{\mathbf{Y}}(k) = \left[\hat{\beta}_0(k),\ \hat{\beta}_1(k),\ \hat{\alpha}_1(k),\ \hat{\beta}_2(k),\ \hat{\alpha}_2(k),\ \dots,\ \hat{\beta}_p(k),\ \hat{\alpha}_p(k)\right]$, $\hat{\mathbf{Y}}(0)$ is an arbitrary initial guess, and $R(0)$ is any symmetric positive definite matrix. The recursive equations for (32) are analogous.

Remark 4.2.4. In a multi-layer neural network, the back propagation algorithm is typically used to update the network weights recursively. This is a gradient-based parameter update algorithm. When expressed in the block diagram form for hardware implementation, the algorithm resembles the forward network except that the signal travels in the opposite direction, leading to the name back propagation. In the present case, because the network is simply one layer of linear neurons, it is not efficient to use the back propagation algorithm to compute the network weights. For on-line implementation, the least squares algorithm given in Eqs. (35-36) or its variants for fast computation are preferred.

5. Predictor Models For Linear Dynamic Systems

In theory, the open-loop model can be used to predict the system response based on current and past input values. However, this is not desirable in practice because of the requirement that both the open-loop model and the initial conditions be known exactly. Such a prediction is also sensitive to noises. On the other hand, an observer which is typically used to estimate the system state based on actual input-output data can also be used to provide an estimate of the system output. This section first discusses the use of an observer as an one-step ahead predictor. This interpretation is important because of its connection to the control problem which will be discussed in later sections. Extensions to the identification and use of multiple-step ahead predictors will then be made. For linear systems, these predictors are simply special single-layer networks of linear neurons.

5.1. An One-Step Ahead Predictor

To express explicitly the observer as an one-step ahead predictor, one simply writes the observer equations as

$$\begin{aligned} \hat{x}(k+1) &= (A+MC)\hat{x}(k)+(B+MD)u(k)-My(k) \\ \hat{y}(k+1) &= C\hat{x}(k+1)+Du(k+1) \end{aligned} \tag{37}$$

As a predictor, the interested quantity is $\hat{y}(k+1)$. One can therefore bypass the state equation by writing

$$\hat{y}(k+1)=\alpha_1 y(k)+\cdots+\alpha_p y(k-p+1)+\beta_0 u(k+1)+\beta_1 u(k)+\cdots+\beta_p u(k-p+1) \tag{38}$$

The following remarks can be made regarding the forms of Eq. (37) and Eq. (38).

Remark 5.1.1. In theory, if the state space model *(A, B, C, D)* is known exactly then one can design an observer gain M such that $A+MC$ is asymptotically stable. To use Eq. (38) as an output predictor, one need to include a sufficient number of terms such that $(A+MC)^i$ is negligible for $i \geq p$. The state space representation in Eq. (37) is a better choice since it involves no such approximation. The above comment no longer holds true if M is such that $A+MC$ is deadbeat, i.e., $(A+MC)^i \equiv 0,\ i \geq p$ since the approximation becomes exact in this case.

Remark 5.1.2. In practice, the system model cannot be known exactly. To identify the system from input-output data using the series-parallel structure, one in fact computes directly the coefficients in Eq. (38) rather than the state space matrices. To obtain a minimal order state space representation from these coefficients, realization is required. As an output predictor, therefore, Eq. (38) should be used directly because conversion to a state space representation is not necessary for this purpose.

Remark 5.1.3. Equation (38) clearly indicates that the one-step ahead predictor takes the form of a single layer network of linear neurons with actual input and output signals entering the network, and the output of the network represents the one-step ahead prediction. Schematically, this is the same as shown in Fig. 10.

5.2. *A Two-Step Ahead Predictor*

This section derives the equations for a two-step ahead predictor for linear systems, and shows that it also has a linear neural network form. First, from Eq. (7), one can write

$$\begin{aligned} x(k+2) &= Ax(k+1)+Bu(k+1) \\ &= A^2x(k)+ABu(k)+Bu(k+1) \\ y(k+2) &= Cx(k+2)+Du(k+2) \end{aligned} \tag{39}$$

Adding and subtracting the term *Gy(i)* to the right hand side of the state equation yields

$$\begin{aligned} x(k+2) &= A^2x(k)+ABu(k)+Bu(k+1)+Gy(k)-Gy(k) \\ &= (A^2+GC)x(k)+\left[AB+GD,\ B\right]\begin{bmatrix} u(k) \\ u(k+1) \end{bmatrix}-Gy(k) \end{aligned} \tag{40}$$

If G is a matrix such that A^2+GC is deadbeat of order p, i.e.,

$$(A^2+GC)^k \equiv 0\ ,\quad k \geq p \tag{41}$$

then the relationship between the input and output of the system can be described as a linear combination of input-output data of the form

$$y(k+2) = g\big(y(k),\ y(k-2),\ y(k-4),\ \ldots,\ u(k+2),\ u(k+1),\ u(k),\ \ldots\big) \tag{42}$$

for $k \geq 2p-2$ so that at sufficiently large time steps, terms involving the states $x(0)$ and $x(1)$ vanish, i.e., $C(A^2+GC)^i x(0) = 0$, $C(A^2+GC)^i x(1) = 0$, for $i \geq p$ due to the imposed deadbeat condition for A^2+GC. Furthermore, there is only a finite number of coefficients that make up the linear combination in $g(.)$, which are the *predictor Markov parameters* of the form

$$\overline{G}(k) = C(A^2+GC)^{k-1}\big[AB+GD,\ B\ ,\ -G\big]\ ,\ k=1,\ 2,\ \ldots,\ p \tag{43}$$

Existence of the matrix G such that $(A^2+GC)^k \equiv 0\ ,\quad k \geq p$ is assured if the pair $(A^2,\ C)$ is observable.

Remark 5.2.1. The above derivation justifies the form of a two-step ahead predictor. In fact, one can identify the coefficients of this predictor from input-output data by minimizing the two-step prediction error. The procedure is similar to that discussed in Section 4.

Remark 5.2.2. To obtain the two-step ahead prediction one can also propagate the observer, which is a one-step ahead predictor, in two successive time steps by treating the estimated output from the first time step as the actual output for the second time step However, such a procedure would amount to performing open-loop prediction in the second time step and is therefore sensitive to noises. On the other hand, if one uses the predictor form with the coefficients directly identified from input-output data then only actual data enter the computation and thus minimizes the errors due to noises.

Remark 5.2.3. Again, the predictor form can be represented by a single layer of linear neurons, and the weights of this network are simply the elements of the predictor Markov parameters shown in Eq. (43). Results presented in this section can be easily generalized to a general multi-step predictor. The relationship between such predictors and the deadbeat control problem will be discussed in a later publication.

6. Control of Linear Systems using Neural Networks

As formulated in previous sections, linear systems can be represented by a single-layer network of linear neurons. The weights of this network can be identified from input-output data. Once identified, the network can be used as a one-step ahead predictor. This section discusses the use of such network directly for control application without requiring the state space model to be extracted from these weights.

6.1. A One-Step Ahead Controller

First, consider the case where the linear system can be expressed in the form,

$$y(k+1) = \alpha_1 y(k) + \ \cdots\ + \alpha_p y(k-p+1) + \beta_0 u(k+1) + \beta_1 u(k) + \ \cdots\ + \beta_p u(k-p+1) \tag{44}$$

where the coefficients are assumed to be known. Let the desired response be denoted by $r(k)$. To obtain a controller directly from the above equation, one simply replaces $y(k+1)$ by its desired value $r(k+1)$ and then solve for the control input $u(k+1)$ to obtain the control law,

$$u(k+1)=\beta_0^{-1}\left[r(k+1)-\alpha_1 y(k)-\cdots-\alpha_p y(k-p+1)-\beta_1 u(k)-\cdots-\beta_p u(k-p+1)\right] \quad (45)$$

If the coefficients in Eq. (44) are not known exactly then Eq. (46) represents an one-step ahead estimate of what the system will produce based on current and past input-output data,

$$\hat{y}(k+1)=\hat{\alpha}_1 y(k)+\cdots+\hat{\alpha}_p y(k-p+1)+\hat{\beta}_0 u(k+1)+\hat{\beta}_1 u(k)+\cdots+\hat{\beta}_p u(k-p+1) \quad (46)$$

The control law then is simply

$$u(k+1)=\hat{\beta}_0^{-1}\left[r(k+1)-\hat{\alpha}_1 y(k)-\cdots-\hat{\alpha}_p y(k-p+1)-\hat{\beta}_1 u(k)-\cdots-\hat{\beta}_p u(k-p+1)\right] \quad (47)$$

The behavior of the closed-loop system using the above control law can be examined as follows. Let $e(k+1)$ denote the error between the actual response and the predicted response,

$$\hat{y}(k+1)=y(k+1)-e(k+1) \quad (48)$$

Substituting Eq. (47) and Eq. (48) into Eq. (46) produces

$$\begin{aligned} y(k+1)-e(k+1) &= \hat{\alpha}_1 y(k)+\cdots+\hat{\alpha}_p y(k-p+1)+\hat{\beta}_1 u(k)+\cdots+\hat{\beta}_p u(k-p+1) \\ &\quad +\hat{\beta}_0\Big\{\hat{\beta}_0^{-1}\big[r(k+1)-\hat{\alpha}_1 y(k)-\cdots-\hat{\alpha}_p y(k-p+1) \\ &\quad -\hat{\beta}_1 u(k)-\cdots-\hat{\beta}_p u(k-p+1)\big]\Big\} \\ &= r(k+1) \end{aligned} \quad (49)$$

or

$$y(k+1)=r(k+1)+e(k+1) \quad (50)$$

Define the tracking error to be the difference between the actual response and the desired response, $\varepsilon(k)=y(k)-r(k)$, Eq. (50) reveals that

$$\varepsilon(k+1)=e(k+1) \quad (51)$$

Therefore, if the predictor is such that its prediction error vanishes in the limit, then the tracking error also vanishes in the limit, i.e.,

$$\lim_{k\to\infty} e(k)=0 \quad \Rightarrow \quad \lim_{k\to\infty}\varepsilon(k)=0 \quad (52)$$

Remark 6.1.1. The one-step ahead control law has the property that the tracking error is the same as the prediction error. The above analysis shows that accuracy of the predictor model governs the accuracy of the tracking response. As long as the predictor can perform a reasonably good one-step ahead prediction of the system response then the control input can be computed to make the system track a desired trajectory. In the ideal case where the system is linear and the data is noise-free, the prediction error and the tracking error will be zero identically. Non-zero prediction and tracking error can only come about during adaptation or when noises are present. This is different from the non-linear case where both the estimation and tracking error are non-zero even when there are no noises in the system.

An important restriction of the one-step ahead controller is that the open-loop system is required to be stably invertible, (i.e., there are no unstable zeros in the linear case). If this condition is not met, it is possible to have the controller producing unbounded input while maintaining zero tracking error.

Remark 6.1.2. To obtain the result in Eq. (51), the controller coefficients must be the same as those of the predictor model. In the event the coefficients of the predictor model are updated at each time step, then the controller coefficients must also match those of the predictor model. Mathematically, if at time step k, the predictor takes the form

$$\begin{aligned}\hat{y}(k+1) = \hat{\alpha}_1(k)y(k) + \cdots + \hat{\alpha}_p(k)y(k-p+1) \\ + \hat{\beta}_0(k)u(k+1) + \hat{\beta}_1(k)u(k) + \cdots + \hat{\beta}_p(k)u(k-p+1)\end{aligned} \tag{53}$$

then the control law will be taken to be

$$\begin{aligned}u(k+1) = \hat{\beta}_0(k)^{-1}\big[r(k+1) - \hat{\alpha}_1(k)y(k) - \cdots - \hat{\alpha}_p(k)y(k-p+1) \\ - \hat{\beta}_1(k)u(k) - \cdots - \hat{\beta}_p(k)u(k-p+1)\big]\end{aligned} \tag{54}$$

Remark 6.1.3. The above controller can be implemented in neural network form. Such a controller simply copies the weights of the feedforward predictor network to generate the control input. This is shown schematically in Fig. 11 below.

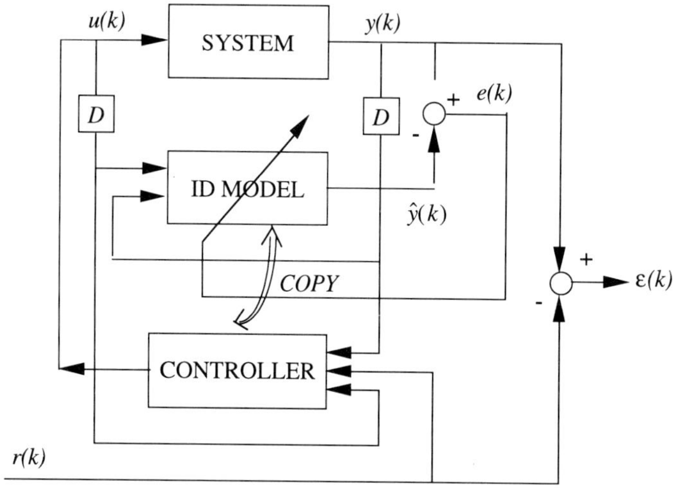

Figure 11: Adaptive one-step ahead controller.

Remark 6.1.4. Since the controller attempts to make the system track the desired trajectory in one step, excessive control efforts are usually required. This makes the approach unattractive in practice. To alleviate this problem, the weighted one-step ahead controller is used, such that at each time step the control input minimizes the following quadratic cost function

$$J(k+1)=\frac{1}{2}\varepsilon(k+1)^T Q\varepsilon(k+1)+\frac{1}{2}u(k+1)^T Su(k+1) \tag{55}$$

where the tracking error $\varepsilon(k+1)=y(k+1)-r(k+1)$. The weighting matrices Q and S are required to be symmetric and positive definite. Substituting the expression for $\varepsilon(k+1)$ and $y(k+1)$ into the cost function and then performing the minimization produces

$$\begin{aligned}\frac{\partial J(k+1)}{\partial u(k+1)}&=\beta_0^T Q\alpha_1 y(k)+\beta_0^T Q\alpha_2 y(k-1)+\cdots+\beta_0^T Q\alpha_p y(k-p+1)+\beta_0^T Q\beta_0 u(k+1)\\&\quad+\beta_0^T Q\beta_1 u(k)+\cdots+\beta_0^T Q\beta_p u(k-p+1)-\beta_0^T Qr(k+1)+Su(k+1)\end{aligned} \tag{56}$$

Setting the result to zero and solving for the control input yields

$$\begin{aligned}u(k+1)=\left(\beta_0^T Q\beta_0+S\right)^{-1}\beta_0^T Q\big[r(k+1)-\alpha_1 y(k)-\cdots-\alpha_p y(k-p+1)\\-\beta_1 u(k)-\cdots-\beta_p u(k-p+1)\big]\end{aligned} \tag{57}$$

The above is known as a weighted one-step ahead controller in adaptive control literature.[6]

6.2. *Model Reference Controller*

A different way to avoid the requirement that the system track a desired trajectory in one step is to use a control scheme known as model reference control. Let the control law in Eq. (47) be modified as

$$\begin{aligned}u(k+1)=\hat\beta_0^{-1}\Big[r(k+1)-\hat\alpha_1 y(k)-\cdots-\hat\alpha_p y(k-p+1)-\hat\beta_1 u(k)-\cdots-\hat\beta_p u(k-p+1)\\-\gamma_1 y(k)-\cdots-\gamma_p y(k-p+1)\Big]\end{aligned} \tag{58}$$

Substituting Eq. (48) and Eq. (58) into Eq. (46) yields

$$\begin{aligned}y(k+1)-e(k+1)&=\hat\alpha_1 y(k)+\cdots+\hat\alpha_p y(k-p+1)+\hat\beta_1 u(k)+\cdots+\hat\beta_p u(k-p+1)\\&\quad+\hat\beta_0\Big\{\hat\beta_0^{-1}\Big[-\hat\alpha_1 y(k)-\cdots-\hat\alpha_p y(k-p+1)-\hat\beta_1 u(k)-\cdots-\hat\beta_p u(k-p+1)\\&\quad-\gamma_1 y(k)-\cdots-\gamma_p y(k-p+1)+r(k+1)\Big]\Big\}\\&=-\gamma_1 y(k)-\cdots-\gamma_p y(k-p+1)+r(k+1)\end{aligned} \tag{59}$$

which can be expressed as

$$y(k+1)+\gamma_1 y(k)+\cdots+\gamma_p y(k-p+1)=r(k+1)+e(k+1) \tag{60}$$

The system response $y(k)$ now no longer follows the reference input $r(k)$ directly as in the case in Eq. (49). Its behavior can be conveniently interpreted in terms of a reference model. Define $y_m(k)$ as the response of a reference model when driven by the reference input $r(k)$,

$$y_m(k+1)+\gamma_1 y_m(k)+\cdots+\gamma_p y_m(k-p+1)=r(k+1) \tag{61}$$

and the tracking error $\varepsilon_m(k)$ as the difference between the system response $y(k)$ and the reference model response $y_m(k)$,

$$\varepsilon_m(k) = y(k) - y_m(k) \tag{62}$$

The equation that governs the behavior of this tracking error is obtained by subtracting Eq. (61) from Eq. (60),

$$\varepsilon_m(k+1) + \gamma_1 \varepsilon_m(k) + \cdots + \gamma_p \varepsilon_m(k-p+1) = e(k+1) \tag{63}$$

Therefore, convergence of the prediction error to zero implies convergence of the tracking error to zero provided that the characteristic equation governing the homogeneous part of the difference equation is asymptotically stable,

$$\lambda^p + \gamma_1 \lambda^{p-1} + \cdots + \gamma_p = 0 \tag{64}$$

This requirement is easily satisfied since the coefficients γ_1, γ_2, ..., γ_p are the design variables to be selected a priori.

Remark 6.2.1. The difference between this case and the previous case is that the desired trajectory is not specified by the reference input $r(k)$, but rather by the response of the reference model. Since the reference model is known, the reference input $r(k)$ that is needed to make the reference model produces the desired response can be easily computed. The introduction of the reference model is to slow down the convergence of the tracking error so that excessive correction during the adaptation process does not occur.

Remark 6.2.2. The model reference control scheme can also be implemented in neural network form. At any time step, the controller network copies the coefficients of the predictor network and uses them in the generation of the control input. The configuration for this control scheme is shown in Fig. 12.

Remark 6.2.3. Equation (63) shows that the prediction error acts as a driving term for the difference equation that governs the behavior of the tracking error. If the reference model coefficients are designed such that the homogeneous solution is asymptotically stable then the steady state tracking error is simply the particular solution of the difference equation. One thus has the ability to affect the steady state tracking error through the reference model coefficients. However, this freedom is constrained by the residual dynamics of the prediction error that the steady state tracking error may be amplified or reduced. Generally speaking, the natural frequencies of the reference model should be placed away from those dominating the residual dynamics.

Remark 6.2.4. If the coefficients of the predictor model are updated at each time step, then the controller coefficients must match those of the predictor model at each time step. The resulting integration between parameter estimation and control computation is known as model reference adaptive control. The adaptive scheme is summarized in the following equations where the ordinary least-squares algorithm is used to perform the parameter estimation step. Again, let $\hat{\mathbf{Y}}(k)$ denote the estimated coefficients of the predictor model at time step k,

$$\hat{\mathbf{Y}}(k) = \left[\hat{\beta}_0(k),\ \hat{\beta}_1(k),\ \hat{\alpha}_1(k),\ \hat{\beta}_2(k),\ \hat{\alpha}_2(k),\ \ldots,\ \hat{\beta}_p(k),\ \hat{\alpha}_p(k)\right] \tag{65}$$

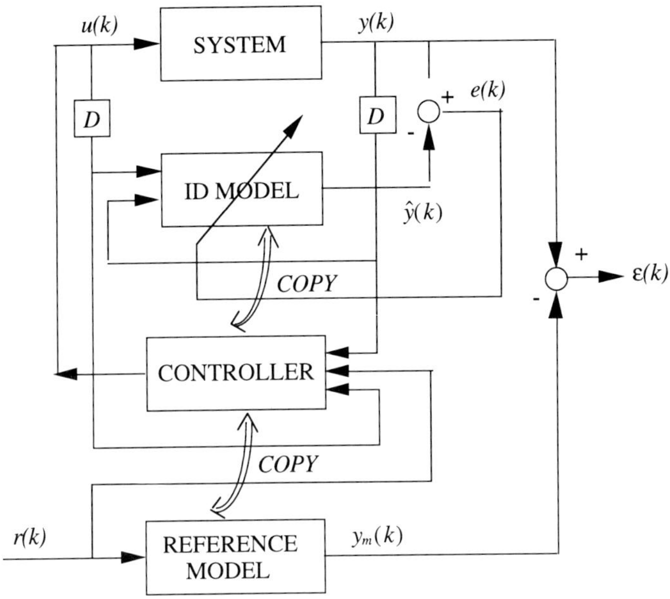

Figure 12: Model reference adaptive control.

starting with $\hat{\mathbf{Y}}(0)$ as an arbitrary initial guess. The control input is computed from

$$u(k+1) = \hat{\beta}_0(k)^{-1}\Big[-\hat{\alpha}_1(k)y(k) - \cdots - \hat{\alpha}_p(k)y(k-p+1) - \hat{\beta}_1(k)u(k) - \cdots$$
$$-\hat{\beta}_p(k)u(k-p+1) - \gamma_1 y(k) - \cdots - \gamma_p y(k-p+1) + r(k+1)\Big] \tag{66}$$

where the reference model coefficients $\gamma_1, \gamma_2, \ldots, \gamma_p$ are time-invariant and chosen a priori. The above control input is applied to the system producing response $y(k+1)$. The predictor coefficients are then updated according to the rule

$$\hat{\mathbf{Y}}(k+1) = \hat{\mathbf{Y}}(k) + \left[y(k+1) - \hat{\mathbf{Y}}(k)\Gamma(k+1)\right]\left\{\frac{\Gamma(k+1)^T R(k)}{1+\Gamma(k+1)^T R(k)\Gamma(k+1)}\right\} \tag{67}$$

$$R(k+1) = R(k) - \frac{R(k)\Gamma(k+1)\Gamma(k+1)^T R(k)}{1+\Gamma(k+1)^T R(k)\Gamma(k+1)} \tag{68}$$

staring with $R(0)$ as any symmetric positive definite matrix. The newly estimated parameters are then used to compute the control input for the next time step $u(k+2)$.

Remark 6.2.5. The control schemes discussed in this section deals with a one-step ahead predictor model of the form shown in Eq. (38). The previous section shows that a two-step ahead predictor or a multi-step ahead predictor has the same linear form. Therefore, the results presented in this section can be easily extended to these predictors. For example, the two-step ahead controller will compute the control $u(k+2)$ requiring the measurements upto $y(k)$ only.

7. Modelling and Control of Non-Linear Systems

Up to this point, the discussion has been restricted to linear systems. It has been shown that for linear systems, it is not necessary to have a complicated neural network for identification and control, but rather a single layer of linear neurons is adequate. This section extends the results to non-linear systems. The significance of linear predictor models for non-linear systems will be discussed. This has an important implication on the extent to which linear techniques can be used to handle non-linear systems. The modelling and control of non-linear systems using non-linear neural networks will then be examined.

7.1. *Linear Predictor Models for Non-Linear Systems*

The predictor model derived in Section 5 is based on the open-loop state space model which is a linear representation. For linear systems, the identified coefficients of an one-step ahead predictor take a particular form, namely, the Markov parameters of an observer model consisting of the open-loop state space model and an observer gain. Recall that a predictor uses actual input and output data to compute the predicted response at each time step. To qualify as a valid open-loop model as well, the predictor must also accurately predicts the system response in an open-loop test using input data alone. As mentioned previously, for linear systems, the predictor is also valid as an open-loop model because it can also produce correct open-loop prediction. For a non-linear system, this is no longer the case. However, when the predicted response is modeled as a linear combination of past input and output data, it turns out that surprisingly good prediction can still be obtained even for non-linear systems. Such predictor models do not qualify as open-loop models of the actual system because they do not predict correct open-loop response using input data alone. This point will be further illustrated by a numerical example in a later section.

7.2. *Control of Non-Linear Systems Using Predictor Models*

In this section, we show that the model reference controller considered in Section 6.2, or its special version, the one-step ahead controller in Section 6.1, can be used to control a class of non-linear systems which can be represented by linear predictors of the form considered in Eq. (38). Suppose that the non-linear system can be represented by an non-linear auto-regressive model of the form

$$y(k+1) = f\big(y(k),\ y(k-1),\ \dots,\ u(k+1),\ u(k),\ u(k-1),\ \dots\big) \tag{69}$$

where $f(.)$ is some non-linear function of past input and output data. First, note that for the response of the system to follow that of a reference model,

$$y_m(k+1) + \gamma_1 y_m(k) + \ \cdots\ + \gamma_p y_m(k-p+1) = r(k+1) \tag{70}$$

we require that the response of the controlled system be described by

$$y(k+1) + \gamma_1 y(k) + \ \cdots\ + \gamma_p y(k-p+1) = r(k+1) \tag{71}$$

so that the tracking error, $\varepsilon_m(k) = y(k) - y_m(k)$, will be governed by

$$\varepsilon_m(k+1) + \gamma_1 \varepsilon_m(k) + \ \cdots\ + \gamma_p \varepsilon_m(k-p+1) = 0 \tag{72}$$

Therefore, at time step k+1, one wishes to determine the control input $u(k+1)$ such that Eq. (69) is satisfied. Since the relationship between $y(k+1)$ and $u(k+1)$ is non-linear and is not known, one cannot solve for $u(k+1)$ directly. However, if the non-linear system is such that there exists a predictor of the form given in Eq. (44) such that $\hat{y}(k+1) \approx y(k+1)$ then one satisfies the following equation,

$$\hat{y}(k+1) + \gamma_1 y(k) + \cdots + \gamma_p y(k-p+1) = r(k+1) \tag{73}$$

instead of Eq. (71). The control law is then determined by substituting Eq. (46) in Eq. (73) and then solve for $u(k+1)$, producing exactly the same control law as given in Eq. (66). The fact that the control law satisfies Eq. (73) instead of Eq. (71) will make the tracking error equation governed by Eq. (63) instead of Eq. (72), where $e(k+1)$ denotes the prediction error defined in Eq. (48).

7.3. Identification and Control of Non-Linear Systems using Neural Networks

The identification and control scheme discussed in this paper can be extended to include non-linear neurons. The basic assumption is that the system response is a non-linear function of previous input and output data which can be represented by a multi-layer feedforward network having a sufficient number of non-linear neurons. Let the non-linear function be denoted by $f(.)$ and its neural network representation by $N(.)$,

$$\begin{aligned} y(k+1) &= f\big(y(k),\ y(k-1),\ \ldots,\ u(k+1),\ u(k),\ u(k-1),\ \ldots\big) \\ &= N\big(y(k),\ y(k-1),\ \ldots,\ u(k+1),\ u(k),\ u(k-1),\ \ldots\big) \end{aligned} \tag{74}$$

When a non-linear network of sufficiently large number of hidden layers is used, then it may also qualify as an open-loop model of the non-linear system besides its being an one-step head predictor. This is the fundamental difference between identification using a linear network versus a non-linear network. Generally speaking, the theoretical advantage of using a non-linear network for non-linear system identification is off-set by the difficulties in finding such a network in practice. Neither the number of hidden layers nor the number of neurons in each layer are known a priori. For a chosen network configuration, the back propagation algorithm is often used to determine the network weights. Typically, the convergence rate is slow and a large amount of data is needed. The back propagation algorithm is well-known and discussed extensively in the literature.

In the model reference control problem, the theoretical advantage of a non-linear network is somewhat diminished because the open-loop model need not be found for purpose of tracking control. The model reference control scheme can accommodate a non-linear network rather easily. Assume that the network representing the non-linear system can be expressed in the form,

$$\begin{aligned} y(k+1) &= N\big(y(k),\ y(k-1),\ \ldots,\ u(k+1),\ u(k),\ u(k-1),\ \ldots\big) \\ &= N_1\big(y(k),\ y(k-1),\ \ldots,\ u(k),\ u(k-1),\ \ldots\big) + \beta_0(k)u(k+1) + e_1(k+1) \end{aligned} \tag{75}$$

where $e_1(k+1)$ denotes the fitting error introduced with the separation of the $u(k+1)$ term from $N(.)$. The control input is computed from

$$u(k+1) = \beta_0(k)^{-1}\big[r(k+1) - \gamma_1 y(k) - \cdots - \gamma_p y(k-p-1) - N_1(.)\big] \tag{76}$$

where γ_1, γ_2,..., γ_p are the coefficients of the reference model representing the desired response. The control input when applied to the system yields the closed-loop response,

$$\begin{aligned} y(k+1) &= N_1(.) + \beta_0(k)\left\{\beta_0(k)^{-1}\left[r(k+1) - \gamma_1 y(k) - \cdots - \gamma_p y(k-p-1) - N_1(.)\right]\right\} + e_1(k+1) \\ &= N_1(.) + r(k+1) - \gamma_1 y(k) - \cdots - \gamma_p y(k-p-1) - N_1(.) + e_1(k+1) \\ &= r(k+1) - \gamma_1 y(k) - \cdots - \gamma_p y(k-p-1) + e_1(k+1) \end{aligned} \tag{77}$$

The tracking error, $\varepsilon_m(k) = y(k) - y_m(k)$, where $y_m(k)$ is the response of the reference model is governed by the difference equation

$$\varepsilon_m(k+1) + \gamma_1\varepsilon_m(k) + \cdots + \gamma_p\varepsilon_m(k-p+1) = e_1(k+1) \tag{78}$$

In practice, one identifies an approximation of $N_1(.)$ denoted by $\hat{N}_1(.)$. The control law is then based on $\hat{N}_1(.)$,

$$u(k+1) = \beta_0(k)^{-1}\left[y(k+1) - \gamma_1 y(k) - \cdots - \gamma_p y(k-p-1) - \hat{N}_1(.)\right] \tag{79}$$

The closed-loop system becomes

$$y(k+1) = N_1(.) + r(k+1) - \gamma_1 y(k) - \cdots - \gamma_p y(k-p-1) - \hat{N}_1(.) + e_1(k+1) \tag{80}$$

Let $e_2(k+1)$ denote the approximation error, $e_2(k+1) = N_1(.) - \hat{N}_1(.)$. The tracking error is now governed by

$$\varepsilon_m(k+1) + \gamma_1\varepsilon_m(k) + \cdots + \gamma_p\varepsilon_m(k-p+1) = e_1(k+1) + e_2(k+1) \tag{81}$$

A schematic diagram of the control scheme is the same as shown in Fig. 12, except that the block representing the identification model is now a non-linear neural network.

8. Numerical Examples

In this section, several examples will be presented to illustrate various concepts discussed in this paper. The case of a linear system is considered first, followed by a non-linear system. Both identification and control aspects of each case will be shown.

8.1. Network Representation of a Linear System

Consider a linear single-input single-output system with three vibration modes at 0.40Hz, 1.37Hz, and 2.21Hz, each with a damping factor of 0.5%. The state space matrices shown represent a discrete model at a sampling rate of 10 Hz.

$$A = \begin{bmatrix} 0.6585 & 0.2100 & 0.0144 & 0.4391 & 0.0382 & 0.0017 \\ 0.2100 & 0.4701 & 0.3149 & 0.0382 & 0.4035 & 0.0573 \\ 0.0144 & 0.3149 & 0.6705 & 0.0017 & 0.0573 & 0.4405 \\ -1.2408 & 0.6925 & 0.1093 & 0.6516 & 0.2119 & 0.0155 \\ 0.6925 & -1.7694 & 1.0387 & 0.2119 & 0.4630 & 0.3179 \\ 0.1093 & 1.0387 & -1.1498 & 0.0155 & 0.3179 & 0.6646 \end{bmatrix}, \quad B = \begin{bmatrix} 0.0050 \\ 0.1124 \\ 0.0075 \\ 0.0382 \\ 0.4035 \\ 0.0573 \end{bmatrix} \tag{82}$$

$$C = [1.0 \quad -0.5 \quad 0.0 \quad 1.0 \quad 0.5 \quad 0.0], \quad D = 1.5$$

The system is excited by random input shown in Fig.13 producing the response shown in Fig.14.

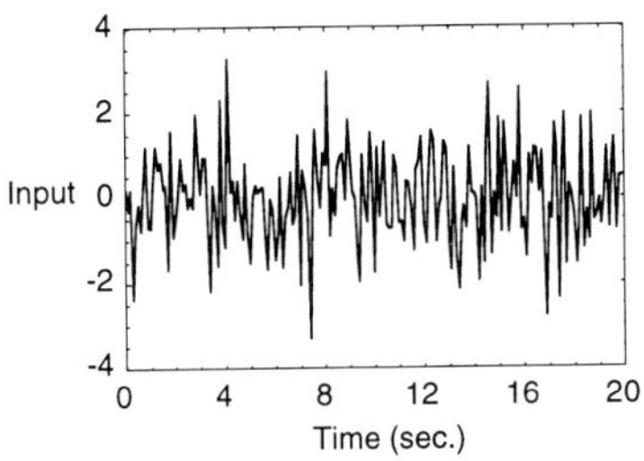

Figure 13: Excitation input time history.

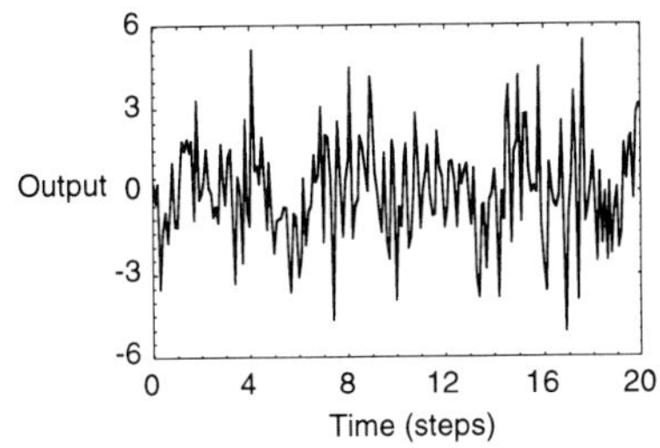

Figure 14: System response time history.

Using the above time histories, the network weights can be identified using Eq. (31). First, consider the case where $p = 6$, the following values for the network weights are obtained:

$$\hat{\alpha}_1 = 8.02 \times 10^{-1},\ \hat{\alpha}_2 = -4.59 \times 10^{-3},\ \hat{\alpha}_3 = -2.31 \times 10^{-2}$$
$$\hat{\alpha}_4 = 4.11 \times 10^{-1},\ \hat{\alpha}_5 = 2.02 \times 10^{-1},\ \hat{\alpha}_6 = -7.24 \times 10^{-1}$$

$$\hat{\beta}_0 = 1.50,\ \hat{\beta}_1 = -1.01,\ \hat{\beta}_2 = -5.48 \times 10^{-4},\ \hat{\beta}_3 = 1.59 \times 10^{-1}$$
$$\hat{\beta}_4 = -5.02 \times 10^{-1},\ \hat{\beta}_5 = -4.27 \times 10^{-1},\ \hat{\beta}_6 = 8.75 \times 10^{-1} \tag{83}$$

The above results are checked against the data by performing an open-loop prediction of the response using the input alone,

$$\hat{y}(k) = \alpha_1 \hat{y}(k-1) + \cdots + \alpha_6 \hat{y}(k-6) + \beta_0 u(k) + \beta_1 u(k-1) + \cdots + \beta_p u(k-6) \tag{84}$$

and an one-step ahead prediction (or observer estimation) using both actual input and output data,

$$\hat{y}(k) = \alpha_1 y(k-1) + \cdots + \alpha_6 y(k-6) + \beta_0 u(k) + \beta_1 u(k-1) + \cdots + \beta_p u(k-6) \tag{85}$$

It can be verified that in both cases, both predicted responses match the actual data exactly. Again, it should be emphasized that the result shown in Eqs. (83) represents a set of weights that can be identified from any feedforward network that uses 6 past values of input and output data to predict the current response. Specifically, if one uses a network consisting of a single neuron, then the values listed in Eqs. (83) are precisely the weights of this neuron. On the other hand, if a feedforward network consisting of several layers of linear neurons is used to identify the system, then the values in Eqs. (83) are the weights of a single neuron representation that is mathematically equivalent to the multi-layer network.

The system in Eqs. (82) in fact contains one uncontrollable mode as revealed by the singular values of the controllability matrix, $\mathbf{C} = [A^5B, A^4B, \ldots, AB, B]$,

$$\sigma_1 = 1.08,\ \sigma_2 = 5.92 \times 10^{-1},\ \sigma_3 = 3.69 \times 10^{-1},\ \sigma_4 = 2.19 \times 10^{-1}$$

$$\sigma_5 = 9.44 \times 10^{-17}, \ \sigma_6 = 1.91 \times 10^{-17} \tag{86}$$

The model in Eq. (84) is therefore an over-parameterized model. The same system can be modeled by using data from only 4 past time steps to predict the current response, i.e., $p = 4$. The corresponding weights are given below:

$$\begin{gathered}\hat{\alpha}_1 = 2.29, \ \hat{\alpha}_2 = -2.67, \ \hat{\alpha}_3 = 2.26, \ \hat{\alpha}_4 = -9.84 \times 10^{-1} \\ \hat{\beta}_0 = 1.50, \ \hat{\beta}_1 = -3.24, \ \hat{\beta}_2 = 3.72, \ \hat{\beta}_3 = -2.98, \ \hat{\beta}_4 = 1.19\end{gathered} \tag{87}$$

Note that the over-parameterization in Eq. (84) is in the form of having more distant past input and output data to predict the current response, corresponding to the case of a neuron having additional input channels. This is in contrast to the case where over-parameterization is in the form of having additional neurons added to the network.

8.2. Model Reference Adaptive Control of A Linear System

Next, we consider the application of the model reference adaptive control of the above system. The goal is to have the system track a desired trajectory prescribed via the reference model,

$$y_m(k+1) = 0.4y_m(k) + 0.5y_m(k-1) - 0.3y_m(k-2) + r(k+1) \tag{88}$$

where $r(k) = \sin(k/2\pi)$. First, consider the ideal case where disturbance and noises are not present. Since the system has a single output, the predictor network consists of only one linear neuron. In this example, 6 past input and output values are used to predict the current response. Recall that this is a case of over-parameterization since the effective order of the system is only 4. The system is assumed to be unknown to the controller at the beginning, and the weights are initially set to zero. Simultaneous prediction and control is carried out producing the results shown in Figs. 15a-d below. Figure 15a shows that the system response (dashed curve) quickly tracks the desired response (solid curve). The time histories of the prediction error and of the tracking error during the process are shown in Fig 15b and 15c, respectively. The control input time history is shown in Fig. 15d revealing that the adaptive mechanism quickly produces the necessary control input to make the system track the desired response.

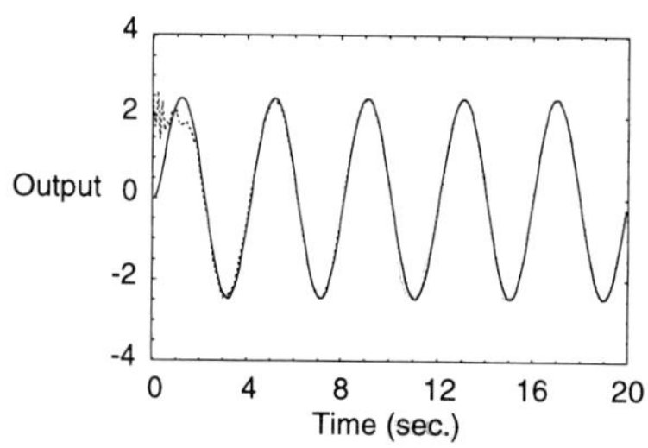

Figure 15a: Tracking response.

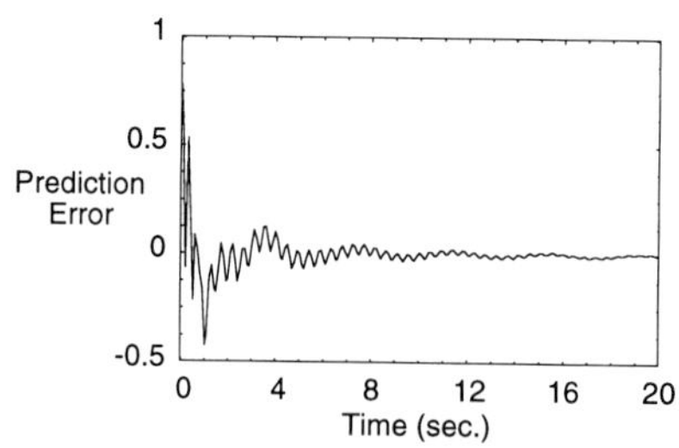

Figure 15b: Prediction error.

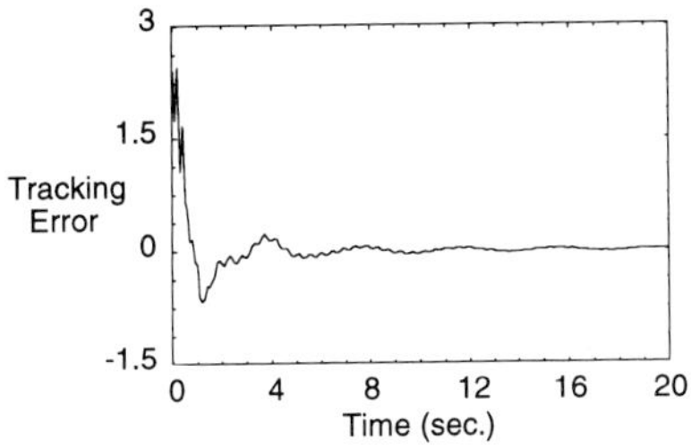

Figure 15c: Tracking error.

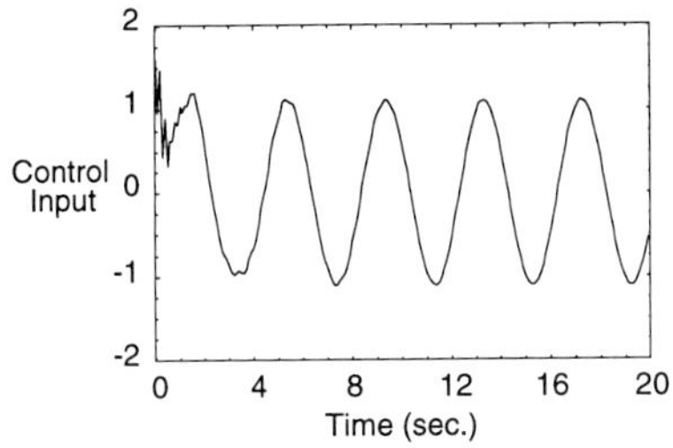

Figure 15b: Control input.

Figures 16 a-c show the adaptation when a disturbance, $d(k) = 0.5\cos(k/2\pi)$, and 5% measurement noise are added to the system. With the same adaptive controller, the system continues to track the desired response as shown in Fig. 16a. The effect of the noises can be seen in the random variation in the prediction error and the tracking error time histories, Figs. 16b and 16c. The new control input history that makes the system track the desired response and accommodate this disturbance is shown in Fig. 16d.

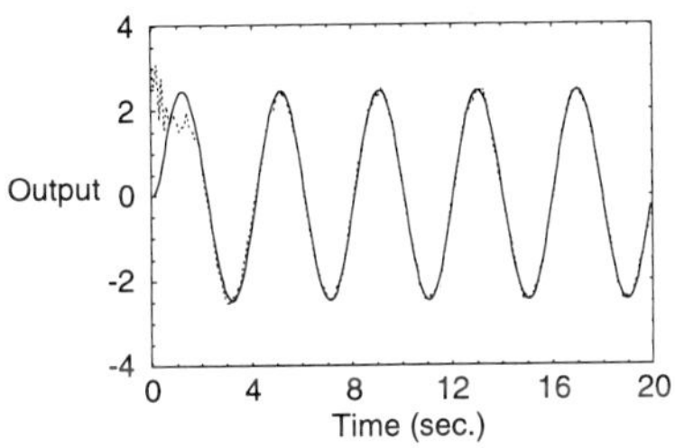

Figure 16a: Tracking response with disturbance and noise present.

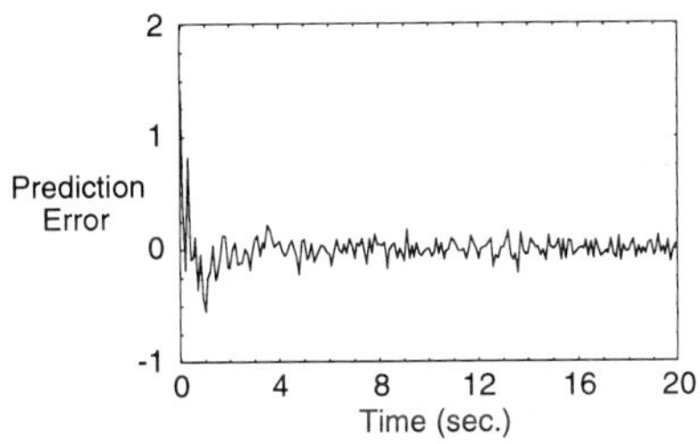

Figure 16b: Prediction error.

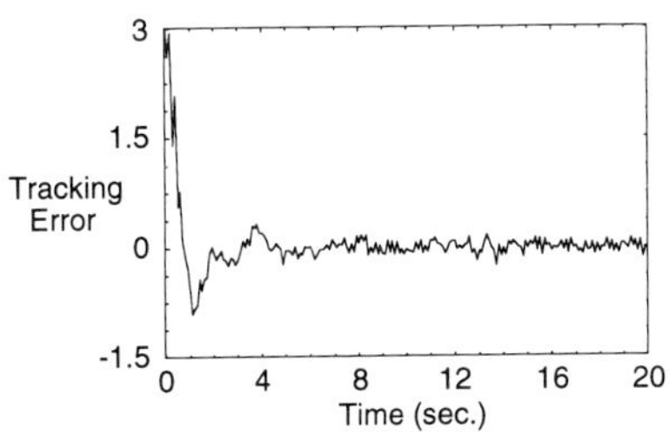

Figure 16c: Tracking error.

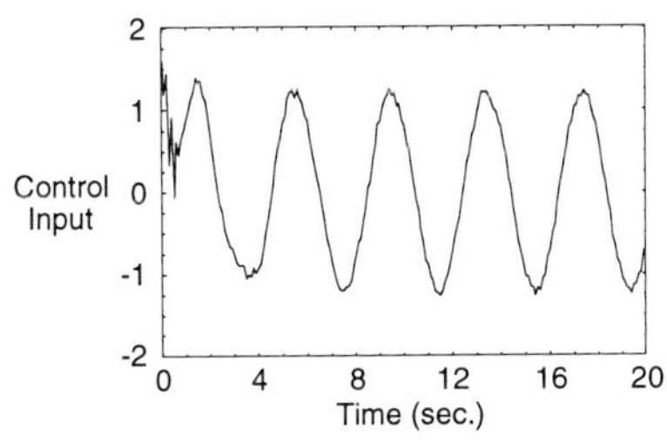

Figure 16b: Control input.

8.3. Identification and Prediction of a Non-Linear System

While it is not possible to have a linear model that can reproduce the open-loop response of a non-linear system, it is possible to have a linear predictor that can reasonably predict the non-linear response. The predictor model uses actual input and output data to compute the predicted response. Consider the system whose state space matrices are shown previously, but the input and output are related by the following non-linear relationship,

$$x(k+1) = A\,f[x(k)] + Bu(k)$$
$$y(k) = g[x(k)] + Du(k) \tag{89}$$

where the non-linear functions $f[x(k)] = |x(k)|^{1/2} sgn[x(k)]$, $g[x(k)] = \sin[Cx(k)]$ operate on each element of the state vector *x(k)*. Note that in this example, the non-linearity affect takes place from one sampling interval to the next. For purpose of identification, the system is excited by a random input sequence and the resulting response is used in the series-parallel identification scheme. For the case $p = 6$, the following model coefficients are identified:

$$\hat{\alpha}_1 = 1.35,\ \hat{\alpha}_2 = -5.40\times10^{-1},\ \hat{\alpha}_3 = 1.39\times10^{-1}$$
$$\hat{\alpha}_4 = -5.95\times10^{-2},\ \hat{\alpha}_5 = 1.33\times10^{-1},\ \hat{\alpha}_6 = -1.62\times10^{-1}$$

$$\hat{\beta}_0 = 1.48,\ \hat{\beta}_1 = -1.89,\ \hat{\beta}_2 = 7.41\times10^{-1},\ \hat{\beta}_3 = -8.74\times10^{-2}$$
$$\hat{\beta}_4 = 9.58\times10^{-2},\ \hat{\beta}_5 = -2.38\times10^{-1},\ \hat{\beta}_6 = 1.93\times10^{-1} \tag{90}$$

Recall that the identified model can be used either as an open-loop model or an one-step ahead predictor. The model is checked against the response of the actual system to a sine-wave input excitation, $u(k) = \sin(0.2\pi k)$. This is shown in Fig. 17 where the solid curve is the actual response of the non-linear system and the dashed curve is the open-loop prediction using the identified linear model. The two curves do not match as expected. However, when the same coefficients are used in the predictor model, then the predicted response follows closely the actual response as shown in Fig. 18. It is precisely this potential ability of the linear predictor model that makes it applicable to the control of non-linear systems. Furthermore, note that the coefficients are time-invariant. Allowing these coefficients to be time-varying as in the case of model reference adaptive control can further enhance the applicability of such models for non-linear systems.

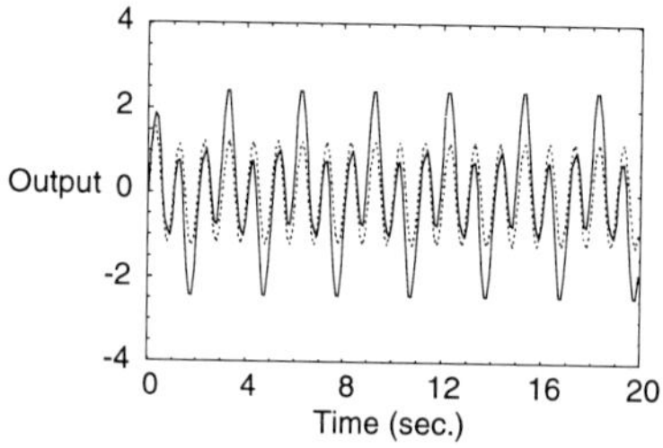

Figure 17: Open-loop prediction. (non-linear system)

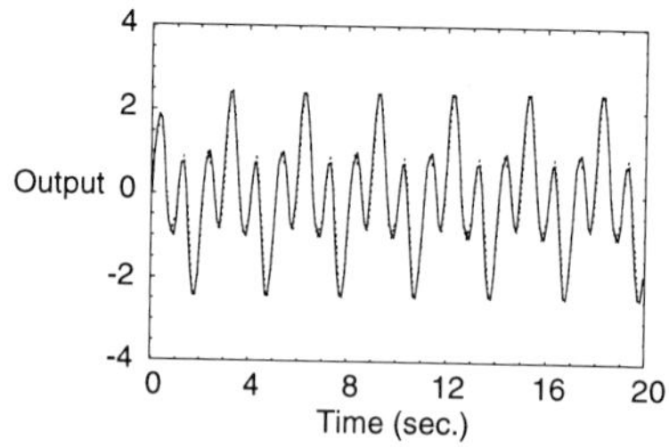

Figure 18: One-step ahead prediction. (non-linear system)

8.4. Model Reference Adaptive Control of a Non-Linear System

Finally, this example illustrates the application of model reference adaptive control to the non-linear system considered in the above section. For comparison purpose, the same adaptive control law is used as in the linear case, except that it is now applied to the non-linear system. Figures 18a-d show the tracking response, the prediction error, the tracking error, and the control input time histories, respectively. Recall that the control method does not require that the open-loop model be identified, but rather the predictor model that can reasonably predict the response, which is the case illustrated in the previous example.

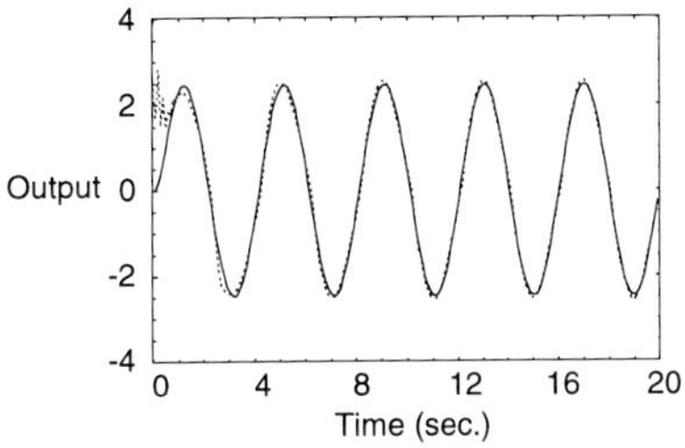

Figure 18a: Tracking response. (non-linear system)

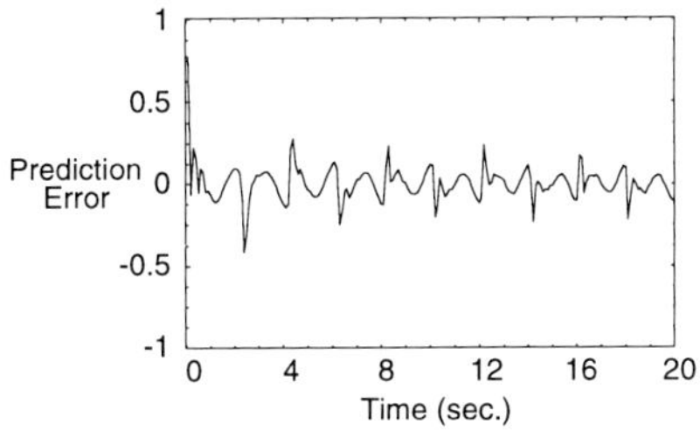

Figure 18b: Prediction error.

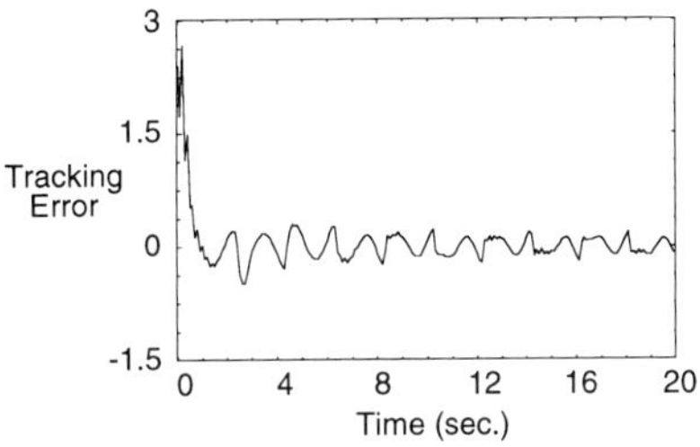

Figure 18c: Tracking error.

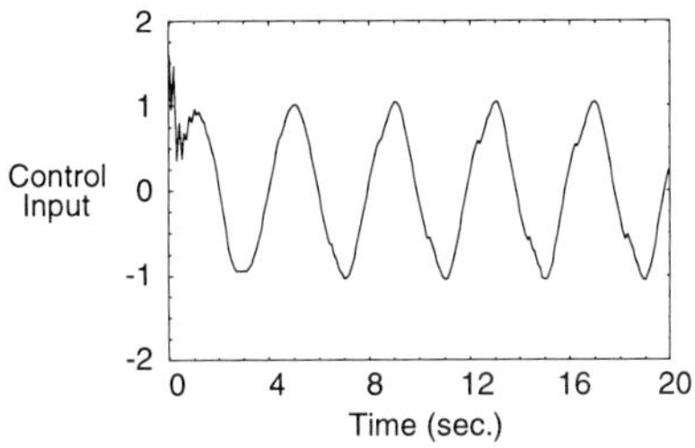

Figure 18b: Control input.

When disturbance and noise are added to the system, the resulting behavior of the system is shown in Figs. 19a-d. Again, this reveals a certain degree of stability robustness of the adaptive scheme to possible disturbance and noises. This is due to the inherent robustness in the ability of linear predictors that can predict the non-linear response.

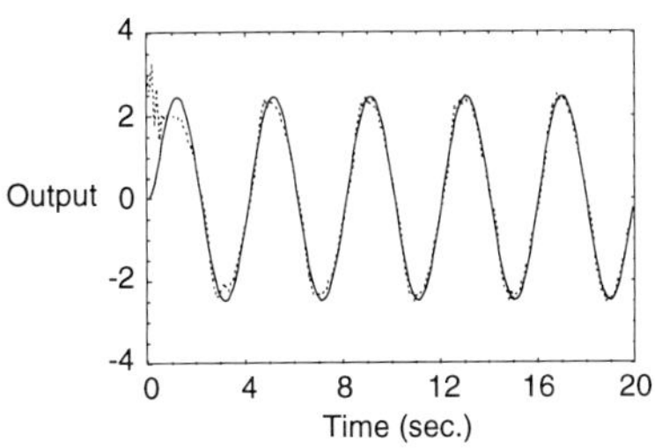

Figure 19a: Tracking response with dist. and noise present (non-linear syst.)

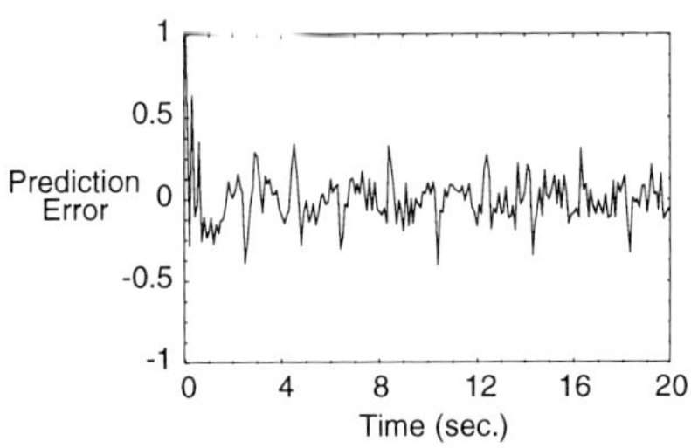

Figure 19b: Prediction error.

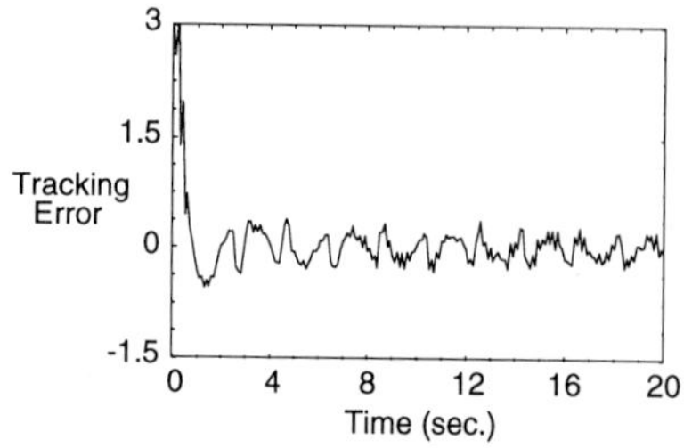

Figure 19c: Tracking error.

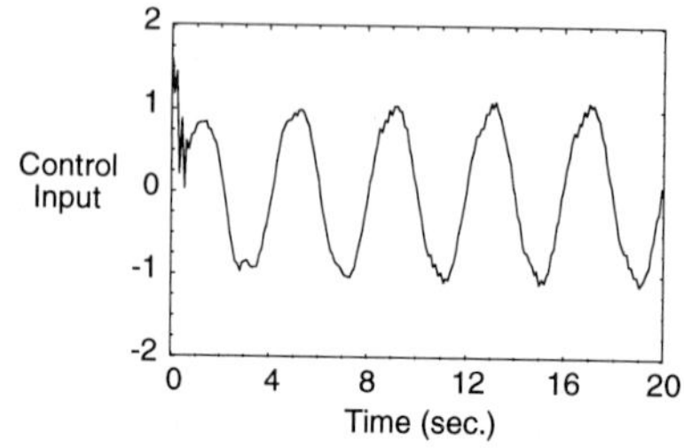

Figure 19b: Control input.

9. Summary and Concluding Remarks

This paper presents the basic concepts of the neural networks as related to the problem of modelling and control of a dynamic system. Two basic forms of the neural networks, the feedforward network and the recurrent network, are discussed. Emphasis is placed on the interpretation of the neural networks in terms of standard linear system theory so that better insight may be gained when these concepts are applied in practice. Relationship between the feedforward neural network and the state space model and between the recurrent network and the observer model is explained. To identify a linear system, the discussion in this paper reveals that it is neither advantageous nor necessary to use a multi-layer network, but rather a single layer of linear neurons is adequate. The resultant simplified network is then equivalent to standard regressive models that are often used in adaptive systems theory. The real benefit of a neural network in system identification is in its capacity to capture non-linearities, in which case the neurons must be non-linear. With respect to the control of both linear and non-linear systems, however, it is shown that it is not the identification of the open-loop system that governs the stability of the tracking behavior, but rather the ability of a mechanism that can predict future response based on actual available input-output data. It is shown that this mechanism can often be provided simply by a linear predictor. A linear predictor, consisting of a single layer of linear neurons, is in fact an optimal choice for a linear system. The same linear predictor can often be adequate for non-linear systems as well, making it directly applicable to the control a non-linear system. The resulting control technique is simply model reference adaptive control, a well-known technique in linear system control. Such a linear predictor can be easily determined from input-output data. If implemented on-line, the method can also adapt to changing dynamics. The recent advent of the Observer/Kalman filter identification (OKID) method has motivated the design of controllers that are based directly on the observer Markov parameters. This paper shows that one such design is model reference control because the observer Markov parameters are precisely the coefficients of an optimal linear predictor. The design takes advantage of the ability of the predictor to handle certain non-linear systems, an often observed fact in practical application of the OKID method.

To make adaptive control truly useful in practice, constraints with respect to sampling and computation speeds must be addressed. Naturally, the adaptive scheme places heavy emphasis on on-line measurements rather than some known model of the system for control. Sensor failure, therefore, becomes an important issue. Practical consideration may dictate a compromise between fixed-gain and adaptive control, thus requiring a mechanism to determine when adaptation should take place. The question of sensor placement and sensor selection are also important ones. This is to avoid the situation where the system can produce bounded output when driven by unbounded input. This case requires additional theoretical treatment than that presented in this paper. Finally, the paper concerned mostly with stability rather than performance robustness issues. Further work is required to assess this aspect of the problem.

10. References

1. Narendra, K.S. and Parthasarathy, K., "Identification and Control of Dynamical Systems Using Neural Networks," *IEEE Transactions on Neural Network,* Vol. 1, No. 1, 1990.
2. Hornik, K., Stinchcombe, M., and White, H., "Multilayer Feedforward Neural Networks Are Universal Approximators," *Neural Networks,* Vol. 2, No. 5, 1989.
3. Billings, S.A. and Leontaritis, I.J., "Input-Output Parametric Models for Non-Linear Systems. Part 1: Deterministic Non-Linear Systems; Part 2: Stochastic Non-Linear Systems, " *International Journal of Control,* Vol. 41, 1985.
4. Chen, S., Billings, and Grant, P.M., "Non-Linear System Identification Using Neural Network," *International Journal of Control*, Vol. 51, No. 6, 1990.
5. Hyland, D.C., "Neural Network Architectures for On-Line System Identification and Adaptively Optimized Control," *Proceedings of the IEEE Conference on Decision and Control,* Brighton, U.K., 1991.
6. Goodwin, G.C. and Sin, K.S., *Adaptive Filtering, Prediction, and Control,* Prentice Hall, Englewood Cliffs, New Jersey, 1984.
7. Ljung, L. and Södertröm, T., *Theory and Practice of Recursive Identification,* The MIT Press, Cambridge, Massachusetts, 1983.
8. Chen, C.-W., Huang, J.-K., Phan, M. and Juang, J.-N., "Integrated System Identification and Modal State Estimation for Control of Large Flexible Space Structures," *Journal of Guidance, Control, and Dynamics,* Vol. 15, No. 1, pp. 88-95, 1992.
9. Juang, J.-N., Phan, M., Horta, L.G., and Longman, R.W., "Identification of Observer/Kalman Filter Markov Parameters: Theory and Experiments," *Journal of Guidance, Control, and Dynamics,* Vol. 16, No. 2, pp. 320-329, 1993.
10. Phan, M., Horta, L.G., Juang, J.-N., and Longman, R.W., "Linear System Identification Via An Asymptotically Stable Observer," *Journal of Optimization Theory and Applications,* Vol. 79, No. 1, pp. 59-86, 1993.
11. Phan, M., Horta, L.G., Juang, J. N., and Longman, R.W., "Improvement of Observer/Kalman Filter Identification (OKID) by Residual Whitening," AIAA Paper No. 92-4385, *Proceedings of the AIAA Guidance, Navigation, and Control Conference,* Hilton Head, South Carolina, 1992; also, *Journal of Vibrations and Acoustics* (to appear).

Wave Motion, Intelligent Structures and Nonlinear Mechanics, pp. 226–238
edited by A. Guran and D. J. Inman
Series on Stability, Vibration and Control of Structures: Vol. 1

AUTOPARAMETRIC SYSTEMS WITH PENDULA

Aleš TONDL
Zborovská 41, 15000 Praha, Czech Republic

and

Radoslav NABERGOJ
Department of Naval Architecture, Ocean and Environmental Engineering, Via Valerio 10, 34127 Trieste, Italy

ABSTRACT

Simple pendula are attached to some masses of a mass-spring system arranged in a chain. The subsystem of masses is periodically excited. It is supposed that each mass of this subsystem can move in the vertical direction only. The stability of the semi-trivial solution, i.e., the solution characterized by the periodic motion of the chain subsystem without oscillation of the pendula, is investigated. The instability of the semi-trivial solution results in autoparametric excitation of one or more pendula.

1. Introduction

Systems of masses connected with springs in a chain belong to the class of autoparametric systems when simple pendula are attached to one or several masses. The masses can move in the vertical direction only and are periodically excited. The autoparametric resonance can be initiated when certain conditions are met.

Little has been given to the investigation of these systems and practically only the simplest system with one mass-spring subsystem and one pendulum has been analyzed[1,2,3]. Most authors have considered directly the vibration in the parametric resonance but for a special tuning only, i.e., when the natural frequency of the pendulum is one half the natural frequency of the excited subsystem (represented by the total mass mounted on the spring) and the excitation frequency is close to the latter natural frequency.

It was shown[4,5] that this special tuning does not represent a single possibility of occurrence of autoparametric resonance. There was proved the advantage of stability investigation of the semi-trivial solution, representing the vibration of the system when the pendulum does not oscillate. The stability investigation of the semi-trivial solution of the excited subsystems, consisting of several masses and springs and with only one pendulum subsystem, has been also considered[6]. As an example, a two-mass subsystem where the pendulum is attached to the upper mass with two alternatives of harmonic excitation acting on the upper or lower mass (with constant amplitude and with amplitude proportional to the square of the excitation frequency) was investigated.

In this paper, the case of several pendulum subsystems is considered. The investigation is limited to the first step only, i.e., to the analysis of the semi-trivial solution. A different approach from that shown in Svoboda, Tondl and Verhulst[6] will be used here.

2. Definition of autoparametric systems

Systems where autoparametric resonance can occur are called autoparametric systems. These consist of one excited subsystem and one or several non-excited subsystems. At particular conditions the excited subsystem is the source of parametric excitation of a certain non-excited subsystem.

The excited subsystem is defined by N coordinates y_i $(i = 1, 2, \ldots, N)$ and the non-excited subsystems by n coordinates φ_j $(j = 1, 2, \ldots, n)$. The motion of the whole system is governed by differential equations having the following properties:

- The non-excited subsystems are coupled with the excited system only by non-linear coupling terms having such a form that they admit the trivial solution for the corresponding coordinates $(y_i = 0, \varphi_j = 0)$. The non-excited subsystems need not be mutually coupled.
- There always exist steady-state semi-trivial solutions $(y_i \neq 0, \varphi_j = 0)$.
- The semi-trivial solution is stable with the exception of some intervals of the vibration frequency.
- In the instability intervals of the semi-trivial solution autoparametric resonances are initiated $(y_i \neq 0, \varphi_j \neq 0)$. The vibration of the excited subsystem acts as parametric excitation for a certain non-excited subsystem.
- The interval of the existence of the stable autoparametric resonant vibration can be broader than that of the semi-trivial solution instability. Outside this interval, both semi-trivial and autoparametric resonant solutions are locally stable and have their own domains of attraction.
- The vibration of the excited subsystem can be done to the action of external excitation, parametric excitation or self-excitation as well.

Furthermore the following has to be noted:

- The plurality in the type of excitation for the excited subsystem implies a much broader concept of autoparametric systems. Up to now the researchers have analyzed systems with external excitation. The exception represents a paper[7] where, in addition to external excitation, also parametric excitation and self-excitation of the excited subsystem have been considered.

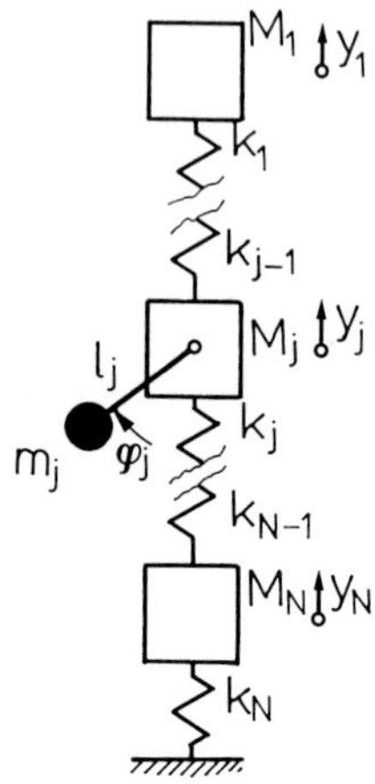

Fig. 1. Example of autoparametric system consisting of a chain of masses and springs (excited subsystem) where, to some of the masses, a pendulum is attached (non-excited subsystem).

- The stability investigation of the semi-trivial solution enables one to determine the domains where autoparametric resonances are initiated. This approach, mostly used by the researchers in their investigations, shows the form of the analytic approximation at a certain autoparametric resonance.

3. General analysis

Let us consider a chain system of N masses $M_1, M_2, \ldots, M_N$ connected with linear springs having stiffnesses $k_1, k_2, \ldots, k_N$ (see Figure 1). The masses can move only in the vertical direction and their motion is damped by linear viscous damping. Some of the masses M_i ($i = 1, 2, \ldots, N$) are supporting simple pendula, characterized by masses m_j and lengths l_j ($j = 1, 2, \ldots, n$), where $1 \leq n \leq N$. Let us denote the coordinates of masses M_i as y_i and the angular deflections of the pendula as φ_j. A harmonic or periodic excitation acts either directly on some of the masses $M_1, M_2, \ldots, M_N$ or by the prescribed motion of the support to which the last mass M_N is connected by a spring having stiffness k_N.

The whole system is governed by a set of differential equations having the following form:

- $(N-n)$ non-homogeneous equations are linear differential equations containing only coordinates y_i;

- $2n$ remaining equations, containing coordinates y_i, y_j and φ_j, can be divided into two groups, each of them having n differential equations.

These two groups of equations have the following form:

$$
\begin{aligned}
&(M_j + m_j)\,\ddot{y}_j + b_j\,\dot{y}_j - b_{j-1,j}(\dot{y}_{j-1} - \dot{y}_j) + b_{j,j+1}(\dot{y}_j - \dot{y}_{j+1}) + \\
&\quad - k_{j-1}(y_{j-1} - y_j) + k_j(y_j - y_{j+1}) + m_j l_j(\ddot{\varphi}_j \sin\varphi_j + \dot{\varphi}_j{}^2 \cos\varphi_j) = P_j(\omega t) \quad ,
\end{aligned}
$$

$$
m_j l_j^2\,\ddot{\varphi}_j + c_j\,\dot{\varphi}_j + m_j l_j(g + \ddot{y}_j)\sin\varphi_j = 0 \quad . \qquad (j = 1, 2, \ldots, n) \tag{1}
$$

The first group of differential equations in (1) is non-homogeneous and linearly coupled with the mentioned set of $(N-n)$ non-homogeneous differential equations. Moreover, they are coupled by non-linear terms only with the corresponding j-th differential equation of the second group of equations (1). The coupling terms are functions of φ_j coordinates and their derivatives. The second group of equations (1) consists of homogeneous differential equations which are coupled only by a non-linear term with the corresponding j-th equation of the first group.

The semi-trivial solution is defined by the excited motion of masses $M_1, M_2, \ldots,$ M_N when the pendula are not oscillating, i.e. $\varphi_j = 0\,(j = 1, 2, \ldots, n)$. For $\varphi_j = 0$ the nonlinear coupling terms in (1) equal zero and the semi-trivial solution is therefore the steady-state solution of N differential equations:

$$
\mathbf{M}\ddot{\mathbf{Y}} + \mathbf{D}\dot{\mathbf{Y}} + \mathbf{K}\mathbf{Y} = \mathbf{P}(\omega t) \quad . \tag{2}
$$

Here $\mathbf{M}$ is the mass matrix (masses m_j must be added to masses M_j), $\mathbf{D}$ is the damping matrix, $\mathbf{K}$ is the stiffness matrix, $\mathbf{Y}$ is the column vector of deflections and $\mathbf{P}(\omega t)$ is the excitation vector.

The solution of (2), i.e. $\mathbf{Y}(\omega t)$, has the same period as $\mathbf{P}(\omega t)$. For each harmonic component of $\mathbf{P}(\omega t)$ there exists the corresponding response component whose amplitude is a function of frequency, having maxima at ω values close to the N eigenvalues of system (2) considered without damping.

Let us now investigate the stability of the semi trivial solution. Inserting in the previous equations the perturbative solutions $y_i = y_{0i} + v_i$ and $\varphi_j = \varphi_{0j} + \psi_j$, where v_i and ψ_j are perturbations and $y_{0i} \neq 0$, $\varphi_{0j} = 0$ represent the semi-trivial solution, the following differential equations in the variations are obtained:

$$
\mathbf{M}\ddot{\mathbf{V}} + \mathbf{D}\dot{\mathbf{V}} + \mathbf{K}\mathbf{V} = 0 \quad ,
$$

$$
m_j l_j^2 \ddot{\psi}_j + c_j \dot{\psi}_j + m_j l_j \left[g + \ddot{y}_{0j}(\omega t)\right] \psi_j = 0 \quad , \qquad (j = 1, 2, \ldots, n) \tag{3}
$$

where $\mathbf{V}$ is the column vector of perturbations. Supposing positive damping in the whole system, it follows that $\lim_{t\to\infty} v_i = 0$, i.e., the stability of the semi-trivial solution is determined only by the second set of equations (3). Their number equals the number of pendula; they are not mutually coupled and can be analyzed separately. These equations are, in general, Hill equations. However, in case of harmonic excitation, $y_{0j}(\omega t)$ is also harmonic and equations (3) reduce to Mathieu equations.

Equations (3) can be arranged by using the time transformation $\tau = \omega_0 t$, where ω_0 is a suitably chosen frequency (e.g., a partial frequency of the system), into the following dimensionless form:

$$\psi_j'' + \delta_j \psi_j' + \left[Q_j^2 + \varepsilon_j f_{0j}''(\eta\tau)\right] \psi_j = 0 \quad , \tag{4}$$

where $\delta_j = c_j/\omega_0 m_j l_j^2$, $Q_j^2 = g/\omega_0^2 l_j$, $\varepsilon_j = a/l_j$, $f_{0j}(\eta\tau) = y_{0j}(\eta\tau)/a$, $\eta = \omega/\omega_0$, and a is a constant having the dimensions of length. From the theory of differential equations of Hill's type, it follows that the instability intervals of (4) lie at prescribed values of the excitation frequency, i.e., for $\eta \approx 2Q_j/L$ (where $L = 1, 2, \ldots$ is the order of the instability interval).

Let us express the dimensionless vibration amplitude of the semi-trivial solution as:

$$f_{0j}(\eta\tau) = \sum_{K=1}^{\infty} \left(A_{jK} \cos K\eta\tau + B_{jK} \sin K\eta\tau\right) \quad . \tag{5}$$

For the K-th harmonic frequency component, the solution on the boundary of the instability interval of the first order ($L = 1$) can be approximated in the form:

$$\psi_j = C_{jK} \cos \frac{1}{2} K\eta\tau + D_{jK} \sin \frac{1}{2} K\eta\tau \quad . \tag{6}$$

Substituting (6) into (4), the following equations are obtained:

$$\left(Q_j^2 - \frac{1}{4}K^2\eta^2 - \frac{1}{2}K^2\eta^2\varepsilon_j A_{jK}\right) C_{jK} + \left(\frac{1}{2}K\eta\delta_j + \frac{1}{2}K^2\eta^2\varepsilon_j B_{jK}\right) D_{jK} = 0 \quad ,$$

$$-\left(\frac{1}{2}K\eta\delta_j - \frac{1}{2}K^2\eta^2\varepsilon_j B_{jK}\right) C_{jK} + \left(Q_j^2 - \frac{1}{4}K^2\eta^2 + \frac{1}{2}K^2\eta^2\varepsilon_j A_{jK}\right) D_{jK} = 0 \quad . \tag{7}$$

From the condition of non-trivial solution of C_{jK} and D_{jK}, one obtains:

$$\left(Q_j^2 - \frac{1}{4}K^2\eta^2\right)^2 + \frac{1}{4}K^2\eta^2\delta_j^2 = \frac{1}{4}K^4\eta^4\varepsilon_j^2 R_{jK}^2 \quad , \tag{8}$$

where $R_{jK} = (A_{jK}^2 + B_{jK}^2)^{1/2}$.

Two different approaches can be used. In the first one the corresponding values A_{jK}, B_{jK} are inserted in (8) from the semi-trivial solution. Then a certain parameter (e.g. the amplitude ε_j of excitation) is expressed both as function of frequency η and of some other parameter of the system (e.g. the tuning coefficient)[4,5,6]. In this way the stability threshold is obtained, yielding information on the influence of certain parameters:

$$\varepsilon_j = \frac{2}{K^2\eta^2 R_{jK}} \left[\left(Q_j^2 - \frac{1}{4}K^2\eta^2\right)^2 + \frac{1}{4}K^2\eta^2\delta_j^2\right]^{1/2} \quad . \tag{9}$$

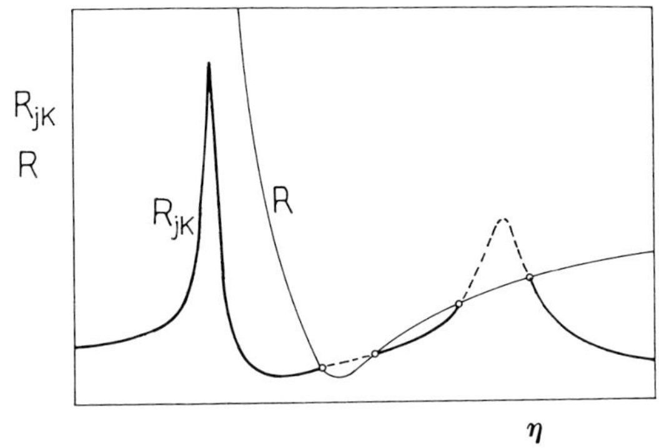

Fig. 2. Schematic representation of the semi-trivial vibration amplitude curve $R_{jK}(\eta)$ (stable solution heavy solid lines, unstable solution dotted lines) and stability boundary curve $R(\eta)$ (light solid line) for a linear excited subsystem.

The second approach is based on the following idea. The oscillation amplitude R_{jK} in (8) is considered as a stability boundary curve defined by the relation:

$$R_{jK} = \frac{2}{K^2\eta^2\varepsilon_j}\left[\left(Q_j^2 - \frac{1}{4}K^2\eta^2\right)^2 + \frac{1}{4}K^2\eta^2\delta_j^2\right]^{1/2} . \tag{10}$$

Plotting the amplitude curve $R_{jK} = (A_{jK}^2 + B_{jK}^2)^{1/2}$, expressed by the semi-trivial solution (5), and the stability boundary curve $R = R_{jK}(\eta)$, calculated from (10), the crossing points of both curves determine the stability boundary of the semi-trivial solution (see Figure 2). In the figure, the vibration amplitude of each mass M_j as a function of η has been considered. The stability boundary curve is marked by light solid line, while the vibration amplitude is marked by a heavy solid line for stable solution and by a dotted line for unstable solution. Index jK means that the diagram corresponds to the pendulum attached to mass M_j and to the excitation harmonic component of K-th order. It is necessary to plot one such diagram for each pendulum and for each component of the excitation, i.e., the number of diagrams equals the product jK.

It is quite evident that — when several pendulum subsystems are applied — one can hardly expect that in a certain interval of the excitation frequency the semi-trivial solution will be unstable for all j cases. Therefore, the autoparametric excitation can relate only to a single pendulum, which will be the most common case. In this autoparametric resonance the vibration of one single pendulum will be initiated.

If the connecting springs of masses M_j are linear, the above approach gives the exact semi-trivial solution. When some springs are non-linear, the same approach can be used, but the semi-trivial solution must be determined by an approximate

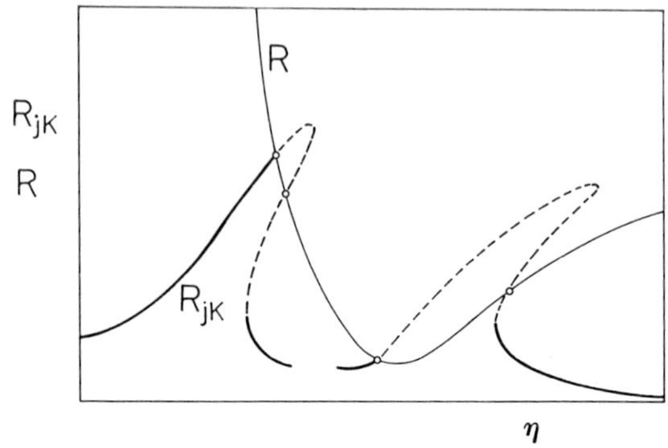

Fig. 3. Schematic representation of the semi-trivial vibration amplitude curve $R_{jK}(\eta)$ (stable solution heavy solid lines, unstable solution dotted lines) and stability boundary curve $R(\eta)$ (light solid line) for a non-linear excited subsystem (hardening spring).

method. The diagrams showing the vibration amplitude and the stability boundary curve for the case corresponding to a non-linear spring with hardening characteristic are presented schematically in Figure 3. Because the approximate methods determine mostly the amplitudes separately in the neighborhood of certain resonances, the response amplitude curve is not continuous. For non-linear systems not only main resonances but also resonances of other types (subharmonic, superharmonic, etc.) must be taken into account.

Note that the first approach of stability investigation presents a more general view, but it is more complicated especially when a system with several masses and pendula has to be analyzed. Therefore, the second approach will be used in the example presented.

4. Illustrative example

A two-mass subsystem with two pendula attached will be presented as an illustrative example (see Figure 4). For simplicity, the harmonic excitation is supposed to be applied to the support.

This kinematically excited system is governed by the following equations:

$$(M_1 + m_1)\,\ddot{z}_1 + b_1(\dot{z}_1 - \dot{z}_2) + k_1(z_1 - z_2) + m_1 l_1(\ddot{\varphi}_1 \sin\varphi_1 + \dot{\varphi}_1^2 \cos\varphi_1) = (M_1 + m_1)a\omega^2 \cos\omega t \quad ,$$

$$(M_2 + m_2)\,\ddot{z}_2 - b_1(\dot{z}_1 - \dot{z}_2) - k_1(z_1 - z_2) + b_2\,\dot{z}_2 + k_2 z_2 + m_2 l_2(\ddot{\varphi}_2 \sin\varphi_2 + \dot{\varphi}_2^2 \cos\varphi_2) = (M_2 + m_2)a\omega^2 \cos\omega t \quad ,$$

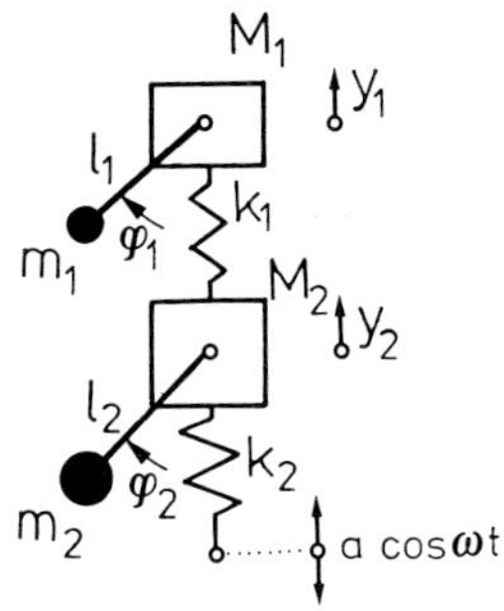

Fig. 4. Scheme of the system considered in the example.

$$m_j l_j^2\, \ddot{\varphi}_j + c_j\, \dot{\varphi}_j + m_j l_j g \sin\varphi_j + m_j l_j (\ddot{z}_j - a\omega^2 \cos\omega t) \sin\varphi_j = 0 \quad , \; (j = 1, 2) \quad (11)$$

where z_1 and z_2 are the mass coordinates relative to the lower spring end with prescribed motion of amplitude a and frequency ω. With this notation, the absolute coordinates of the masses are $y_j = z_j + a\cos\omega t$.

Using the time transformation $\tau = \omega_1 t$, equations (11) can be transformed into the following dimensionless form:

$$w_1'' + \kappa_1(w_1' - w_2') + w_1 - w_2 + \mu_1(\varphi_1'' \sin\varphi_1 + \varphi_1'^2 \cos\varphi_1) = \eta^2 \cos\eta\tau \quad ,$$

$$w_2'' - \mu\left[\kappa_1(w_1' - w_2') + w_1 - w_2\right] + \kappa_2 w_2' + q^2 w_2 \\ + \mu_2(\varphi_2'' \sin\varphi_2 + \varphi_2'^2 \cos\varphi_2) = \eta^2 \cos\eta\tau \quad ,$$

$$\varphi_j'' + \delta_j \varphi_j' + \left[Q_j^2 + \varepsilon_j(w_j'' - \eta^2 \cos\eta\tau)\right] \sin\varphi_j = 0 \quad , \qquad (j = 1, 2) \qquad (12)$$

where $w_j = z_j/a$, $\omega_1^2 = k_1/(M_1 + m_1)$, $\kappa_j = b_j/\omega_1(M_j + m_j)$, $\mu_j = m_j l_j / a(M_j + m_j)$, $\mu = (m_1 + M_1)/(m_2 + M_2)$, $\delta_j = c_j/\omega_1 m_j l_j^2$, $Q_j^2 = g/\omega_1^2 l_j$, and $\varepsilon_j = a/l_j$ $(j = 1, 2)$.

The semi-trivial solution of system (12) is the steady-state solution of the subset of differential equations:

$$\mathbf{E}\mathbf{W}'' + \mathbf{D}\mathbf{W}' + \mathbf{K}\mathbf{W} = \eta^2 \mathbf{P} \cos\eta\tau \quad , \qquad (13)$$

where $\mathbf{E} = \begin{vmatrix} 1 & 0 \\ 0 & 1 \end{vmatrix}$, $\mathbf{D} = \begin{vmatrix} \kappa_1 & -\kappa_1 \\ -\mu\kappa_1 & \mu\kappa_1 + \kappa_2 \end{vmatrix}$, $\mathbf{K} = \begin{vmatrix} 1 & -1 \\ -\mu & \mu + q^2 \end{vmatrix}$, $\mathbf{P} = \begin{vmatrix} 1 \\ 1 \end{vmatrix}$, $\mathbf{W} = \begin{vmatrix} w_1 \\ w_2 \end{vmatrix}$. According to the general theory of linear differential equations, the solution of system (13) has the form:

$$w_j(\tau) = a_j \cos\eta\tau + b_j \sin\eta\tau \quad , \qquad (j = 1, 2) \qquad (14)$$

where the coefficients a_j, b_j can be determined by solving the algebraic equations:

$$\begin{vmatrix} 1-\eta^2 & \kappa_1\eta & -1 & -\kappa_1\eta \\ -\kappa_1\eta & 1-\eta^2 & \kappa_1\eta & -1 \\ -\mu & -\mu\kappa_1\eta & \mu+q^2-\eta^2 & \mu\kappa_1+\kappa_2 \\ \mu\kappa_1\eta & -\mu & -(\mu\kappa_1+\kappa_2) & \mu+q^2-\eta^2 \end{vmatrix} \begin{vmatrix} a_1 \\ b_1 \\ a_2 \\ b_2 \end{vmatrix} = \eta^2 \begin{vmatrix} 1 \\ 0 \\ 1 \\ 0 \end{vmatrix} \quad . \tag{15}$$

The stability of the semi-trivial solution leads, using the approach presented in the previous Section, to the following perturbative equations:

$$\psi_j'' + \delta_j\psi_j' + \left[Q_j^2 - \varepsilon_j\eta^2 \left(A_j \cos\eta\tau + B_j \sin\eta\tau\right)\right]\psi_j = 0 \quad , \qquad (j=1,2) \tag{16}$$

where $A_j = 1 + a_j$, $B_j = b_j$. Thus, in the first instability region, the stability boundary curve can be determined from the relation:

$$R_j = \frac{2}{\eta^2\varepsilon_j}\left[\left(Q_j^2 - \frac{1}{4}\eta^2\right)^2 + \frac{1}{4}\eta^2\delta_j^2\right]^{1/2} \quad , \tag{17}$$

where $R_j = (A_j^2 + B_j^2)^{1/2}$. Thus, the absolute coordinates of the masses are:

$$y_j(\tau) = a(A_j \cos\eta\tau + B_j \sin\eta\tau) = aR_j \cos(\eta\tau + \alpha_j) \quad . \qquad (j=1,2) \tag{18}$$

In all further diagrams the following parameter values are used: $\kappa_1 = \kappa_2 = 0.10$, $\delta_1 = \delta_2 = 0.05$. Different alternatives are taken for parameters μ, q, ε_j, Q_j $(j = 1,2)$. However, it is necessary to keep in mind that ε_j and Q_j depend on $1/l_j$, i.e. when choosing a certain value of Q_j then ε_j is no more arbitrary and vice versa. In all the diagrams the values of μ, q, ε_j, are marked. Because there are two pendula attached to the masses, two diagrams are always necessary for each alternative.

Three alternatives are presented in Figures 5 to 7. The vibration amplitude curves $R_1(\eta)$ and $R_2(\eta)$ are marked with heavy solid lines for stable solution and with dotted lines for unstable solution. Light solid lines denote again the stability boundary curves. In all the examples considered an appropriate tuning of the pendula has been chosen: Q_1 is close to one half of the η at lower resonance, Q_2 is close to one half of η at the higher resonance. We can see that for such a tuning the autoparametric resonance can be initiated at a certain resonance of the excited subsystem due to the action of only a single pendulum. This means that, at a certain resonance of the excited subsystem, part of the excitation energy will flow into the pendulum subsystem, which results in diminishing the vibration amplitude of the excited subsystem.

5. Conclusions

The presented approach, using the stability investigation of the semi-trivial solution of an autoparametric system consisting of several masses and springs arranged in a chain, enables us to find the optimal tuning of pendulum subsystems attached

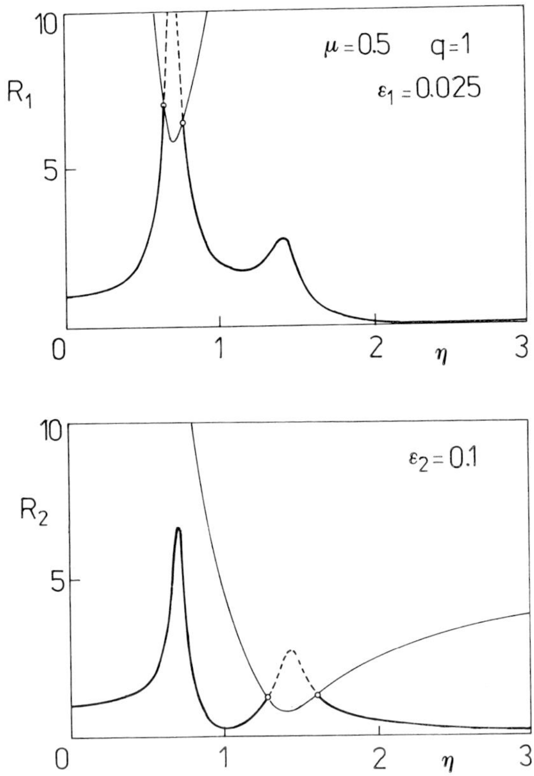

Fig. 5. Vibration amplitude curves R_1 and R_2 (stable solution heavy solid lines, unstable solution dotted lines) and stability boundary curves R (light solid lines) as a function of the excitation frequency η. The following values have been used for the system parameters: $\kappa_1 = \kappa_2 = 0.10$, $\delta_1 = \delta_2 = 0.05$, $\mu = 0.5$, $q = 1.0$, $\varepsilon_1 = 0.025$, $\varepsilon_2 = 0.100$.

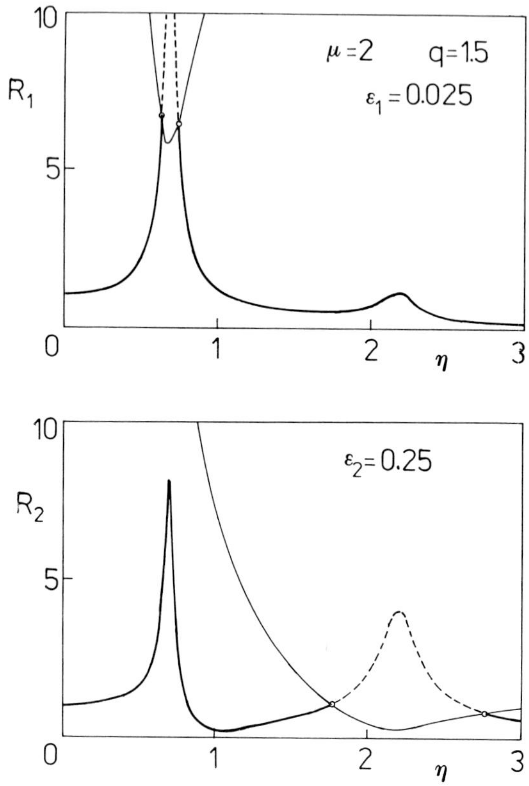

Fig. 6. Vibration amplitude curves R_1 and R_2 (stable solution heavy solid lines, unstable solution dotted lines) and stability boundary curves R (light solid lines) as a function of the excitation frequency η. The following values have been used for the system parameters: $\kappa_1 = \kappa_2 = 0.10$, $\delta_1 = \delta_2 = 0.05$, $\mu = 2.0$, $q = 1.5$, $\varepsilon_1 = 0.025$, $\varepsilon_2 = 0.250$.

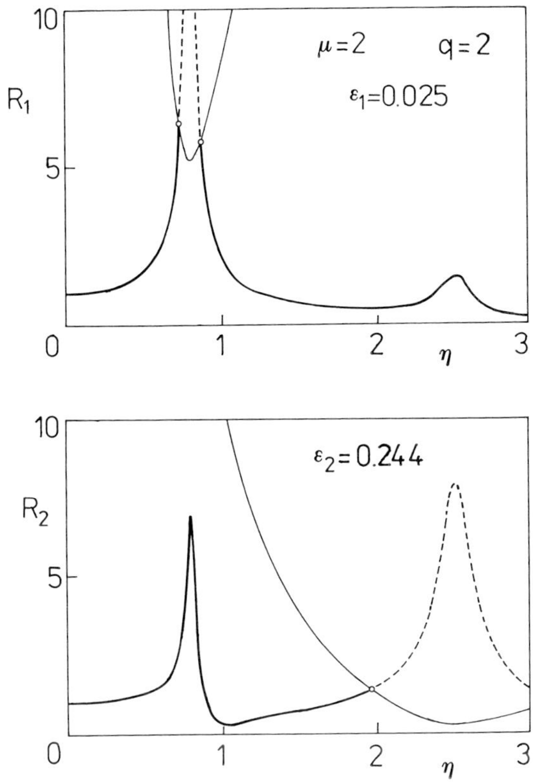

Fig. 7. Vibration amplitude curves R_1 and R_2 (stable solution heavy solid lines, unstable solution dotted lines) and stability boundary curves R (light solid lines) as a function of the excitation frequency η. The following values have been used for the system parameters: $\kappa_1 = \kappa_2 = 0.10$, $\delta_1 = \delta_2 = 0.05$, $\mu = 2.0$, $q = 2.0$, $\varepsilon_1 = 0.025$, $\varepsilon_2 = 0.244$.

to the masses of the excited subsystem. At a certain resonance, for a convenient tuning of the pendulum (the natural frequency of the pendulum subsystem is close to one half of the excitation frequency), the mass subsystem can diminish the vibration amplitude due to the flow of a part of the excitation energy into the pendulum subsystem. In most cases only one resonance of the excited subsystem can be influenced with one pendulum.

6. Acknowledgements

The authors would like to thank Professors A. Guran and R.H. Plaut for useful comments that improved this work.

7. References

1. A. H. Nayfeh and D. T. Mook, *Nonlinear Oscillations* (John Wiley & Sons, New York, 1979).
2. G. Schmidt and A. Tondl, *Non-Linear Vibrations* (Cambridge University Press, Cambridge, 1986).
3. M. Cartmell, *Introduction to Linear, Parametric and Nonlinear Vibrations* (Chapman and Hall, London, 1990).
4. A. Tondl and R. Nabergoj, *Nonlinear Dynamics* **1** (1990) 131.
5. A. Tondl, *Festschrift zum 80. Geburtstag von Prof. K. Magnus* (Lehrstuhl B für Mechanik, Technische Universität München, 1992), 291.
6. R. Svoboda, A. Tondl and F. Verhulst, *International Journal of Non-Linear Mechanics* **29** (1994) 225.
7. A. Tondl, *Acta Technica ČSAV* **37** (1992) 735.

Wave Motion, Intelligent Structures and Nonlinear Mechanics, pp. 239–279
edited by A. Guran and D. J. Inman
Series on Stability, Vibration and Control of Structures: Vol. 1

A SURVEY OF GEOMETRIC STIFFENING IN MULTIBODY DYNAMICS FORMULATIONS

INNA SHARF
Department of Mechanical Engineering
University of Victoria
Victoria, British Columbia, Canada
V8W 3P6

Abstract

In this chapter, the issue of geometric stiffening as it arises in the context of multibody dynamics is discussed. This topic has been treated in a number of previous publications and appears to be a debated subject. The controversy revolves primarily around the "correct" methodology for incorporating the stiffening effect into dynamics formulations. The main goal of this work is to present the different approaches that have been developed for this problem through an in-depth review of several publications dealing with this subject. The contribution is a precise understanding of the existing methodologies for modelling the stiffening effects in multibody systems. Thus, the material is presented so as to illuminate the key characteristics of the various methods and how they relate to each other. In addition, the chapter offers some novel insights and clarifying interpretations of these schemes. It is concluded with a general classification and comparison of the different approaches.

1. Introduction

The issue of geometric stiffening, also referred to as dynamic, centrifugal, or rotational stiffening and foreshortening has been a topic of many recent publications dealing with the dynamics of flexible bodies for applications to multibody systems. Kane *et al.*[1] were the first to observe that majority of existing multibody dynamics formulations and accordingly the dynamics simulation packages do not incorporate the geometric stiffening effect. They have attributed this flaw to the "conventional approach" for describing the deformation of elastic bodies, which yields a set of dynamics equations that inherently lack the geometric stiffening terms. Kane *et al.* propose an alternative approach which corrects this flaw and apply it to develop a set of equations for the deformation of a beam attached to a moving base.

Eke and Laskin[2] took up the issue raised in Ref. 1 and investigated regimes of validity of existing formulations with the simulation package DISCOS on a spin-up beam example. They qualified the error in the conventional approach as a "premature linearization" of the displacement field. This was later supported by Padilla and von Flotow[3] and Banerjee and Dickens.[4]

Shortly after Kane *et al.*'s publication, two commentaries appeared on the material presented in Ref. 1. In particular, London[5] pointed out that geometric stiffening has been previously considered by many researchers in a number of various applications. Furthermore, London observed that several approaches have been employed to include this effect in the dynamics equations and compiled a table characterizing the various methods. In a technical note,[6] Hanagud and Sarkar state that, contrary to the claim made in Ref. 1, the conventional method for modelling the kinematics of elastic deformation *can* be used.

The existing controversy over the nature of geometric stiffening, the debate on "the correct" approach to model it and the seeming incongruity of the methods used to include the effect in the motion equations—all of these have motivated us to review several of the works on this subject. In doing so, we have attempted to understand precisely how geometric stiffening is incorporated into the dynamics equations in different approaches, what assumptions and approximations are made in the derivation, what motivated these and whether they are justified. This chapter contains the main results of our review.

Our starting point will be the landmark paper by Kane *et al.*[1] and the subsequent commentaries.[5,6] Following that, we give a thorough treatment of the works by Likins *et al.*,[7] Vigneron[8] and Kaza and Kvaternik[9] and a summary of the relevant material from the publications by Lips and Modi[10] and Hughes and Fung.[11] Section 4 contains the main results from Laskin *et al.*,[12] Simo and Vu-Quoc,[13] Meirovitch,[14,15] and the recent works by Banerjee *et al.*[4,16] and Wallrapp and Schwertassek.[17] In reviewing the works of these researchers, we do not simply repeat their derivations, nor do we include the dynamics equations developed in these publications. Instead, we concentrate on the fundamental assumptions made in formulating the basic elements necessary for deriving these equations, where the "formulation" ends when the development becomes a purely mechanical process. For instance, in the cases where dynamics equations are derived *via* Hamilton's principle, we limit ourselves to stating kinetic and potential energy functions, and do not go through the procedure of applying the variational principle. This allows us to compare the various approaches based on the fundamental physical assumptions.

In addition to presenting the key features of different procedures, making comparisons and establishing relationships between them, we also provide clarifications and give some new insights. A classification of the approaches and comments on their suitability for multibody dynamics simulation culminate the review. We conclude the chapter with a numerical comparison based on a spin-up of a flexible beam problem.

2. Kane *et al.* and Commentaries

2.1. Main Results of Kane, Ryan and Banerjee

In Ref. 1, Kane, Ryan and Banerjee develop the dynamics equations of a general flexible beam built into a rigid base. The base body can undergo arbitrary, but prescribed translational and rotational motion. The generality of the beam refers to the fact that its geometric and material properties are not assumed to be constant, but can vary along the length of the beam. In addition, Kane *et al.* do not make the common assumption that the elastic and centroidal axes coincide. As a result, their motion equations contain terms dependent on the components of the eccentricity vector, e_2 and e_3.

The formulation of equations in Ref. 1 differs from many existing procedures in several respects. First, it incorporates the effect of the transverse displacement on the axial displacement in the kinematic description of the deformation. This is achieved *indirectly* by expressing the distance along the deformed elastic axis as a nonlinear function of the transverse displacements with:

$$x + s(x,t) = \int_0^{x+u_1} \left[1 + \left(\frac{\partial u_2}{\partial \sigma}\right)^2 + \left(\frac{\partial u_3}{\partial \sigma}\right)^2\right]^{1/2} d\sigma \tag{1}$$

The above equation is the same as Eq. (19) of Ref. 1, where the variable s denotes "the stretch in the beam along the elastic axis."

The second major difference relates to the choice of elastic deformations that are employed to describe kinematics of the deformed beam. The standard procedure is to use three orthogonal elastic displacements u_1, u_2 and u_3 to represent the displacement field in a deformable body. Kane *et al.* employ the *stretch* s with the transverse translations u_2 and u_3 as a set of generalized coordinates. Thus, when discretizing the continuous displacement field, they discretize the stretch variable *instead* of the axial displacement u_1. This is expressed by Eqs. (25) and (26) of Ref. 1, which we rewrite here for convenience as:

$$\begin{aligned} s(x,t) &= \sum_{j=1}^{\nu} \phi_{1j}(x)\, q_j(t) \\ u_i(x,t) &= \sum_{j=1}^{\nu} \phi_{ij}(x)\, q_j(t), \qquad i = 2,3 \end{aligned} \tag{2}$$

Contrary to the statement made by London[5] that Kane *et al.* "choose to represent the three elastic degrees of freedom by five variables $\{u_1, u_2, u_3, s, \zeta\}$," we emphasize that the stretch s is introduced to replace the axial displacement u_1. The variable ζ is employed by Kane *et al.* as a short-hand for the combination $x + u_1$.

In addition to equations (2), Kane *et al.* discretize the rotations $\{\theta_1, \theta_2, \theta_3\}$, which represent the orientation of the beam cross-sections. This implies that the rotation field is treated independently from the displacement field, unlike the common practice in modelling flexible beams. Accordingly, the dynamics equations based on (2) and the discretized rotations represent a model for the time-evolution of $\{s, u_2, u_3, \theta_1, \theta_2, \theta_3\}$, or rather, the corresponding discrete elastic coordinates.

The "conventional approach" involves discretizing the orthogonal set of elastic displacements $\{u_1, u_2, u_3\}$ with:

$$u_i = \sum_{j=1}^{\nu} \phi_{ij}\, q_j, \qquad i = 1, 2, 3 \tag{3}$$

or in matrix form as:

$$\mathbf{u}(\mathbf{r}, t) = \mathbf{\Phi}(\mathbf{r})\mathbf{q}(t) \tag{4}$$

where $\mathbf{r}$ contains the undeformed (material) coordinates of a point. Kane *et al.* argue that in the standard procedure (3), the three elastic deformations cannot account for the fact that every transverse displacement gives rise to an axial displacement, because the form (3) inherently precludes such an interdependence. Although we do not share this conviction and, therefore, do not believe that the dynamics equations *must be* formulated in terms of the stretch variable in order to predict stiffening of the beam, this particular feature of their procedure provides, at the least, an interesting alternative to the conventional approach.

The general methodology employed in Ref. 1 to derive an explicit (literal) set of motion equations for the elastic coordinates is that presented in Kane and Levinson. The procedure requires one to construct the *generalized inertia* and *generalized active* forces. The former are developed in Ref. 1 according to the algorithm outlined by Kane and Levinson. The generalized *active* forces, which for the particular system considered result from internal forces, are derived from the strain energy function. The expression for this function U is given in terms of the components of the force and torque vectors which act on a cross-section of the beam (see Eq. (51) in Ref. 1). Thus, it takes the form of a sum of six integrals, three for each of the force and torque, where the integrand in each term is a *quadratic* function of the appropriate load. To determine the generalized internal force by using Castigliano's theorem (Eq. (50) of Ref. 1), one needs to formulate the strain energy as a function of the generalized coordinates. To this end, Kane *et al.* express each of the six loads as a *linear* function of elastic deformations or their spatial derivatives.

For reasons that will become apparent §2.3, we draw attention to one particular term in the strain energy function. This term represents the contribution of the axial load and will be denoted here by U_P. It corresponds to the first term in Eq. (51) of Ref. 1, which we rewrite, omitting the subscripts as:

$$U_P = \int_0^L \frac{P^2}{2EA}\, dx \tag{5}$$

Usually, one approximates the axial load as a linear function of the axial displacement gradient with $P = EA\frac{\partial u}{\partial x}$. The expression for P given in Ref. 1 (Eq. (58)) takes the same form, but with the axial translation u_1 replaced with the stretch variable s. With that, the axial load contribution to the strain energy function becomes:

$$U_P = \frac{1}{2}\int_0^L EA\left(\frac{\partial s}{\partial x}\right)^2 dx \tag{6}$$

To complete the presentation of the key features of the dynamics formulation as put forward by Kane, Ryan and Banerjee,[1] we observe that the equations of motion presented in Ref. 1 are *linear* in the elastic coordinates, and they incorporate the geometric stiffening. In fact, one of the crucial points made by these authors is that the linear motion equations in other works do not contain all of the linear terms, in particular the stiffening term. As implied by Kane *et al.*,[1] and later stated by Eke and Laskin[2] and Padilla and von Flotow,[3] this occurs because of premature linearization implicit in the "conventional approach." We will further comment on this issue in §2.3. In the following subsection, we summarize the main points of an engineering note, which appeared shortly after the publication of Kane *et al.*

2.2. Hanagud and Sarkar's Note

Contrary to that claimed in Ref. 1, Hanagud and Sarkar[6] believe that the axial and transverse motions can be treated independently with the standard discretization procedure and the stiffening effect can be accounted for if "the nonlinear effects are properly included in the formulation."[6]

Hanagud and Sarkar present a formulation for an elastic beam where, as in the conventional approach, they discretize the axial displacement u_1, not the stretch variable s. Hanagud and Sarkar derive the differential equations for the corresponding discrete elastic coordinates by using the same general methodology as employed in Ref. 1, which as they point out is sometimes referred to as Kane's method. Similar to the development of Kane *et al.*, they also determine the generalized elastic forces from the strain energy function. However, Hanagud and Sarkar formulate the strain energy as a *quartic* function of the spatial derivatives of u_1, u_2 and u_3. This is accomplished by employing the *nonlinear* strain-displacement relations, through which the aforementioned nonlinear effects are introduced into the formulation. Lastly, Hanagud and Sarkar do not linearize the final equations of motion, but retain terms of second and third order.

A notable contribution of Hanagud and Sarkar's work is an observation that the expression for the stretch variable presented by Kane *et al.* (Eq. (19) in Ref. 1) is inconsistent with the rest of their development. The inconsistency results from the fact that this relation for s is applicable if one expresses the transverse displacements u_2 and u_3 as a function of the *deformed* coordinate X which corresponds to the axial projection of a generic point in the deformed configuration of the beam. (We have chosen to follow the traditional notation employed in the theory of elasticity, where one distinguishes the deformed and undeformed coordinates by different-case letters.) Since in their formulation, Kane *et al.* express the translations u_2 and u_3 in terms of the undeformed coordinate x (see Eq. (2)), the consistent expression for the stretch variable is:

$$x+s(x,t) = \int_0^x \left[\left(1 + \left.\frac{\partial u_1(x,t)}{\partial x}\right|_{x=\sigma}\right)^2 + \left(\left.\frac{\partial u_2(x,t)}{\partial x}\right|_{x=\sigma}\right)^2 + \left(\left.\frac{\partial u_3(x,t)}{\partial x}\right|_{x=\sigma}\right)^2 \right]^{1/2} d\sigma \tag{7}$$

The above equation, although looks different, is equivalent to Eq. (2) in Ref. 6. We also observe that it embodies the nonlinear formulation of the strain. In (7), we have used the notation $\left.\frac{\partial(\cdot)}{\partial x}\right|_{x=\sigma}$ to emphasize that elastic translations must be expressed as a function of the undeformed coordinate x, while σ in this case is a dummy integration variable. We also point out a change in the upper limit of the integral from $x + u_1$ in Eq. (1) to x in Eq. (7) which, of course, is a consequence of the transformation from deformed to undeformed axial coordinate.

London[5] pointed out that the use of $x + u_1$ as an upper limit of the integral in Eq. (1) creates problems in computing the modal integrals β_{ij} and γ_{ij}, which are required in the evaluation of motion equations in Ref. 1 (please refer to Eq. (41) in Ref. 1). According to their definition in Ref. 1, these modal integrals are indeed time-dependent through $u_1 = u_1(x,t)$ in their upper limit of integration. This would require that they be evaluated at each time step in the numerical integration of the motion equations. (Although that would add to the computational cost of the simulation, it should not otherwise pose a problem.) The situation changes, however, if one corrects the formulation in Ref. 1 in accordance with observations made by Hanagud and Sarkar. This can be achieved by replacing the inconsistent expression for the stretch, given by (1), with a consistent form of Eq. (7). With this correction, the aforementioned modal integrals become *independent* of time, and can therefore be evaluated prior to the numerical integration of the motion equations.

2.3. Commentaries

In the following, we offer additional clarification on the points made in the commentaries of London[5] and Hanagud and Sarkar.[6] We hope that these will serve to elucidate the more subtle features of the formulation proposed by Kane *et al.* A few of the issues raised in the commentaries will be further addressed in section 5 of the chapter.

London questions why the foreshortening is not considered during evaluation of the strain energy in Ref. 1. This comment has motivated us to consider closely the corresponding expression, and in the process we have singled out the "axial" strain energy, cited in Eq. (5), as the only contribution which may possibly comprise the foreshortening effect. The following brief development shows that it does indeed.

The function U_P, as can be seen from (6), is quadratic in the spatial derivative of the stretch, $\frac{\partial s}{\partial x}$. Thus, in order to illuminate the nature of this strain energy, we need to obtain an explicit expression for the stretch gradient. As noted previously, Eq. (7) provides the proper form for s that should be employed in Ref. 1. Abbreviating the notation, we rewrite (7) for s as:

$$s(x,t) = \int_0^x \left[\left(1 + \frac{\partial u_1}{\partial \sigma}\right)^2 + \left(\frac{\partial u_2}{\partial \sigma}\right)^2 + \left(\frac{\partial u_3}{\partial \sigma}\right)^2 \right]^{1/2} d\sigma - x \tag{8}$$

Differentiating the above with respect to x gives:

$$\frac{\partial s}{\partial x} = \left[\left(1 + \frac{\partial u_1}{\partial x}\right)^2 + \left(\frac{\partial u_2}{\partial x}\right)^2 + \left(\frac{\partial u_3}{\partial x}\right)^2 \right]^{1/2} - 1 \tag{9}$$

and upon expansion of the first term, the required gradient takes the form:

$$\frac{\partial s}{\partial x} = \left[1 + 2\frac{\partial u_1}{\partial x} + \left(\frac{\partial u_1}{\partial x}\right)^2 + \left(\frac{\partial u_2}{\partial x}\right)^2 + \left(\frac{\partial u_3}{\partial x}\right)^2 \right]^{1/2} - 1 \tag{10}$$

Let us now introduce the axial component of the Lagrangean strain tensor, denoted by $\varepsilon_{0,xx}$, where the 0 subscript signifies that it refers to the elastic axis. (Note that u_1, u_2 and u_3 are defined in Ref. 1 as translations of points along the elastic axis only.) The strain $\varepsilon_{0,xx}$ can be expressed in terms of the elastic displacements with a well-established strain-displacement relation. It has the following *exact* and *nonlinear* form:

$$\varepsilon_{0,xx} = \frac{\partial u_1}{\partial x} + \tfrac{1}{2} \left[\left(\frac{\partial u_1}{\partial x}\right)^2 + \left(\frac{\partial u_2}{\partial x}\right)^2 + \left(\frac{\partial u_3}{\partial x}\right)^2 \right] \tag{11}$$

With the above, the stretch gradient of Eq. (10) can be succinctly written as:

$$\frac{\partial s}{\partial x} = [1 + 2\varepsilon_{0,xx}]^{1/2} - 1 \tag{12}$$

Before we continue, it is worthwhile to point out that in all formulations dealing with the subject of geometric stiffening in the context of multibody dynamics, it is always assumed, although not always stated, that the strains are small, and specifically, $\varepsilon_{0,xx} << 1$. Therefore, we can make use of the Binomial Theorem to simplify Eq. (12). Retaining the first two terms in the binomial expansion we get:

$$\frac{\partial s}{\partial x} \approx [1 + \varepsilon_{0,xx}] - 1 = \varepsilon_{0,xx} \tag{13}$$

Finally, substituting for $\frac{\partial s}{\partial x}$ from (13), the axial contribution to the strain energy function employed by Kane *et al.* takes the form:

$$U_P = \frac{1}{2} \int_0^L EA \left(\varepsilon_{0,xx}\right)^2 dx \tag{14}$$

with the axial strain given by the *nonlinear* Eq. (11).

At this point, we comment on the form of the strain energy employed by Hanagud and Sarkar in their formulation (Eq. (8) in Ref. 6). This expression can be derived by using Eq. (14) in conjunction with the following simplified expression for the axial strain:

$$\varepsilon_{0,xx} \approx \frac{\partial u_1}{\partial x} + \tfrac{1}{2}\left[\left(\frac{\partial u_2}{\partial x}\right)^2 + \left(\frac{\partial u_3}{\partial x}\right)^2\right] \tag{15}$$

The above form is obtained from the exact relation (11) by omitting the $\left(\frac{\partial u_1}{\partial x}\right)^2$ term—an approximation often made for beams and in the context of moderate-deformation theories.[19] The third- and fourth-order terms in the strain energy of Hanagud and Sarkar result from the nonlinear terms $\left(\frac{\partial u_2}{\partial x}\right)^2$ and $\left(\frac{\partial u_3}{\partial x}\right)^2$ in the strain-displacement relation (15). It is these terms that lead to the geometric stiffening and foreshortening of the beam.

The development presented here demonstrates that, contrary to London's comment, the geometric nonlinearity *is* included in the strain energy expression given in Ref. 1. This is not apparent because Kane *et al.* employ the stretch s instead of the axial translation u_1 as a generalized coordinate in their formulation. As a consequence, they do not need to expand $\frac{\partial s}{\partial x}$ when evaluating U so that their strain energy function remains quadratic in the discrete generalized coordinates and does not explicitly contain higher-order terms. Indeed, this is precisely the reason why the final equations of Kane *et al.* are linear in elastic coordinates, but yet include the stiffening term. As the above analysis shows, it is through the use of stretch instead of axial displacement, that Kane *et al.* incorporate foreshortening in their formulation, without introducing nonlinearity explicitly into the motion equations. By contrast, Hanagud and Sarkar demonstrated that one *can* obtain the stiffening effect with the conventional approach. This is achieved by employing the nonlinear strain-displacement relations in constructing the strain energy function in which case the resulting stiffening term is a *nonlinear* function of elastic coordinates.

To conclude the discussion of the article by Kane *et al.* and the related commentaries, we would like to state that, in agreement with London, the theory proposed in Ref. 1 is *not* fundamentally new; a number of formulations which incorporate the foreshortening effect and the resultant stiffening of an elastic beam during its rotation, have been previously published. Indeed, Kane *et al.* refer to some of these works, but only in a superficial and in some cases misleading manner. Nevertheless, we feel that the formulation developed in Ref. 1 does have several novel features. Most notably, these are the generality of the flexible beam model and use of the stretch variable as a generalized coordinate in deriving the dynamics equations.

3. References Contended by London

London's comments on the literature review and Kane *et al.*'s reply,[20] motivated us to investigate the material presented in several of the references in question. The main objective of this section is to summarize these findings.

3.1. Likins et al., Vigneron, and Kaza and Kvaternik

Our choice to discuss the contributions of Likins *et al.*,[7] Vigneron,[8] and Kaza and Kvaternik[9] in the same section is based on several reasons. First, these articles appeared within a time-span of four years and therefore belong to the same "era." Second, all three publications contain formulations of the dynamics equations for a flexible beam spinning in a plane at a constant speed. Finally, there is a logical relationship between these works, since that by Vigneron is a comment on Likins *et al.*, while Kaza and Kvaternik extend Vigneron's approach to obtain the nonlinear equations of motion of the aforementioned system.

Our presentation is not a plain copy of Refs. 7, 8 and 9, as it is structured to make apparent the key features of the approaches taken in the three works, establish a relationship between them, as well as identify the particular contributions of each one. Towards this end, we present the results of these works in a common notation, which will also be employed throughout the rest of the chapter. This notation is similar to that used in Ref. 1 with one major difference. We choose to denote the three orthogonal elastic displacements with symbols u, v and w. Thus, u now represents the axial elastic displacement measured along the x-axis of the floating reference frame, while v and w are the transverse elastic displacements. Furthermore, these symbols are not restricted to the elastic axis, but represent elastic displacements of any point in the beam. This convention follows that of Kaza and Kvaternik.[9]

As already mentioned, the system considered in all three publications[a] is an Euler-Bernoulli beam with a symmetric cross-section, spinning at a constant angular speed Ω about the z-axis of the reference frame. This system represents a special case of that treated by Kane *et al.*, defined with: $\omega_3 = \Omega$, $\omega_1 = \omega_2 = \mathrm{v}_1 = \mathrm{v}_2 = \mathrm{v}_3 = 0$, $e_2 = e_3 = 0$, as well as constant geometric and material properties. The common features of the formulations[7,8,9] are listed below.

(i) The dynamics equations are constructed *via* Hamilton's principle.

(ii) The position of a generic point located at $\mathbf{r}^T = [x, y, z]^T$ in the undeformed beam is given by $[x + u, y + v, z + w]^T$, where v and w are functions of x and time only, that is $v(x, y, z, t) = v_0(x, t)$ and similarly for w. This implies a linearization of the displacement field and precludes torsional deformation.

(iii) The kinetic energy is calculated with:

$$T = \tfrac{1}{2} \iiint \left[\dot{u}^2 + \dot{v}^2 + \dot{w}^2 + \Omega^2(x+u)^2 + \Omega^2(y+v)^2 + \right. \\ \left. 2\Omega(x+u)\dot{v} - 2\Omega(y+v)\dot{u}\right] \frac{\rho}{A}\, dx\, dy\, dz \qquad (16)$$

where, like in Ref. 1, the symbol ρ denotes the mass per unit length of the beam and A is the cross-sectional area. The above equation is identical to Eq.

[a] Likins *et al.* also consider "Axial Beams."

(6) of Likins *et al.* and applies to any elastic body spinning as specified before. Indeed, the form of T remains the same for nonconstant Ω.

(iv) The potential energy is calculated with

$$U = \frac{E}{2} \iiint (\varepsilon_{xx})^2 \, dx \, dy \, dz \tag{17}$$

where ε_{xx} denotes the axial strain at any point in the beam, and is given by a nonlinear expression:

$$\varepsilon_{xx} = \frac{\partial u}{\partial x} + \tfrac{1}{2}\left[\left(\frac{\partial u}{\partial x}\right)^2 + \left(\frac{\partial v}{\partial x}\right)^2 + \left(\frac{\partial w}{\partial x}\right)^2\right] \tag{18}$$

(v) The dynamics equations are formulated for the *continuous* displacement variables, and accordingly take the form of partial differential equations.

There are two major differences between formulations presented in Refs. 7, 8 and 9. The first one relates to the form of the assumed *axial* displacement field. Likins *et al.* "expand" the axial displacement in the beam with:

$$u = u_0 - y\frac{\partial v}{\partial x} - z\frac{\partial w}{\partial x} \tag{19}$$

Vigneron, followed by Kaza and Kvaternik adopt a different form. Their axial displacement is given by Eqs. (2) and (1a) in Refs. 8 and 9 respectively, which we write as:

$$u = u_s - y\frac{\partial v}{\partial x} - z\frac{\partial w}{\partial x} - u_f \tag{20}$$

In the above, u_f is the "displacement associated with the foreshortening effect."[8] In both references, this component of the axial displacement is specified as an explicit function of the transverse displacements:

$$u_f = u_f(x,t) = \tfrac{1}{2}\int_0^x \left[\left(\frac{\partial v}{\partial x}\right)^2 + \left(\frac{\partial w}{\partial x}\right)^2\right] dx \tag{21}$$

We note that expression (21) is a second-order approximation for the foreshortening of the beam. Clearly, u_f corresponds to the axial displacement which results from the transverse deformation. Comparing Eqs. (19) and (20), we observe that u_0 must equal $u_s - u_f$ and hence u_0 and u_s represent different physical quantities. We have employed the subscript s in u_s to signify that it corresponds to axial displacement resulting strictly from *extension* or *stretch* of the elastic axis. By comparison, u_0 represents the total axial displacement on the elastic axis which may be due to *both* stretch and bending.

The second difference between the derivations of Likins *et al.*, Vigneron, and Kaza and Kvaternik lies in the approximations made in deriving the final equations of motion for the elastic coordinates u_0 or u_s, v and w. To be specific, these are approximations made in the kinetic and strain energies that are used to construct the Lagrangian of the system.

Likins *et al.* substitute their expansion for u from Eq. (19) into the kinetic and strain energy expressions (16) and (17), the latter combined with (18). Expanding the result and simplifying for a beam with a symmetric cross-section yields:

$$\begin{aligned} T &= \tfrac{1}{2}\rho \int_0^L \left[(\dot{u}_0 - \Omega v)^2 + [\dot{v} + \Omega(x + u_0)]^2 + \dot{w}^2\right] dx \\ &+ \tfrac{1}{2} \int_0^L \left[\left(\frac{\partial \dot{v}}{\partial x} + \Omega\right)^2 + \Omega^2 \left(\frac{\partial v}{\partial x}\right)^2\right] \frac{\rho I_z}{A} dx \end{aligned} \tag{22}$$

where we have used the standard definition $I_z = \iint y^2\, dy\, dz$. Likins *et al.* simplify the results by making the following assumptions.

I. The vibration of the beam takes place in the meriodinal direction only. Accordingly, the in-plane deformation is ignored based on the argument that it is present only because of the Poisson effect.

- **(a)** This amounts to dropping all terms in kinetic energy T which involve v or its derivatives.
- **(b)** The strain energy U is derived by omitting $\left(\frac{\partial v}{\partial x}\right)^2$ in the definition (18) for the strain.

II. Likins *et al.* neglect all terms that involve u_0 and $\dot{u}_0$ in T, thus effectively eliminating the axial equation of motion from the model.

III. They retain only one term in the strain energy U among the additional third- and fourth-order terms which arise from the remaining nonlinearities in the strain-displacement relation.

With the above assumptions, the kinetic and potential energy functions take the form:

$$T = \tfrac{1}{2}\rho \int_0^L \dot{w}^2\, dx + \rho\Omega^2 L \left(\frac{1}{2A} I_z + \frac{1}{6} L^2\right) \tag{23}$$

$$U = \frac{EI_z}{2} \int_0^L \left(\frac{\partial^2 w}{\partial x^2}\right)^2 dx + \frac{EA}{2} \int_0^L \frac{\partial u_0}{\partial x} \left(\frac{\partial w}{\partial x}\right)^2 dx \tag{24}$$

where in accordance with the assumption **II** above, we have also dropped the term in the strain energy which involves u_0 only. These correspond to Eqs. (39) and (45) in Ref. 7. Note that the second term in (23) includes the rotary inertia contribution. Being constant, it does not contribute to the motion equation for w. Also, the second term in (24) is the additional term mentioned in **III** and is of third order. It represents the coupling between axial and transverse displacements which leads to the stiffening of the beam in bending. This term is identical to the corresponding third-order term in the strain energy function of Hanagud and Sarkar (Eq. (8) in Ref. 6.)

At this point in their development, Likins *et al.* establish a connection between their approach, as we have just outlined, and "the textbook derivation for the transverse vibrations of beams subject to an external axial force P." As noted by them, the axial load to first approximation is given by:

$$P = EA\frac{\partial u_0}{\partial x} \tag{25}$$

so that Eq. (24) can be rewritten as:

$$U = \frac{EI_z}{2}\int_0^L \left(\frac{\partial^2 w}{\partial x^2}\right)^2 dx + \tfrac{1}{2}\int_0^L P\left(\frac{\partial w}{\partial x}\right)^2 dx \tag{26}$$

To proceed with the application of Hamilton's variational principle, Likins *et al.* assume that P is time-independent and can be approximated by its steady-state value. In fact, for the particular problem of a beam rotating at a constant speed, the axial load P is the centrifugal load on the beam. Furthermore, since the latter is a known function of the prescribed Ω, P can be calculated with:

$$P(x) = \tfrac{1}{2}\rho\Omega^2(L^2 - x^2) \tag{27}$$

With equations (23), (26) and (27), Likins *et al.* derive the motion equation for the elastic displacement w. Due to the assumptions made in evaluating the strain energy and the axial load P, geometric stiffening appears as a *linear* term in this equation.

Let us now proceed with the developments presented by Vigneron, and Kaza and Kvaternik. As we had already mentioned, both derivations are premised on expansion (20) for the axial displacement. As well, the kinetic energy is constructed by ignoring the effects of rotary inertia. This can be achieved by neglecting the "off-elastic-axis" contribution, $(-y\frac{\partial v}{\partial x} - z\frac{\partial w}{\partial x})$, in the expression for u used in Eq. (16) or by dropping the second integral (I_z-term) in Eq. (22). Then, kinetic energy can be written as a function of u_s, v, w and u_f as:

$$\begin{aligned} T = \quad & \tfrac{1}{2}\rho \int_0^L \left(\dot{u}_s^2 + \dot{v}^2 + \dot{w}^2 - \underline{2\dot{u}_s\dot{u}_f} + \underline{\underline{\dot{u}_f^2}}\right) dx \\ & + \tfrac{\Omega^2}{2}\rho \int_0^L \left(x^2 + u_s^2 + v^2 + \underline{\underline{u_f^2}} + 2xu_s - [2xu_f] - \underline{2u_s u_f}\right) dx \\ & + \rho\Omega \int_0^L \left(-\dot{u}_s v + v\dot{u}_f + x\dot{v} + \dot{v}u_s - \underline{u_f\dot{v}}\right) dx \end{aligned} \tag{28}$$

Recall that according to (21), the foreshortening u_f is a quadratic function of the transverse displacement gradients and, therefore, the above expression for T includes terms of third and fourth order. These terms are underlined in (28) with single and double lines, respectively. We have also singled out by enclosing in square brackets the term which is *linear* in u_f. This term gives rise to the stiffening effect in the motion equation. We also emphasize that Eq. (28) originates from the same expression for the kinetic energy (Eq. (16)) as used by Likins *et al.* The higher order terms in it are a consequence of dividing the axial displacement u_0 into two components, u_s and u_f, and explicitly assuming a second-order form for the latter.

Vigneron approximates T by keeping only second-order terms in (28) (including the term in the square brackets) as well as, setting the axial displacement u_s and its derivatives to zero. The latter approximation corresponds to *inextensibility* assumption, which implies that the beam is modelled as axially rigid.[9] Note that dropping the u_s-terms in Vigneron's formulation is *not* equivalent to dropping the u_0- terms in Likins *et al.*'s formulation, although in both cases this achieves elimination of the independent axial equation of motion. Vigneron's approximation yields:

$$T = \tfrac{1}{2}\rho \int_0^L \left(\dot{v}^2 + \dot{w}^2\right) dx + \frac{\Omega^2}{2}\rho \int_0^L \left(x^2 + v^2 - [2xu_f]\right) dx + \rho\Omega \int_0^L (x\dot{v})\, dx \tag{29}$$

which, without the last term, corresponds to Eq. (9) of his paper. Unlike Vigneron, Kaza and Kvaternik do not make the inextensibility assumption and therefore keep the axial equation of motion in the final model. They also retain third-order terms in (28). The result is given by Eq. (7) in Ref. 9 which is identical to our Eq. (28) without the double-underlined terms.

To determine the strain energy according to (17), Vigneron and Kaza and Kvaternik first evaluate the strain ε_{xx} of (18) by substituting for the axial displacement u from (20) in conjunction with (21). Upon dropping the third- and fourth-order terms, this produces the following second-order expression for the axial strain:

$$\varepsilon_{xx} = \frac{\partial\left(u_s - y\frac{\partial v}{\partial x} - z\frac{\partial w}{\partial x}\right)}{\partial x} - \frac{\partial u_f}{\partial x} + \tfrac{1}{2}\left[\left(\frac{\partial\left(u_s - y\frac{\partial v}{\partial x} - z\frac{\partial w}{\partial x}\right)}{\partial x}\right)^2 + \left(\frac{\partial v}{\partial x}\right)^2 + \left(\frac{\partial w}{\partial x}\right)^2\right] \tag{30}$$

where we have intentionally separated the contribution of u_f. However, according to (21), we have

$$\frac{\partial u_f}{\partial x} = \tfrac{1}{2}\left[\left(\frac{\partial v}{\partial x}\right)^2 + \left(\frac{\partial w}{\partial x}\right)^2\right] \tag{31}$$

so that (30) is simplified to:

$$\varepsilon_{xx} = \frac{\partial\left(u_s - y\frac{\partial v}{\partial x} - z\frac{\partial w}{\partial x}\right)}{\partial x} + \tfrac{1}{2}\left(\frac{\partial\left(u_s - y\frac{\partial v}{\partial x} - z\frac{\partial w}{\partial x}\right)}{\partial x}\right)^2 \tag{32}$$

or

$$\varepsilon_{xx} = \frac{\partial u_s}{\partial x} - y\frac{\partial^2 v}{\partial x^2} - z\frac{\partial^2 w}{\partial x^2} + \tfrac{1}{2}\left[\frac{\partial u_s}{\partial x} - y\frac{\partial^2 v}{\partial x^2} - z\frac{\partial^2 w}{\partial x^2}\right]^2 \tag{33}$$

The above is equivalent to Eq. (5) in Ref. 8 and Eq. (4) in Ref. 9, although neither publication includes a description of the intermediate step (30). It is interesting to note that ε_{xx} evaluated with (32) without the second-order term, can be derived from the axial strain of (18) without the $\left(\frac{\partial u}{\partial x}\right)^2$ term, but with the "transverse" nonlinear terms. As is demonstrated by Eq. (30), the latter is cancelled when foreshortening u_f is explicitly defined in the axial displacement u, which results in a linear expression for the strain. Vigneron's strain energy can be constructed by using this linearized form of (32), since he retains only quadratic terms in the strain energy. In fact, the strain energy function given in Ref. 8 has a standard form used in the linear theory of elasticity. Kaza and Kvaternik formulate their strain energy with the second-order strain (32), but drop the resultant fourth-order terms.

Because of the differences in the approximations made, Vigneron and Kaza and Kvaternik derive different sets of motion equations. Vigneron obtains two *linear* equations of motion for the elastic variables v and w (Eqs. (10) and (11) in Ref. 8), the latter being identical to the equation derived by Likins *et al.* Kaza and Kvaternik present a set of nonlinear dynamics equations for the three elastic deformations, u_s, v and w (Eq. 8 in Ref. 9). They are nonlinear, in particular second-order, since Kaza and Kvaternik retain third-order terms in their kinetic and strain energies. It is important to emphasize, that the term responsible for the stiffening of the beam derived in Refs. 8 and 9 is a first-order term and appears *via* kinetic energy. This occurs because of the particular form assumed for the axial displacement (Eq. (20)) as well as, the expression adopted for the foreshortening (Eq. (21).) If, as is done by Likins *et al.*,[7] one does not explicitly identify u_f in the axial displacement, then the stiffening appears in the motion equations through the strain energy (see Eq. (24)) and is fundamentally a nonlinear term. In the special case of a beam rotating at *constant* Ω, where Eq. (27) holds, the appoaches in Ref. 8 and Ref. 7 will yield identical results.

To conclude this section, we draw attention to some of the observations made by Kaza and Kvaternik. They identify four different approaches for deriving linear or nonlinear equations of motion for a rotating flexible beam. They are: (1) "the effective applied load artifice;" (2) the use of Newton's second law applied to the deformed configuration; (3) an approach in which nonlinear strain-displacement relations and a first-degree displacement field are used; (4) Vigneron's approach which uses nonlinear strain-displacement relations and a second-degree displacement field. Kaza and Kvaternik show that all four appraches "make use of geometric nonlinear theory of elasticity either implicitly or explicitly." They state that for developing the equations of motion for a rotating beam, "the geometric nonlinear theory is necessary to obtain even the correct linear equations." In their paper, Kaza and Kvaternik also discuss whether foreshortening must be explicitly included in the axial displacement field. They conclude that although it is not necessary, the approaches where the foreshortening effect is accounted for otherwise, require special considerations.

3.2. Lips and Modi, Hughes and Fung

Lips and Modi[10] investigate the dynamics of satellite systems composed of a central rigid body with flexible appendages. The base body is allowed to undergo general rotational motion and therefore, their dynamics model includes the rotational rigid-body equations. To illustrate the procedure for modelling deformation of appendages, Lips and Modi present an explicit form of the elastic equations for a rotating beam. In their treatment of this system, they take into account: (i) the offset between the attachment point of the appendage and the center of mass of the rigid body; (ii) variable flexural rigidity and variable density. The basic assumptions made in deriving the equations for elastic displacements are similar to those made by Kaza and Kvaternik.[9] In particular, Lips and Modi explicitly separate the foreshortening component from what they refer to as the oscillation component in the assumed form of the axial displacement. As well, the kinetic and potential energy functions employed to construct the Lagrangian contain terms up to third order.

The main subject of the work by Hughes and Fung[11] is the stability of spinning satellites with long flexible appendages. Therefore, they formulate the dynamics equations with a view to addressing this issue. The system is modelled as a spinning rigid body with appended beams. The rigid-body equations are formulated for small perturbations from the nominal spin configuration. The development of the elastic motion equations presented in Ref. 11 is different from the previously considered works in two respects. First, Hughes and Fung employ *deformed* coordinates to describe the kinematics of the deformed beam, a fact which they do not state explicitly. Thus, the position of a point on the elastic axis is defined by $[X, v, w]^T$ where $X = X(t)$, $v = v(X,t)$ and $w = w(X,t)$. (Note, the corresponding undeformed description is $[x+u_0, v, w]^T$ where $u_0 = u_0(x,t)$, etc.) Another distinct feature of Hughes and Fung's formulation is that they evaluate the kinetic and potential energies by integrating the appropriate integrands over the volume, which in the case of a slender beam reduces to the arc length $\hat{s}$, of the *deformed* configuration. Thus, the energy functions are defined by means of the line integral $\int(\cdot)d\hat{s}$. In general, the integration necessary to determine kinetic and potential energies of an elastic body can be performed over the deformed or undeformed configurations. However, the former is the standard choice when the kinematic description is given in terms of the deformed coordinates. (This representation corresponds to the Eulerian or spatial description of the problem.)

Hughes and Fung incorporate the geometric nonlinearity into their formulation by expressing the differential arc length $d\hat{s}$ with

$$d\hat{s} = \sqrt{1 + v'^2 + w'^2}\, dX \tag{34}$$

The prime in the above denotes differentiation with respect to the deformed coordinate X. We note that the integral of Eq. (34) is equivalent to Eq. (4) of Hanagud and Sarkar.[6] It defines the distance along the beam as a function of the transverse displacement gradients, when these are expressed in terms of the deformed coordinates. Hence, it is also equivalent to Eq. (1) of this chapter, with $\hat{s} = x + s$. To simplify the derivation, Hughes and Fung use a second-order approximation of $d\hat{s}$ so that

$$\int_0^L (\cdot)\, d\hat{s} = \int_0^{L^\star} (\cdot)\, dX + \tfrac{1}{2} \int_0^{L^\star} (\cdot) \left(v'^2 + w'^2\right) dX \tag{35}$$

where $L^\star$ denotes the projection of the tip on the axial coordinate axis. We also note that by using L in the upper limit of the integral $\int(\cdot)d\hat{s}$, Hughes and Fung implicitly assume that the beam is inextensible.

The kinetic and potential energies derived in Ref. 11 contain second-order terms only. That, combined with the inextensibility assumption makes the formulation of Hughes and Fung similar to Vigneron's. In fact, the kinetic energy of a single beam rotating at a nominal speed, deduced from Eq. (9) of their paper, is equivalent to Vigneron's kinetic energy given by our Eq. (29). The form of the strain energy function is also the same.

This completes our overview of selected references from London's and Kane *et al.*'s commentaries.[5,20] It is clear that these works incorporate geometric stiffening into the dynamics equations of a rotating flexible beam, although through different approaches. A discussion of these will be given in §5 of the chapter. At this point, we only note that they all account for the coupling between the transverse and axial deformations. Indeed, it is exactly this phenomenon that causes stiffening of an elastic body under certain conditions. The differences between the approaches lie in what we view as the *mechanism* for introducing the coupling effect into the formulation and accordingly, the stage in the derivation at which it is introduced. In the following section of the chapter we discuss some of the other approaches that have been employed to account for the geometric stiffening in the dynamics equations of multibody systems. These are presented in the chronological order of their publication in open literature.

4. Laskin *et al.*, Simo and Vu-Quoc, Meirovitch, Banerjee *et al.*, Wallrapp and Schwertassek

4.1. Laskin et al.

Similar to Vigneron[8] and Kaza and Kvaternik,[9] Laskin *et al.*[12] assume a form for the axial displacement u in which the displacement of points along the elastic axis of the beam, u_0, is divided into two parts. We rewrite their Eq. (6) as:

$$u = u_{qs} + u_t - y\frac{\partial v}{\partial x} - z\frac{\partial w}{\partial x} \tag{36}$$

which can be directly compared to (20). Laskin *et al.* refer to u_{qs} (v_0 in Ref. 12) as a "quasi-steady" component. This appropriately reflects the fact that this axial displacement is present even when there is no axial vibration, and furthermore, it exists even under static loading. The "transient" component u_t (their $v^\star$) accounts for longitudinal vibrations. Laskin *et al.* justify this arrangement by arguing that it allows one to consider u_t as a small, more precisely, infinitesimally small displacement which is of the order of the transverse displacements v and w. The quasi-steady component u_{qs} may be comparatively large.

To discretize the elastic deformation field, Laskin *et al.* use modal expansion similar in form to Eq. (3) but written for the transient axial displacement u_t rather than u_0 (u_1 in (3)). By doing so, they implicitly choose u_t as a generalized coordinate in their formulation. Laskin *et al.* explain this choice by saying that because u_t is small, its modal coordinates can also be regarded as small. By contrast, if one discretizes $u_0 = u_{qs} + u_t$, its modal expansion cannot adequately account for both large (u_{qs}) and small (u_t) components without having to include a large number of terms in the expansion.

Laskin *et al.* employ Kane's method to derive the motion equation of the beam. Thus, the next step in the formulation is to develop the generalized inertia and active forces. Starting with the latter, they consider two contributions—the elastic forces and the controller forces. In keeping with the subject of this chapter, discussion is limited to the derivation of the generalized *elastic* forces.

Let us recall that Kane *et al.*, and Hanagud and Sarkar derive the elastic force from the strain energy function. Laskin *et al.* take a different approach and express the elastic force $\mathbf{F}$ as a function of the stresses in the beam. This is actually a traditional way of writing the elastic force contribution to the equilibrium equations of the theory of elasticity. As pointed out,[12] one can express $\mathbf{F}$ in *linear* elasticity with:

$$\mathbf{F} = \boldsymbol{\nabla} \cdot \boldsymbol{\sigma} \tag{37}$$

where $\boldsymbol{\sigma}$ is referred to as the "spatial stress tensor," which we deduce to be the Cauchy stress. We note that this form is also valid in nonlinear theory, provided the divergence operator is defined with respect to the deformed coordinates and the strains are small. If $\boldsymbol{\nabla}$ is defined with respect to the undeformed coordinates, then Eq. (37) must be modified to take into accout the nonlinear nature of the deformation gradient. The resultant expresssion for $\mathbf{F}$ is given by Eq. (21) in Ref. 12 which we do not repeat here. In agreement with Kaza and Kvaternik's comments on the Newton's second law approach, we point out that this form allows for arbitrary rotations and implicitly assumes nonlinear strain-displacement relations.

Since the generalized elastic forces given by (22)-(25)[12] are expressed in terms of the generalized coordinates, Laskin *et al.* must take an intermediate step of substituting for the stresses in their (21) from stress-strain and then strain-displacement relations, in order to reformulate $\mathbf{F}$ in terms of u_{qs}, u_t, v and w. What is most notable about their approach is that, by using the "nonlinear" relation for $\mathbf{F}$ in terms of the stresses, they are able to derive the geometric stiffening component of elastic forces with a *linear* form of the strain-displacement relations. Laskin *et al.* emphasize that if one were to use strain energy to derive the elastic forces $\mathbf{F}$, one would have to retain nonlinear terms in the strain-displacement relations. To add to their interpretation of this "paradoxical situation", we offer this observation. The nonlinear terms in the strain-displacement relations give rise to third- and fourth-order terms in the strain energy. As was shown in §3.1, it is the third-order terms that are responsible for what is usually referred to as geometric stiffening. (In fact, the fourth-order terms also contribute to stiffening of the beam in bending.) Since the "nonlinear" formulation of **F** in terms of stresses is one-order higher then the "linear" one, it essentially provides a mechanism for incorporating only third-order terms in the strain energy.

We note that geometric stiffening is represented by G_{ij}- and a_{i0}- terms in Eqs. (23)-(25) of Ref. 12. Although these are linear in the discrete elastic coordinates, they are also dependent on the spatial derivative of the quasi-steady axial displacement u_{qs} (see the definitions of G_{ij} and a_{i0} on p. 515 of Ref. 12, with $v_0 = u_{qs}$).

The derivation of generalized inertia forces in Ref. 12 is similar to the procedure in Ref. 1 with the main difference being the addition of six rigid generalized inertia forces. These appear because the rigid-body motion of the beam is not prescribed, but is unknown. Accordingly, the final equations of motion include six equations for the position and orientation of the floating reference frame and the differential equations for the discrete elastic coordinates. Since Laskin *et al.* do not substitute for u_{qs} in terms of v and w as in Eq. (21) of this chapter, nor any other expression, their motion equations are explicitly dependent on u_{qs}, and therefore require this axial displacement as an input. Furthermore, u_{qs} appears in *both* generalized inertia and elastic forces.

Laskin *et al.* apply their dynamics equations to a number of special cases. They specify u_{qs} as the stretch that would occur if the beam were executing its nominal or intended motion. This stretch is defined as a solution of a linear second-order differential equation which we deduce from examples discussed in Ref. 12 to be:

$$EA\frac{\partial^2 u_{qs}}{\partial x^2} = -\frac{\partial P}{\partial x} = p \tag{38}$$

Here, P is nominal axial load on the beam and we have introduced the bymbol p to denote the axial load density. In the case of general *rotational* motions, Laskin *et al.* propose to approximate p by the axial component of the *centripetal* acceleration (multiplied by an appropriate inertia) of a point on the elastic axis of the beam. This yields, in accordance with Eq. (78) in Ref. 12:

$$p = \rho(x + u_{qs})\left(\omega_2^2 + \omega_3^2\right) \tag{39}$$

They suggest that the above should provide a good approximation for the steady-state axial load in the case of rotational motions at low angular acceleration rates. Moreover, it allows for a closed-form solution of (38) for u_{qs} which can then be used as input to their dynamics model.

4.2. Simo and Vu-Quoc

In Ref. 13, Simo and Vu-Quoc describe the application of the *geometrically nonlinear* theory to the dynamic analysis of flexible beams and plates. In contrast to the formulations presented in the previous works, with the exception of Ref. 1, they employ an *exact* kinematic description of beams and plates ("exact" within the plane sections remain plane assumption.) This involves using an exact form of the displacement field, and not the linearized form described in §3.1(ii) of this chapter. Accordingly, the rotation field is treated independently, as in Ref. 1. Another significant feature of the formulation developed by Simo and Vu-Quoc is the use of nonlinear strain measures that are invariant under superposed rigid-body motion. As demonstrated in their earlier work,[21] geometrically nonlinear theory allows one to formulate the dynamics equations of a flexible body directly in the inertial frame, rather than the traditionally employed floating frame.

To illustrate the procedure of Ref. 13, we once again consider as an example a planar beam rotating at a prescribed, but not necessarily constant angular speed, $\omega_3 = \Omega(t)$. In addition, we drop the meriodinal motion apriori, as in Ref. 13. Then, the position of a generic point in the beam expressed in the floating frame is given by:

$$\mathbf{r} + \mathbf{u} = \begin{bmatrix} x + u_0 - y \sin\alpha \\ v_0 + y \cos\alpha \end{bmatrix} \tag{40}$$

where α is the angle between the x-axis of the floating reference frame and the local normal to the beam cross-section and is equivalent to θ_3 of §2.1. Note that the linearized description of the beam kinematics is obtained by assuming that $\alpha << 1$. This, coupled with the additional assumption of negligible shear deformation, that is $\alpha \approx \frac{\partial v_0}{\partial x}$, yields the kinematic description of the beam employed in Refs. 6–12.

The kinetic energy expression that emanates from the position field (40) is:

$$T = \tfrac{1}{2}\int_0^L \rho\left[(\dot{u}_0 - \Omega v)^2 + (\dot{v}_0 + \Omega(x + u_0))^2\right] dx + \tfrac{1}{2}\int_0^L (\dot{\alpha} + \Omega)\,\frac{\rho I_z}{A}\,dx \tag{41}$$

which is the analogue of (22). The aforementioned nonlinear strain measures for a beam are defined in Eq. (6) of Ref. 13 which we present here as:

$$\begin{bmatrix} \Gamma_1 \\ \Gamma_2 \\ \kappa \end{bmatrix} = \begin{bmatrix} \cos\alpha(1 + \frac{\partial u_0}{\partial x}) + \sin\alpha\frac{\partial v_0}{\partial x} - 1 \\ -\sin\alpha(1 + \frac{\partial u_0}{\partial x}) + \cos\alpha\frac{\partial v_0}{\partial x} \\ \frac{\partial\alpha}{\partial x} \end{bmatrix} \tag{42}$$

The corresponding strain energy is written as:

$$U = \frac{1}{2}\int_0^L \left(EA\Gamma_1^2 + GA_s\Gamma_2^2 + EI_z\kappa^2\right) dS \tag{43}$$

where S is the coordinate along the underformed line of centroids of the beam.[22] The second term in the integrand of (43) represents the strain energy due to shearing stresses.

With expressions for the kinetic and strain energies as given by Eqs. (41) and (43), Simo and Vu-Quoc use Hamilton's principle to construct the motion equations of a beam. The successive approximations of these equations are derived by approximating the fully nonlinear strain measures of (42). Thus, Simo and Vu-Quoc define the first-order and second-order strains by using the power series expansion in terms of a small parameter. Based on these and the exact definition of the strain measures, they present the corresponding linear, third-order and fully nonlinear equations of motion for the beam. The stiffness term in these equations is expressed in terms of the internal (elastic) forces.

Of particular interest here are the motion equations derived using the second-order approximation for the strain measures, given by:

$$\begin{bmatrix} \overline{\overline{\Gamma}}_1 \\ \overline{\overline{\Gamma}}_2 \\ \overline{\overline{\kappa}} \end{bmatrix} = \begin{bmatrix} \frac{\partial u_0}{\partial x} + \alpha \frac{\partial v_0}{\partial x} - \frac{1}{2}\alpha^2 \\ \frac{\partial v_0}{\partial x} - \alpha(1 + \frac{\partial u_0}{\partial x}) \\ \frac{\partial \alpha}{\partial x} \end{bmatrix} \tag{44}$$

We observe that for the case of negligible shear, $\overline{\overline{\Gamma}}_1$ reduces to:

$$\overline{\overline{\Gamma}}_1 \approx \frac{\partial u_0}{\partial x} + \frac{1}{2}\left(\frac{\partial v_0}{\partial x}\right)^2 \tag{45}$$

which coincides with the approximate Lagrangean strain (for the planar case) of Eq. (15). The strain energy derived with the above $\overline{\overline{\Gamma}}_1$, $\overline{\overline{\kappa}} \approx \frac{\partial^2 v_0}{\partial x^2}$ and neglecting contribution from the shearing stresses is of the same form as the strain energy expression in Hanagud and Sarkar.[6] Therefore, the third-order motion equations derived in the two references are equivalent.

Finally, Simo and Vu-Quoc develop what they refer to as the consistent linear motion equations. This is carried out by linearizing the third-order equations of motion with the axial internal force approximated by its zeroth-order steady-state value. For a rotating beam, this force is equal to the centrifugal load given in Eq. (27) of this chapter. When extending the procedure to the case of a rotating plate, they generalize this approximation by evaluating the axial load from the lowest-order component of the in-plane inertia force. For a plate undergoing general rotation at $\boldsymbol{\omega}(t)$, the latter is given by:

$$\mathbf{F}_{inertia} \approx \rho\left[\dot{\boldsymbol{\omega}} \times \mathbf{r}_0 + \boldsymbol{\omega} \times (\boldsymbol{\omega} \times \mathbf{r}_0)\right] \tag{46}$$

where ρ is taken as the mass per unit of undeformed area of the plate, and $\mathbf{r}_0$ denotes the position of a point on the mid-surface. We observe that for a beam rotating at $\boldsymbol{\omega}(t)$, this procedure yields the following equation for the axial internal force P:

$$\frac{\partial P}{\partial x} = -p = -\rho x(\omega_2^2 + \omega_3^2) \tag{47}$$

Comparing the above with expression (39) proposed by Laskin *et al.* to approximate p for a beam in general rotation, we observe that the two expressions give identical values to zeroth order. Moreover, we conclude that the value of Laskin *et al.* applies to motions at *any* angular acceleration rates since the first term in Eq. (46) does not contribute to the axial internal force in the beam.

4.3. Meirovitch

In his 1967 book, Meirovitch[14] includes a section on the effect of axial forces in the bending vibration of a bar, which as he states, cannot be ignored in some cases. In this section, Meirovitch derives an equation of motion for the transverse displacement of the beam by means of extended Hamilton's principle. Thus, expressions for kinetic energy T and work function W are developed. The former takes the simple form used in planar bending vibration problems without the axial force. Rewritten in our notation, T specified in Ref. 14 is:

$$T = \tfrac{1}{2} \int_0^L \rho(x) \left(\frac{\partial v(x,t)}{\partial t} \right)^2 dx \tag{48}$$

In evaluating the work function, Meirovitch proposes to include the effect of the bending moment, the transverse (external) load and axial force. The first two are formulated in the same way as for the case without the axial force. To determine the "axial" work, the change in the horizontal projection of an element $d\hat{s}$ is calculated. This differential of the foreshortening is expressed with:

$$d\hat{s} - dx = \sqrt{1 + \left(\frac{\partial v}{\partial x} \right)^2} dx - dx \approx \tfrac{1}{2} \left(\frac{\partial v}{\partial x} \right)^2 dx \tag{49}$$

where the approximation results from retaining two terms in the binomial expansion of $\sqrt{1 + \left(\frac{\partial v}{\partial x}\right)^2}$. Then, the work done by the axial force is:

$$W_P = -\tfrac{1}{2} \int_0^L P(x,t) \left(\frac{\partial v}{\partial x} \right)^2 dx \tag{50}$$

It is worth pointing out that in adopting the above formulation, Meirovitch makes a tacit assumption that the axial force is given as a known function of x and t.

In the more recent publication,[15] Meirovitch derives an exact set of motion equations for an unconstrained *flexible body*[b] in general motion. The stiffness term in these equations is represented with the homogeneous stiffness differential operators which "include all the terms that can be accounted for in a strain energy function."[15] The motion equations are written in terms of the rigid-body quasi-coordinates and the continuous elastic coordinates u, v and w. Meirovitch illustrates their application to a system made up of a rigid hub and a beam-like flexible appendage in general translation and rotation. He begins by assuming that the axial displacement can be

[b] All previous works deal with flexible beams only.

ignored and therefore sets $u = 0$ apriori. The kinetic energy is derived in the standard manner and contains terms that are of second degree in the elastic variables. The strain energy includes the standard second-order contributions due to bending in two directions as well as the contribution due to "shortening of the projection." The latter is expressed as

$$U_P = \frac{1}{2}\int_0^L \left[\int_x^L p(\zeta,t)\,d\zeta\right]\left[\left(\frac{\partial v}{\partial x}\right)^2 + \left(\frac{\partial w}{\partial x}\right)^2\right] dx \tag{51}$$

where $p(x,t)$ is the axial component of the internal force density. The above can be directly compared to (50).

Meirovitch proposes to determine the force density p from the motion equation for the axial displacement u, which as we had mentioned is excluded from the dynamics model of the beam. Thus, he defines p as a sum of the terms in the u-equation, omitting terms that involve elastic displacements, as well as the control force density. The resultant expression for p is given by Eq. (29) in Ref. 15, which we rewrite here as:

$$p = \rho\left[-\dot{\mathrm{v}}_1 - \omega_2\mathrm{v}_3 + \omega_3\mathrm{v}_2 + x\left(\omega_2^2 + \omega_3^2\right)\right] \tag{52}$$

with $\mathrm{v}_i, i = 1,2,3$ as components of the absolute velocity of the hub in the floating frame.

Let us compare expression (52) for the internal axial force density, which is employed by Meirovitch to evaluate the strain energy due to the shortening of the beam with the corresponding expression of Laskin *et al.*, which they use to determine the quasi-steady component of the axial displacement. We recall that the expression for p proposed in Ref. 12, as given by Eq. (39) of this chapter, originates from inertial acceleration of a mass element on the beam's elastic axis. In fact, it is defined as the centripetal component of the acceleration evaluated with $u = u_{qs}$. It can be shown that Meirovitch's expression for p can also be derived from the inertial acceleration distribution, but neglecting the elastic contributions. Indeed, this is precisely the method used by Simo and Vu-Quoc to evaluate the axial internal force. This is not surprising since an equation of motion in the absence of external forces and the stiffness term is essentially a linear homogeneous relation for the inertial acceleration. Thus, the "centripetal component" of p, given by $\rho x\left(\omega_2^2 + \omega_3^2\right)$ in Eq. (52), is identical to p of Eq. (39) if one drops u_{qs} in the latter. By comparing (52) to (39), we also observe that Laskin *et al.*'s approximation for the axial force density should apply to a beam in *general* motion if the translational velocity and acceleration are small relative to the angular velocity. Finally, we note that in the case of pure rotation, Eq. (52) reduces to Simo and Vu-Quoc's axial force density from (47), as expected.

4.4. Banerjee et al.

Banerjee and Dickens[4] present a formulation for the dynamics of an arbitrary flexible body in large rigid-body motion. The formulation is based on the standard description of the deformation field where the elastic translations u, v and w are discretized by means of a modal expansion. The equations of motion are derived by employing the same general methodology as in Refs. 1 and 6. Therefore, as was done in reviewing other works, the following analysis focuses on the procedure to derive the stiffening terms only.

Banerjee and Dickens introduce the notion of "motion stiffness" as a special case of the geometric stiffness or "initial stress stiffness"[16] caused by the inertia loading on the body due to its large rigid motion. As is noted by Banerjee and Lemak,[16] this motion-induced stiffness has its origin in the strain energy term,

$$U_{NL} = \int_V \boldsymbol{\varepsilon}_{NL}{}^T \boldsymbol{\sigma}_0 \, dV \tag{53}$$

where $\boldsymbol{\sigma}_0$ and $\boldsymbol{\varepsilon}_{NL}$ are 6x1 columns. The former is defined in Ref. 16 as the "reference" state of stress in the body, while $\boldsymbol{\varepsilon}_{NL}$, collects the nonlinear (second-order) components of the Lagrangian strains. Thus, one can view U_{NL} as the "nonlinear" part of U, since it arises from the nonlinear terms in the strain-displacement relations.

The key to the method proposed Ref. 4 is the observation that the stress state of the deformable body undergoing large translations and rotations results from the inertia loading on it. With that, Banerjee and Dickens determine the distributions of the inertia force and torque in the body from the inertial linear and angular accelerations, respectively, in accordance with Newton's Second Law and Euler equations. In applying this procedure, they omit elastic terms in the velocity and acceleration distributions. Therefore, the resultant inertia loads are expressed as functions of rigid velocities and accelerations.

In the next step of their development, Banerjee and Dickens rewrite the inertia load as a product of a matrix and a column vector. For the inertia *force*, the matrix is 3x12 and is dependent on the spatial coordinates, x, y and z. The column vector contains 12 parameters A_i, $i = 1, \ldots, 12$, which are dependent strictly on the velocities and accelerations of the body. The inertia *torque* is factorized into a constant 3x9 matrix and a time-dependent 9x1 column vector of A_i, $i = 13, \ldots, 21$, which are also functions of rigid velocities and accelerations. Although not stated in Ref. 4, the motivation for this "matrix-vector factorization" is to separate the time-dependent component of the inertia forces from the space-dependent or constant part. By doing so, Banerjee and Dickens are able to construct the geometric stiffness term in the motion equations in two stages. The first stage produces 21 geometric stiffness matrices denoted by $S^{(i)}$.[4] These can be assembled with the standard finite-element procedure from the constant or space-dependent matrix factors of the inertia loads, *prior* to evaluation of the motion equations. The second stage involves calculating the geometric stiffness term from $S^{(i)}$, A_i and the discrete elastic coordinates.

We will comment on the relationship between the approach of Banerjee and Dickens and the other methods. First, it is observed that the "nonlinear" strain energy of Eq. (53) accounts for the same geometric nonlinearity as that treated in the previous works in the context of a flexible beam. In fact, the first component of $\boldsymbol{\varepsilon}_{NL}$ comprises the second-order terms in equation (18) of this chapter. Moreover, U_{NL} can be rewritten in terms of displacement variables if one expresses stresses in terms of strains and then substitutes for strains in terms of displacements. Following this procedure, one will obtain U_{NL} in the form of third- and fourth-order terms in the displacement gradients. These were mentioned in our discussion of the works by Hanagud and Sarkar and Likins *et al.*

It can be shown that Banerjee and Dickens' expression for the *axial* inertia force is equivalent to Meirovitch's axial component of the internal force density, which is employed to derive the motion equations for a beam on a rigid hub. The latter is given by our Eq. (52) and the former can be obtained from Eqs. (28) and (29) of Ref. 4. Substituting for $x_1 = x$, $x_2 = y = 0$, $x_3 = z = 0$ and $u_i = \mathrm{v}_i$, $u_{i+3} = \omega_i$ for $i = 1, \ldots, 3$, the axial inertia force $f_1^\star$ of Banerjee and Dickens takes the form:

$$\begin{aligned} f_1^\star &= -\left[A_1 + xA_4\right] dm \\ &= -\left[\dot{\mathrm{v}}_1 + \omega_2 \mathrm{v}_3 - \omega_3 \mathrm{v}_2 - x\left(\omega_2^2 + \omega_3^2\right)\right] dm \end{aligned} \tag{54}$$

Since Meirovitch's internal force density must be interpreted as the internal force per unit length, the equivalence between (54) and (52) follows if one substitutes for dm with $\rho\, dx$. Then, we have

$$f_1^\star = p\, dx \tag{55}$$

where p is as defined by Meirovitch (Eq. (52)). Noting that the axial stress in the beam is

$$\sigma_{xx}(x) = \frac{1}{A}\int_x^L p\, dx \tag{56}$$

and if the nonlinear part of the axial strain (18) is simplified to

$$\varepsilon_{NL,xx} = \tfrac{1}{2}\left[\left(\frac{\partial v}{\partial x}\right)^2 + \left(\frac{\partial w}{\partial x}\right)^2\right], \tag{57}$$

Eq. (53) applied to the beam reduces to:

$$\begin{aligned} U_{NL} &= \int_V \varepsilon_{NL,xx}\sigma_{xx}\, dV \\ &= \tfrac{1}{2}\int_0^L \int_x^L p\, d\zeta \left[\left(\frac{\partial v}{\partial x}\right)^2 + \left(\frac{\partial w}{\partial x}\right)^2\right] dx \end{aligned} \tag{58}$$

The above, which we derived from U_{NL} of Ref. 4 is identical to Meirovitch's strain energy due to shortening of the projection. In this light we propose to interpret Meirovitch's formulation of the strain energy due to shortening of an elastic beam as a particular application of Eq. (53) from Banerjee and Dickens' formulation.

To conclude this section, we would like to draw attention to a fundamental approximation made in the development of Banerjee and Dickens. It is that the stresses which contribute to stiffening of the body are only those that are caused by the inertia loading. We believe that the stress in (53) must represent the *complete* stress state in the body which arises from the total loads—inertia and external. Indeed, only then can Eq. (53) be consistent with the general formulation of the strain energy. The approximation for $\boldsymbol{\sigma}_0$ suggested in Ref. 4 is certainly valid in the absence of external forces on the body. However, it may not be appropriate for multibody systems in which each body is acted upon by the "external" (to it) interbody forces, even in the absence of external forces on the whole system. Indeed, this is confirmed by Banerjee in Ref. 23, where he extends the formulation to include stiffening due to interbody forces.

4.5. Wallrapp and Schwertassek

In the introduction to their article,[17] Wallrapp and Schwertassek note that in Ref. 1 "the stresses in a flexible body are tacitly assumed to be zero when the body is in an 'undeformed' or reference configuration." They refer to this as a paradox because in this case "the variable external loads upon a flexible body are balanced by internal forces or stresses, which are zero." The loads alluded to are the inertial forces arising due to the "rigid-body" motion. Motivated by this deduction, Wallrapp and Schwertassek develop a formulation where they explicitly account for the reference stresses in the body. In the following, we briefly outline the key steps in their procedure, which is presented under *Approach C* in Ref. 17.

Wallrapp and Schwertassek derive the motion equations for a *general* flexible body by using the virtual power principle. In formulating this principle they account for the inertial, gravitational, applied surface forces, the constraint forces at the joints and the internal forces. The latter are included with the term of the form:

$$\delta\dot{U} = \int_V \delta\dot{\boldsymbol{\varepsilon}}^T \boldsymbol{\tau}\, dV \tag{59}$$

where $\boldsymbol{\varepsilon}$ contains the components of the Lagrangian strain tensor and $\boldsymbol{\tau}$ collects the components of the second Piola-Kirchhoff stress tensor. Noting that to lowest order, the second Piola-Kirchhoff stresses are the same as the Cauchy stresses, we will use our previous notation for stress, $\boldsymbol{\sigma}$, which also coincides with the notation in Ref. 17.

In §4.4, we have implicitly factored $\boldsymbol{\varepsilon}$ into a linear and nonlinear parts, $\boldsymbol{\varepsilon}_L$ and $\boldsymbol{\varepsilon}_{NL}$. For convenience, let us express the linear strain as:

$$\boldsymbol{\varepsilon}_L = \mathbf{L}^0\mathbf{u} \tag{60}$$

where following the notation employed in Ref. 17, $\mathbf{L}^0$ denotes a constant operator matrix. Differentiating the exact Lagrangian strains gives:

$$\dot{\boldsymbol{\varepsilon}} = \left(\mathbf{L}^0 + \mathbf{L}^1\right)\dot{\mathbf{u}} \tag{61}$$

where $\mathbf{L}^1$ is an operator matrix, the elements of which are linear in the displacements. The product $\mathbf{L}^1\dot{\mathbf{u}}$ gives the nonlinear contribution to $\dot{\boldsymbol{\varepsilon}}$ which of course originates from the nonlinear strains $\boldsymbol{\varepsilon}_{NL}$. As alluded to in the beginning of this section, the state of stress used by Wallrapp and Schwertassek to evaluate the virtual power in (59) is partitioned explicitly into the reference stress and the stress due to the deformation, the latter being a linear function of strain. Thus, Wallrapp and Schwertassek write

$$\boldsymbol{\sigma} = \boldsymbol{\sigma}^{\mathrm{r}} + \mathbf{H}\boldsymbol{\varepsilon} \tag{62}$$

where $\boldsymbol{\sigma}^{\mathrm{r}}$ is their reference stress and is analogous to $\boldsymbol{\sigma}_0$ of Banerjee and Lemak.[16] We note that Eq. (62) corresponds to the standard method for incorporating the initial stress, in this case the reference stress, into the model. To derive the stiffness term, however, Wallrapp and Schwertassek use an approximation for $\boldsymbol{\sigma}$ based on the linearized strain, that is:

$$\boldsymbol{\sigma} = \boldsymbol{\sigma}^{\mathrm{r}} + \mathbf{H}\mathbf{L}^0\mathbf{u} \tag{63}$$

This approximation is motivated by the small displacement assumption. Thus, the stiffness term in the motion equations is obtained by formulating the virtual power of the internal forces defined in (59) with the virtual strain from (61) and the stress specified by (63). With the displacement field discretized in the conventional manner, as in Eq. (4) of this paper, the resulting virtual power takes the following form:

$$\begin{aligned} \delta\dot{U} &= \delta\dot{\mathbf{q}}^T \left\{ \int_V \left(\mathbf{L}^0\boldsymbol{\Phi}\right)^T \mathbf{H} \left(\mathbf{L}^0\boldsymbol{\Phi}\right) \, dV \, \mathbf{q} + \int_V \left[\left(\mathbf{L}^0\boldsymbol{\Phi}\right)^T \boldsymbol{\sigma}^{\mathrm{r}} + \left(\mathbf{L}^1\boldsymbol{\Phi}\right)^T \boldsymbol{\sigma}^{\mathrm{r}} \right] dV \right. \\ &\quad \left. + \int_V \left(\mathbf{L}^1\boldsymbol{\Phi}\right)^T \mathbf{H} \left(\mathbf{L}^0\boldsymbol{\Phi}\right) \, dV \, \mathbf{q} \right\} \end{aligned} \tag{64}$$

We immediately note that the last term in $\delta\dot{U}$ is of second order in $\mathbf{q}$ and is omitted in the linearized motion equations presented in Ref. 17. The first term represents the standard linear stiffness of the body. Wallrapp and Schwertassek point out the two new terms that result from the nonzero reference stress—these are collected in the second integral of (64). They propose to compute the reference stress from the motion equations obtained by neglecting all terms which involve the elastic coordinates and their derivatives. This procedure is similar in concept to that employed by Banerjee and Dickens[4] to determine the reference stress $\boldsymbol{\sigma}_0$ in (53) and the method of Meirovitch[15] to compute the internal force density. However, unlike Ref. 4, Wallrapp and Schwertassek do not restrict their $\boldsymbol{\sigma}^{\mathrm{r}}$ to be due to nominal inertia forces only, but include in its definition the terms in the motion equations due to external forces. Recall that in their formulation, Wallrapp and Schwertassek account for gravitational, applied surface forces, and the constraint forces at the joints. In this light, we suggest that the technique for evaluating the reference stress for computing the stiffening terms in Ref. 17 corrects the deficiency of the approximation for the reference stress employed by Banerjee and Dickens.[4]

Finally, we would like to point out an even more significant difference between the formulations of Refs. 4 and 17. As noted, Wallrapp and Schwertassek derive *two* additional terms which involve the reference stress $\boldsymbol{\sigma}^{\mathrm{r}}$. We repeat these below, using their designations for the two terms. They are:

$$\mathbf{k}_\mathbf{q}^\mathrm{r} = \int_V \left(\mathbf{L}^0\boldsymbol{\Phi}\right)^T \boldsymbol{\sigma}^\mathrm{r}\, dV \tag{65}$$

$$\mathbf{N}_\mathbf{q}\mathbf{q} = \int_V \left(\mathbf{L}^1\boldsymbol{\Phi}\right)^T \boldsymbol{\sigma}^\mathrm{r}\, dV \tag{66}$$

It can be shown that the second of the above terms, which is linear in $\mathbf{q}$, is identical to the stiffening term derived by Banerjee and Dickens from Eq. (53). Indeed, Wallrapp and Schwertassek suggest that this term be evaluated using the two-stage procedure proposed in Ref. 4, which we briefly outlined in §4.4 of this chapter. However, the term given in Eq. (65) is not included in the formulation of Ref. 4. This naturally raises the question as to the correct formulation. In the opinion of the author, the notion of treating the reference stress as an initial stress that exists in the undeformed configuration of the body, as is done in Eqs. (62) and (63), is incorrect. The inertial and external forces are all included in the motion equations of the body and they produce the internal stresses and the corresponding strains in it. These are accounted for *via* the stiffness term in the dynamics equations. Hence, the complete state of stress in an elastic body undergoing large motions can be directly expressed as a function of strains, and subsequently deformations, as these result from its motion and application of external forces. The separation of stress into a reference stress and the stress due to linear strains, proposed in Ref. 17, gives rise to an extra term in the motion equations. This term is not present in any of the previous formulations reviewed in this chapter.

5. Discussion

Based on our review, we suggest that to account for the geometric stiffening of an elastic body undergoing large motions, it is necessary to introduce the geometric nonlinearity into the model. As demonstrated in the previously considered references, this must be done through the use of Lagrangian strains or stated differently, the nonlinear strain-displacement relations. These are used, explicitly or implicitly in all the works considered in this chapter. We have seen, however, that there are different approaches to incorporate the geometric stiffening into the dynamics equations. To distinguish between them, we propose a classification according to the following two criteria: (i) kinematic description of the deformation field (criterion DF); (ii) the formulation of the strain energy function (criterion SE). We feel that these represent two most general criteria that can fully characterize a particular approach. The first one is related to the assumed displacement field, which in turn determines the generalized coordinates employed in deriving the motion equations. The second criterion defines the form of the elastic force term in the motion equations.

According to each criterion, we identify the following possible cases.

Criterion DF. It is suggested that the deformation field can be described by *any* of the following sets of variables:

DF-1: Three independent elastic translations $\{u, v, w\}$. This description is employed by Likins *et al.*, Hanagud and Sarkar, Simo and Vu-Quoc, Meirovitch, Banerjee *et al.* and Wallrapp and Schwertassek.

DF-2: Three independent elastic translations $\{u_s, v, w\}$, where recall u_s is the axial displacement which results strictly from the stretching of the deformable body. This description is used by Vigneron, Kaza and Kvaternik, Lips and Modi, and Laskin *et al.*

DF-3: Three independent but non-orthogonal elastic deformations $\{s, v, w\}$ where s denotes the stretch. These coordinates are employed by Kane *et al.*

It must be observed that classifications DF-2 and DF-3 as stated above have been used for and apply to "beam-like" bodies only. The notions of "axial" displacement and "stretch" require appropriate generalizations for other types of elastic bodies.

Criterion SE. We distinguish two basic formulations of the strain energy:

SE-1: The total strain energy and the corresponding stiffness term in the motion equations are formulated as a function of strains (or displacements) only, where strains are defined with the *nonlinear* strain-displacement relations. This formulation is employed by Likins *et al.*, Vigneron, Kaza and Kvaternik, Lips and Modi and Hanagud and Sarkar as well as, Hughes and Fung, and Kane *et al.* It can be viewed as a *displacement* formulation.

SE-2: The strain energy is subdivided into a "standard linear" contribution and a "nonlinear" part, where the latter is constructed from the stresses or internal forces in the body and the nonlinear components of the Lagrangian strains. This formulation is used by Meirovitch, Banerjee *et al.* and Wallrapp and Schwertassek and can be considered as a *force* formulation.

Note that Laskin *et al.* do not compute the strain energy, but derive the elastic forces directly from the stress state. However, since they eventually reformulate these in terms of strains, their method is fundamentally a displacement formulation. Simo and Vu-Quoc write down their strain energy as a function of strains, but express the stiffness term in the motion equations in terms of internal forces. Hence, we classify their approach as a force formulation. We also point out that Wallrapp and Schwertassek do not use the strain energy directly, but rather the variation of its rate.

Below, we comment on the three options for the kinematic description of the deformation field.

DF-1. The set $\{u, v, w\}$ is the standard set of elastic displacements used in both linear and nonlinear theories of elasticity as well as, structural analysis. It requires no apriori decisions on the "foreshortening" of the body and in that sense is general.

Since u, v and w represent deformations along three orthogonal axes, they clearly represent *independent* degrees of freedom. However, contrary to what has been previously stated, that does not preclude the fact that in case of a beam, part of u, in particular the foreshortening component, is dependent on v and w. The total axial displacement will remain an independent variable as long as it includes the axial displacement u_s. Furthermore, it is not necessary to either separate u into these two components, nor to explicitly account for the coupling between the axial and transverse displacements. In addition, we believe that discretizing u with a standard expansion does not imply "premature linearization." To support this point, let us consider the foreshortening part of u, which for the present purpose we simplify to:

$$u_f = \frac{1}{2} \int_0^x \left(\frac{\partial v}{\partial x} \right)^2 dx \tag{67}$$

If the transverse deformation v is expanded as in (3), then upon substitution, u_f becomes:

$$u_f = \frac{1}{2} \sum_{j=1}^{\nu} \sum_{k=1}^{\nu} \left[\int_0^x \left(\frac{\partial \phi_{2j}(x)}{\partial x} \right) \left(\frac{\partial \phi_{2k}(x)}{\partial x} \right) dx \right] q_j(t)\, q_k(t) \tag{68}$$

The above clearly represents a summation of terms $\phi_{f,i}(x)\, q_{f,i}(t)$ where $q_{f,i}(t) = q_j(t)\, q_k(t)$ and the space-dependent basis functions are defined accordingly. This summation is of the same form as the standard expansion for u.

As exemplified by the formulations of Likins *et al.* and Hanagud and Sarkar, geometric stiffening *can* be modelled with this option, provided one incorporates the nonlinear strain-displacement relations in the evaluation of strain energy and, hence, the elastic forces.

DF-2. In this case, the foreshortening component of u is explicitly separated from u_s in the axial displacement field. It may be specified in terms of v and w or left as a parameter to be defined by the user. Either case, however, involves making an approximation for u_f, although in the first option it is "known" and quantified prior to deriving the motion equations. Furthermore, as alluded to earlier, the concept of "foreshortening" is only meaningful for long flexible bodies.

DF-3. Employment of the stretch variable is unconventional, and certainly is not a common choice, if made at all. The main advantage of using this variable instead of u is that the strain energy retains its ("linear") quadratic form. However, the resulting expression for kinetic energy, and as a consequence the generalized inertia forces is more complicated than that based on DF-1 and DF-2 descriptions.

As was demonstrated in the previous sections (at least we hope it was), it is a particular combination of the coordinates and the strain energy formulation that determines how the geometric stiffening is incorporated into the motion equations as well as, what form it appears in. Thus, we will now comment on the two strain energy formulations taken in combination with the different DF options and point out some of the advantages and/or disadvantages of the resulting approaches.

The main advantage of the SE-1 formulation of the strain energy for any description of the deformation field is that one is not required to make any additional assumptions or approximations. In particular, one does not have to ponder on what is the state of stress in the body and how to compute it. Furthermore, since the motion equations are traditionally formulated in terms of displacements—the use of complementary (force) formulations in dynamics is very scarce—the SE-1 approach fits naturally into the framework of dynamics simulation.

SE-1/DF-1. In this approach, taken by Hanagud and Sarkar, the stiffening term appears in the motion equations through the strain energy and is a nonlinear function of elastic coordinates. In particular, it can be factored into an (elastic)coordinate-dependent second-order stiffness matrix and a column vector of elastic coordinates.

As pointed out by Banerjee and Lemak, evaluation of this stiffness term requires that the stiffness matrix be updated at each time step in the simulation, which may be computationally costly. However, the update does not involve iterations, but is a simple functional evaluation. It should also be pointed out that for a flexible beam, this approach requires that the axial elastic equation be included in the dynamics model. As demonstrated by our numerical results in the following section, this has an adverse effect on the computational efficiency of the simulation.

Finally, we note that the present method corresponds to the third approach identified by Kaza and Kvaternik and as they comment is "the one usually employed for a general three-dimensional rotating body." However, contrary to Kaza and Kvaternik's conclusion, we believe that it does *not* require special consideration, but is the most general. Moreover, this approach can be employed to extend the existing "small" deflection multibody dynamics formulations to incorporate "large" deflection theories.

SE-1/DF-2. With this procedure, applicable to beam-like bodies only, the foreshortening term is always present in the kinetic energy expression. It may or may not appear in the strain energy, depending on the additional approximations made. In the simplest case, as in Vigneron's formulation, the resultant stiffening term takes a linear form, and its discrete form (not available in any of the publications reviewed in this chapter) is relatively simple. If one wishes to retain higher-order terms, as in Ref. 9, then the discrete form of the motion equations becomes fairly complex.

From the computational standpoint, this method of determining the stiffening term allows one the option of dropping the axial deformation from the model—

not a trivial advantage, as it can considerably improve the efficiency of the simulation.

SE-1/DF-3. This approach produces a linear geometric stiffening term *via* kinetic energy (or generalized inertia force). Compared to the SE-1/DF-1 option, it can be just as accurate, but not as general since the stretch variable can be defined only for a particular type of elastic bodies.

As was shown in §4.4 and §4.5, the SE-2 formulation of the strain energy requires an approximation for the stress field in the body, and hence, in contrast to SE-1, is inherently approximate. Among the works presented in this chapter, this formulation has only been used with the DF-1 description of the deformation field. In this case, the geometric stiffening term results from the "nonlinear" strain energy, as in Refs. 4, 15, 16 and 17. With appropriate assumptions, it is linear in elastic variables, but involves rigid-body accelerations and velocities. As for its efficiency, the algorithm presented in Ref. 16 requires an update of the geometric stiffness matrix (see Eq. (37) in Ref. 4), because of the time-dependent quantities A_i. Moreover, the final motion equations no longer have a symmetric coefficient matrix (because of acceleration dependency of the stiffening term), thereby making evaluation of accelerations computationally more costly.

To summarize, we believe that a description of the deformation field in terms of u, v and w, combined with the displacement formulation of the strain energy, is the most accurate and general approach. It does not require an approximation of foreshortening, nor the stress state of the body—two critical advantages for applications to general multibody systems. These approximations, required in other approaches, introduce additional error which is not easily quantified for multibody systems. By contrast, the only source of error in the SE-1/DF-1 formulation is the approximation, if any, made in the Lagrangian strains.

A definitive statement on the efficiency of the different procedures can only be made through implementation of all methods. Moreover, we would expect their relative efficiency to vary depending on the complexity of the system and the number of elastic degrees of freedom in the model. The following section contains some results towards identifying the numerical performance of the four formulations described above.

6. Numerical Comparison

This section presents a series of numerical results for a benchmark problem—a single flexible beam undergoing a spin-up maneuver. This system has been considered in a number of previous works dealing with the subject of the paper. However, there is not a single beam example for which all four approaches identified in §5 can be compared from the published results. Therefore, the following procedure is adopted towards that goal. First, a detailed numerical comparison of two of the aforementioned approaches, SE-1/DF-1 and SE-1/DF-2 is made. Both of these have been implemented by the author which allows us to comment on the more subtle issues, such as ease of implementation and convergence of the discretization scheme. As well, in implementing the SE-1/DF-2 approach, inertial terms up to third-order are included in the motion equations and their significance is assessed. This in itself represents a novel contribution. The second part of the comparison makes use of a different flexible-beam system, one for which results for the SE-1/DF-3 and SE-2/DF-1 formulations are available in literature. A comparison of responses predicted by all four approaches is then made for this beam with the results generated by the author and those cited from the literature.

6.1. SE-1/DF-1 vs. SE-1/DF-2

Equations of Motion and Discretization

The continuous dynamics equations for a flexible beam rotating about the z-axis at a prescribed speed $\Omega(t)$ are presented below. The equations obtained with both formulations are based on the approximate nonlinear strain:

$$\varepsilon_{xx} \approx \frac{\partial u}{\partial x} + \tfrac{1}{2}\left[\left(\frac{\partial v}{\partial x}\right)^2 + \left(\frac{\partial w}{\partial x}\right)^2\right] \tag{69}$$

With the exception of the rotary inertia terms, all of the terms resulting from the respective kinematics descriptions (Eqs. (19) and (20)) are retained.

SE-1/DF-1 dynamics equations:

$$\begin{aligned}
&\rho\left[\ddot{u}_0 - \Omega^2 x - \left(\Omega^2 u_0 + 2\Omega\dot{v} + v\dot{\Omega}\right)\right] \quad + EAu_0'' - EAv'v'' &&= 0\\
&\rho\left[\ddot{v} + \dot{\Omega}x - \left(\Omega^2 v - 2\Omega\dot{u}_0 - \dot{\Omega}u_0\right)\right] \quad + EI_z v'''' - EA\left(v'u_0'' + u_0'v'' + \tfrac{3}{2}(v')^2 v''\right) &&= 0
\end{aligned} \tag{70}$$

SE-1/DF-2 dynamics equations:

$$\begin{aligned}
&\rho\left[\ddot{u}_s - \ddot{u}_f - \Omega^2 x - \left(\Omega^2 u_s + 2\Omega\dot{v} + v\dot{\Omega}\right) + \underline{\Omega^2 u_f}\right] && + AEu_s'' &&= 0\\
&\rho\left[\ddot{v} + \dot{\Omega}x - \left(\Omega^2 v - 2\Omega\dot{u}_s - \dot{\Omega}u_s\right) - \underline{\left(2\Omega\dot{u}_f - \dot{\Omega}u_f\right)} - (Tv')'\right] && + EI_z v'''' &&= 0
\end{aligned} \tag{71}$$

where

$$T = \int_x^L T_I\, dx, \qquad T_I = -\ddot{u}_s + \underline{\underline{\ddot{u}_f}} + \left[\Omega^2 x\right] + \left(\Omega^2 u_s + 2\Omega\dot{v} + v\dot{\Omega}\right) - \underline{\underline{\Omega^2 u_f}} \tag{72}$$

The $()'$ notation in the above denotes partial differentiation $\frac{\partial}{\partial x}$; different terms have been grouped and listed according to their order. The last terms in Eqs. (70) are of second- and third-order and are derived from the third- and fourth-order terms in the strain energy. The markings of the terms in Eqs. (71) and (72) refer to their counterparts in the kinetic energy of (28), where recall the term in the square brackets is responsible for the stiffening of the beam. It should be noted that Eqs. (71) are closest to the dynamics equations derived in Ref. 9. They differ by the third-order inertial terms (underlined with double lines) which, as noted in §3, are omitted in that reference. The second difference is that the equations here do not contain terms of third-order in the spatial derivatives of the elastic displacements. This is a consequence of the approximation adopted for the nonlinear strain. Lastly, the present equations accommodate motions at a nonconstant angular speed.

The discrete form of Eqs. (70), derived with the standard expansion of the displacement field $\{u_0, v, w\}$ is given in Ref. 6. The discrete linearized Eqs. (71), including the stiffening term, have been recently employed in Ref. 25. In the present implementation, both sets of equations are discretized according to the Galerkin procedure with the shape functions for a beam element. The discretization was carried out with Maple symbolic computation program.

It is worthwhile to comment on the structure of the resulting discretized equations for the two formulations. In the SE-1/DF-1 approach, the discrete form of the stiffening terms can be expressed as a product of an (elastic)coordinate-dependent second-order stiffness matrix and a column vector of elastic coordinates. The corresponding discrete equations have a standard form, but with the constant stiffness matrix replaced by a configuration-dependent stiffness matrix. If the beam is discretized with more than one element, the latter can be assembled using the established finite-element assembly procedure. Banerjee and Lemak[16] pointed out that evaluation of the stiffness term in the SE-1/DF-1 formulation requires that this stiffness matrix be updated at each time step in the simulation, which may be computationally costly. This is supported by our numerical results.

The discretization of the continuous equations resulting from SE-1/DF-2 approach is relatively simple if one retains only the linear stiffening term. If higher-order terms are included, then the process is complicated by the spatial integration of these time-dependent terms in the calculation of T in (72). In particular, the time-derivative and the u_f-terms, make the discretization of these equations a non-trivial task, not likely to be feasible without a symbolic manipulation software. Moreover, the aforementioned time-derivative terms in T_I, as well as the $\ddot{u}_f$-term in the *axial* motion equation modify the usual symmetric structure of the mass matrix which is also configuration-dependent. The assembly of this matrix requires a special procedure and adds to the computational cost.

Approximate Models

Numerical results are presented here for several approximations of equations (70) and (71). The two simpler approximate models are common to both formulations. In particular, the first of these does not contain any inertial terms which involve elastic displacements or their rates and has been coined by Padilla and von Flotow[3] as *ruthlessly linearized* (RL). The second model includes terms that are linear in elastic coordinates with the exception of the linear stiffening term in the SE-1/DF-2 formulation. The response of the beam predicted by this model is unstable and is not presented here. Further approximations of (70) and (71) must be considered separately for the two formulations.

One approximation that has been used by some researchers[24] in conjunction with the SE-1/DF-1 formulation can be derived by retaining only the third-order "nonlinear" contribution to the strain energy and treating the $EAu_0{}'$ as "constant" when deriving the corresponding stiffness term. This amounts to keeping only the $EAu_0'v''$ term in the transverse equation of motion and dropping the rest of the nonlinear terms in (70). The resulting model will be refered to as second-order and denoted by S-2, the letter S to signify that the nonlinearity appears in the *stiffness* term. The last model considered for the SE-1/DF-1 approach contains all terms in (70) and is of third order (S-3).

For the dynamics equations derived with the SE-1/DF-2 approach, one can identify the following models. The consistently linearized model (I-1), as the name suggests is a linear approximation of Eqs. (71) and it includes the stiffening term. It is obtained by dropping the underlined terms and setting $T_I = \Omega^2 x$. The consistent second-order model (I-2) is derived from the third-order expression for the kinetic energy. It is defined with T_I of Eq. (72) without the underlined terms. The final model (I-3), which is of third-order, includes all terms in Eqs. (71). In addition to these, an approximate linear model (I-1a) is defined, which is similar to I-1 but excludes the $\ddot{u}_f$-term in the axial equation. This model generates results indistinguishable from the I-1 approximation, but is considerably more efficient.

To compare the performance of different models defined for the two sets of the dynamics equations, a flexible beam example of Ref. 6 is used. Fig. 1(a) shows the transverse deflection v of the beam obtained with one-element discretization for the RL, S-2 and S-3 models. The S-2 response has been compared to the response presented in Ref. 24 (for a beam with slightly different properties.) The maximum transverse deflection shown in that reference is -0.59 m while the corresponding value predicted with the present implementation is -0.57 m. Hanagud and Sarkar present the axial and transverse displacements obtained with the S-3 model for a beam discretized with one axial and three transverse eigenfunctions. The results of Fig. 1(a) are in excellent agreement with theirs. As noted below, however, the solution predicted with one-element discretization of the S-3 model greatly underestimates the deformation of the beam. Fig. 1(b) displays the responses for the I-1, I-2, I-3 and I-1a models defined for the SE-1/DF-2 dynamics equations. Models I-1 and I-1a produce identical results. Also, it can be seen that the second- and third-order inertial terms derived with this formulation have negligible impact on the response. It is noted that the transverse deflection of Fig. 1(b) is identical to that presented in Ref. 13.

Another aspect of the two formulations investigated here is convergence of the finite-element discretization scheme. Fig. 2 illustrates this for the I-1a, S-2, and S-3 models (results for the I-1, I-2 and I-3 models are similar to those of I-1a). These

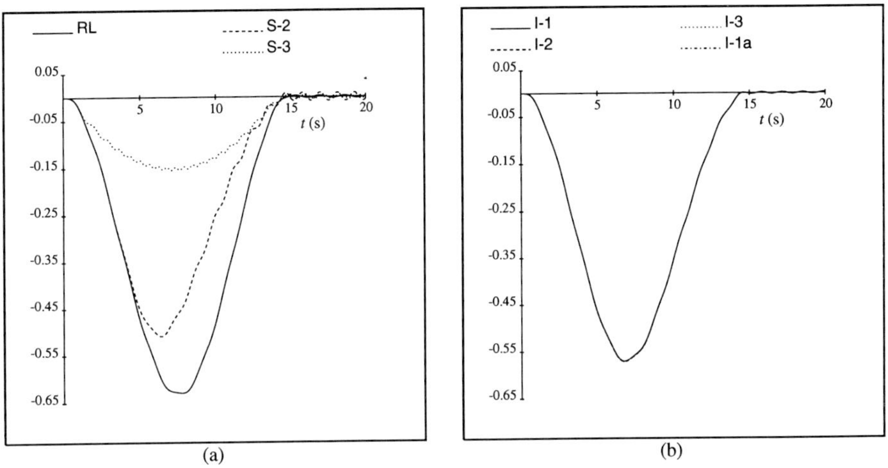

Figure 1: *Tip Displacement v (m) for* SE-1/DF-1 *and* SE-1/DF-2 *Models*

converge in two, four and over 30 elements, respectively! Clearly, the SE-1/DF-2 formulation yields a superior performance in terms of the convergence of the discretization scheme. By contrast, the exact SE-1/DF-1 approach has very poor convergence characteristics. This supports the remarks made by Laskin *et al.*[12] to the effect that a large number of modes is required if one does not explicitly separate the stretch from the foreshortening of the beam (as in the SE-1/DF-1 formulation.)

To summarize the comparison of the responses predicted by different models, the converged axial and transverse displacements are plotted in Fig. 3. It is evident that the results predicted by different formulations are in good agreement. It is also noted that for a beam spin-up problem, the S-2 approximation of the SE-1/DF-1 approach provides a very accurate solution.

It is fitting to comment on the efficiency of the computation of the various models. The CPU times and the number of function evaluations incurred during the numerical integration for responses of Fig. 1 are summarized in Table 1. Also included are the data for the I-1a model without the axial deformation. These results demonstrate the superior efficiency of the SE-1/DF-2 formulation, particularly due to the fact that it allows one to drop the axial motion equation. They are also in support of the previous observation regarding the high cost associated with the assembly of the configuration-dependent stiffness or mass matrices. The differences in the computational performance of the two formulations become particularly pronounced when one compares the corresponding data for the converged solutions.

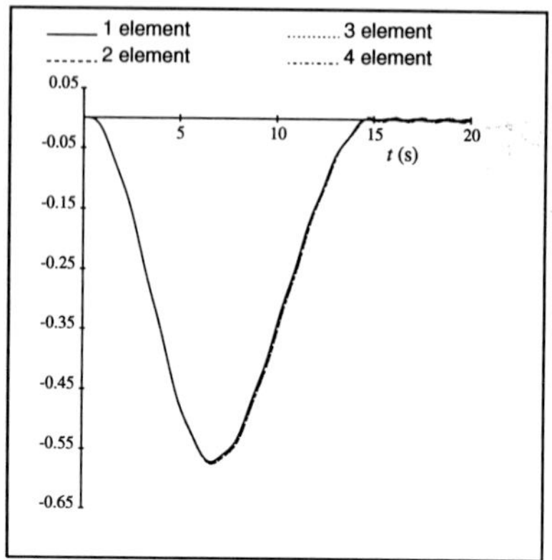

I-1a Model

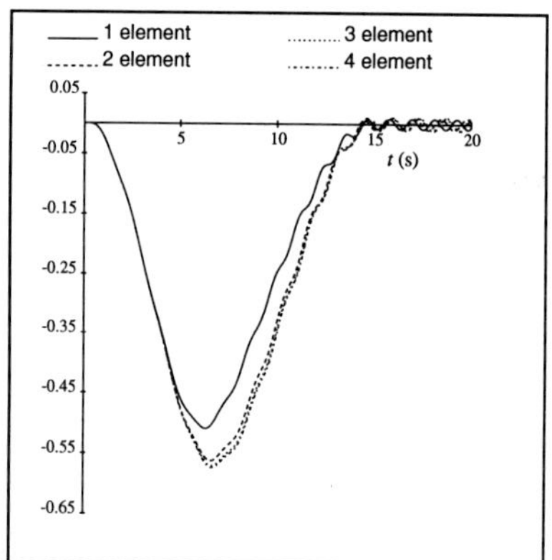

S-2 Model

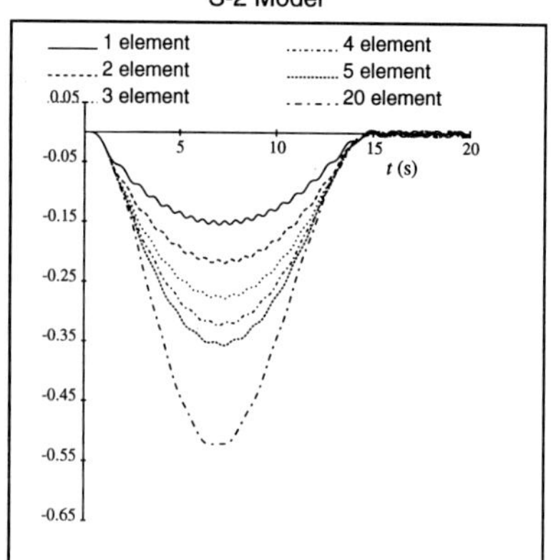

S-3 Model

Figure 2: *Convergence Results for Tip Displacement v (m)*

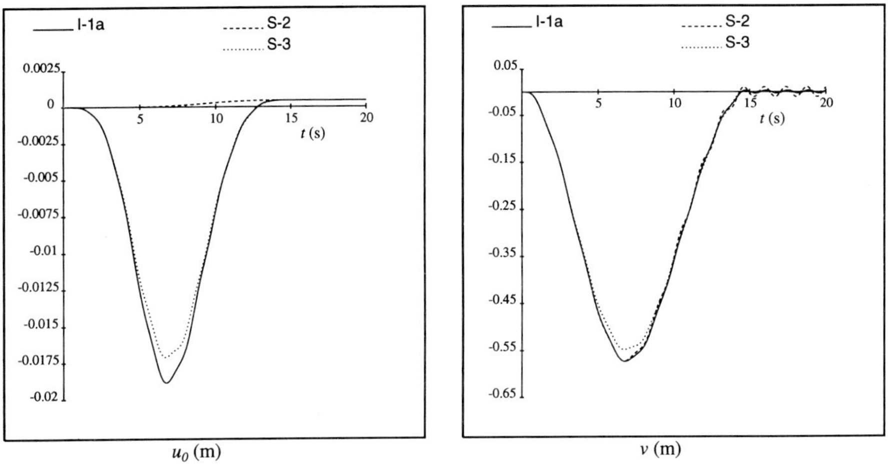

Figure 3: *Axial and Transverse Tip Displacements for Different Models*

6.2. Comparison with SE-1/DF-3 and SE-2/DF-1 Approaches

The flexible beam considered here is a 150-m-long antenna tube of Ref. 16. Fig. 4(a) contains the converged transverse deflection at the tip calculated with the S-2, S-3 and I-1a models defined for the SE-1/DF-1 and SE-1/DF-2 formulations. Fig. 4(b) displays the corresponding response reproduced from Ref. 16. It contains the deflection predicted with the formulation proposed in that reference (thick line) and the response obtained with the theory developed in Ref. 1 (lighter line). These were classified here as SE-2/DF-1 and SE-1/DF-3 approaches. It can be seen that the responses obtained with the SE-1/DF-1 and SE-1/DF-2 formulations are identical to the results of Fig. 4(b).

Table 1: *Performance Statistics for Different Models*

Model	CPU on HP 730 (s)	# of function evaluations
RL	5.	46091
S-2	7.	47068
S-3	10.	46221
I-1	10.	46814
I-1a	7.	47069
I-1a, without u_s	0.3	1838

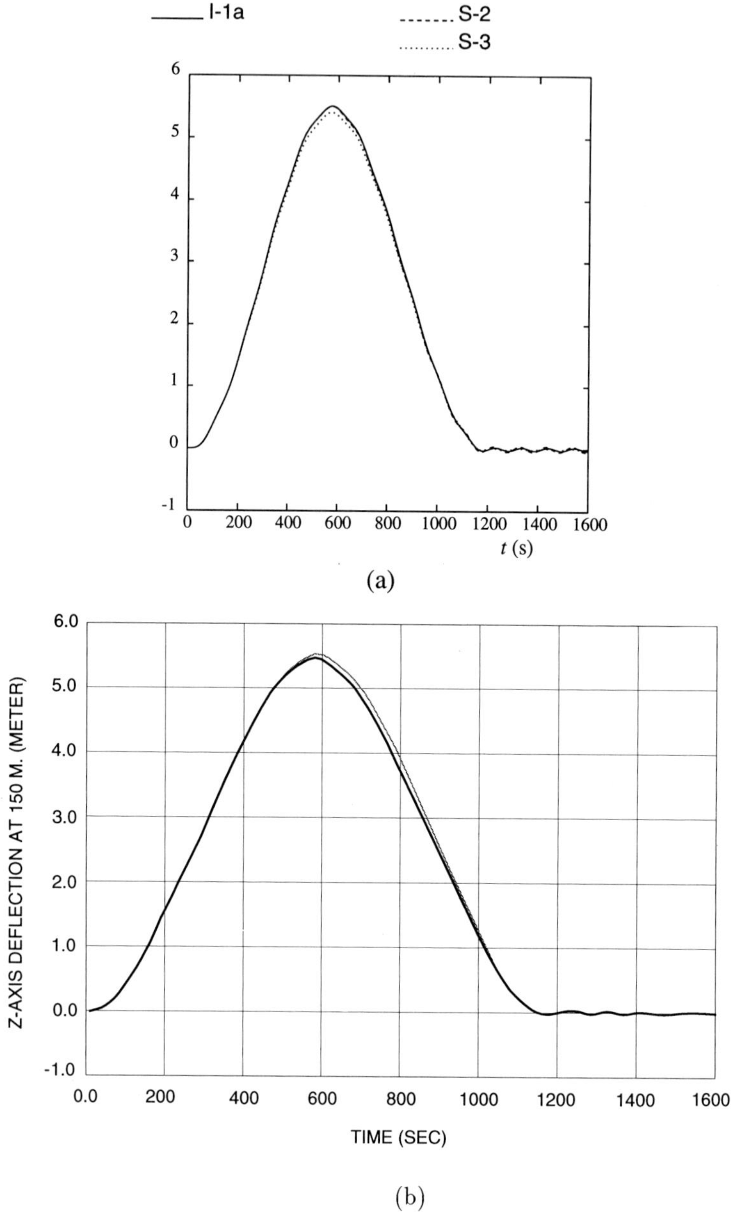

Figure 4: *Comparison of Different Approaches*

7. Concluding Remarks

In this chapter, we presented an exposition of several approaches to model the geometric stiffening effect for dynamics simulation of flexible-body systems. Our review included 13 papers published in the period from 1973 to 1991. Although it does not represent a complete literature review of the works that have addressed this issue, it covers a wide range of formulations developed for the problem.

In reviewing these works, we demonstrated that all formulations incorporate the geometric nonlinearity, which in case of beam-like bodies gives rise to foreshortening, through the nonlinear Lagrangian strains. Two key characteristics of the different methods were identified, based on which a general classification was put forward. We have also established the interrelationships between the various approaches, provided a number of clarifications and new interpretations and offered our opinions on their benefits. A numerical example of a flexible beam was considered which brought to light the differences in their performance. It is hoped that this work will contribute to a better understanding of the origin of geometric stiffening and how this effect can be incorporated into the dynamics model of a flexible body undergoing large rigid-body motion.

Acknowledgements

I would like to thank Dr. C. J. Damaren of Royal Roads Military College and Dr. J. Haddow for the help they have given me in tackling the subject of this chapter. This research has been funded by the Natural Sciences and Engineering Research Council of Canada.

References

1. Kane, T. R., Ryan, R. R., and Banerjee, A. K., "Dynamics of a Cantilever Beam Attached to a Moving Base," *J. Guidance, Control, and Dynamics*, Vol. 10, No. 2, 1987, pp. 139-151.

2. Eke, F. O., and Laskin, R. A., "On the Inadequacies of Current Multi-Flexible Body Simulation Codes," AIAA Paper No. 87-2248, 1987, pp. 79-89.

3. Padilla, C. E., and von Flotow, A. H., "Nonlinear Strain-Displacement Relations and Flexible Multibody Dynamics," *J. Guidance, Control, and Dynamics*, Vol. 15, No. 1, 1992, pp. 128-136.

4. Banerjee, A. K., and Dickens, J. M., "Dynamics of an Arbitrary Flexible Body in Large Rotation and Translation," *J. Guidance, Control, and Dynamics*, Vol. 13, No. 2, 1990, pp. 221-227.

5. London, K. W., "Comment on 'Dynamics of a Cantilever Beam Attached to a Moving Base'," *J. Guidance, Control, and Dynamics*, Vol. 12, No. 2, 1989, pp. 284-286.

6. Hanagud, S., and Sarkar, S., "Problem of the Dynamics of a Cantilever Beam Attached to a Moving Base," *J. Guidance, Control, and Dynamics*, Vol. 12, No. 3, 1989, pp. 438-441.

7. Likins, P. W., Barbera, F. J., and Baddeley, V., "Mathematical Modeling of Spinning Elastic Bodies for Modal Analysis," *AIAA Journal*, Vol. 11, 1973, pp. 1251-1258.

8. Vigneron, F. R., "Comment on 'Mathematical Modeling of Spinning Elastic Bodies for Modal Analysis," *AIAA Journal*, Vol. 13, 1975, pp. 126-127.

9. Kaza, K. R., and Kvaternik, R. G., "Nonlinear Flap-Lag-Axial Equations of a Rotating Beam," *Acta Astronautica*, Vol. 17, No. 12, 1980, pp. 1349-1360.

10. Lips, K. W. and Modi, V. J., "General Dynamics of a Large Class of Flexible Satellite Systems," *Acta Astronautica*, Vol. 17, No. 12, 1980, pp. 1349-1360.

11. Hughes, P. C. and Fung, J. C., "Lyapunov Stability of Spinning Satellites with Long Flexible Appendages," *Journal of Celestial Mechanics*, Vol. 4, 1971, pp. 295-308.

12. Laskin, R. A., Likins, P. W., and Longman, R. W., "Dynamical Equations of a Free-Free Beam subject to Large Overall Motions," *Journal of the Astronautical Society*, Vol. XXXI, 1983, pp. 507-528.

13. Simo, J. C. and Vu-Quoc, L., "The Role of Non-linear Theories in Transient Dynamic Analysis of Flexible Structures," *Journal of Sound and Vibration*, Vol. 119, 3, 1987, pp. 487-508.

14. Meirovitch, L., *Analytical Methods in Vibrations*, Macmillan, New York, 1967, pp. 440-453.

15. Meirovitch, L., "Hybrid State Equations of Motion for Flexible Bodies in Terms of Quasi-Coordinates," *J. Guidance, Control, and Dynamics*, Vol. 14, No. 5, 1991, pp. 1008-1013.

16. Banerjee, A. K., and Lemak, J. M., "Multi-Flexible Body Dynamics Capturing Motion-Induced Stiffness," *ASME J. Applied Mechanics*, Vol. 58, 1991, pp. 766-775.

17. Wallrapp, O. and Schwertassek, R., "Representation of Geometric Stiffening in Multibody System Simulation," *International Journal for Numerical Methods in Engineering*, Vol. 32, 1991, pp. 1833-1850.

18. Kane, T. R., and Levinson, D. A., *Dynamics, Theory and Applications*, McGraw-Hill, New York, 1985.

19. Brush, D. O., and Almroth, B. O., *Buckling of Bars, Plates, and Shells*, McGraw-Hill, New York, 1975.

20. Kane, T. R., Ryan, R. R., and Banerjee, A. K., "Reply by Authors to K. W. London," *J. Guidance, Control, and Dynamics*, Vol. 12, No. 2, 1989, pp. 286-287.

21. Simo, J. C. and Vu-Quoc, L., "On the Dynamics of Flexible Beams under Large Overall Motions—The Plane Case: Part I," *Journal of Applied Mechanics*, Vol. 53, No. 4, 1986, pp. 849-854.

22. Vu-Quoc, L. and Simo, J. C., "Dynamics of Earth-Orbiting Flexible Satellites with Multibody Component," *J. Guidance, Control, and Dynamics*, Vol. 10, No. 6, 1987, pp. 549-558.

23. Banerjee, A.K. "Block-diagonal Equations for Multi-body Elastodynamics with Geometric Stiffness and Constraints," *J. Guidance, Control, and Dynamics*, Vol. 16, No. 6, 1993, pp. 1092-1100.

24. Ider, S.K. and Amirouche, F.M.L., "Nonlinear Modeling of Flexible Multibody Systems Dynamics Subjected to Variable Constraints," *J. Applied Mechanics*, Vol. 56, 1989, pp. 444–450.

25. Boutaghou, Z.E., Erdman, A.G. and Stolarski, H.K., "Dynamics of Flexible Beams and Plates in Large Overall Motions," *J. Applied Mechanics*, Vol. 59, 1992, pp. 991–999.

Wave Motion, Intelligent Structures and Nonlinear Mechanics, pp. 280–291
edited by A. Guran and D. J. Inman
Series on Stability, Vibration and Control of Structures: Vol. 1

FLOW INSTABILITY BETWEEN COAXIAL ROTATING CYLINDERS WITH A FLEXIBLE SUPPORT

VOLKER MEHL and JÖRG WAUER
Institut für Technische Mechanik, Universität Karlsruhe, Kaiserstraße 12
Karlsruhe, D–76128, Germany

ABSTRACT

The flow instability between coaxial rotating cylinders with a flexible support is analyzed. The fluid is assumed to be incompressible and viscous and the film thickness is not restricted to be small, in general. The rotational speed is chosen to be sufficiently low so that laminar–turbulent transition or even fully developed turbulent flow can not appear. Then, a semianalytical approach is applied to discuss the stability behavior. First, the governing nonlinear boundary value problem is generated. Next, the simplified boundary value problem describing the stationary motion is deduced. Subsequently, the linearized equations of motion for small superimposed perturbations are derived. Finally, solutions of the formulated eigenvalue problem are given. Attention is focussed to the stability boundary in form of a threshold speed for varying characteristic parameters.

1. Introduction

The stability behavior of an incompressible viscous fluid flow between two coaxially arranged rotating cylinders is a classical problem in hydrodynamics [1,2]. Interesting secondary effects are related to an eccentric mounting of the cylinders [3], the presence of an axial flow [4] or the influence of the compressibility of the fluid [5]. A further important modification concerns the support itself which is assumed to be rigid in all available references. In the present work, a flexible bedding is assumed and the dependence of the critical Taylor number on the stiffness property of the suspension is the main objective. Reviewing the literature, at first sight, also this problem seems to be discussed [6]. But a critical examination shows that this paper has nothing to do with the above mentioned topic which will be analyzed here.

First of all, a correct formulation of the governing boundary value problem in the sense of continuum mechanics — including an accurate description of all boundary and transition conditions — is given. Next, the simplified boundary value problem representing the stationary motion in form of a laminar Couette flow is deduced. Subsequently, the linearized equations of motion for small superimposed perturbations are derived. Finally, the procedure is described to solve the formulated equation set approximately. It can be recognized that two independent stability problems appear:

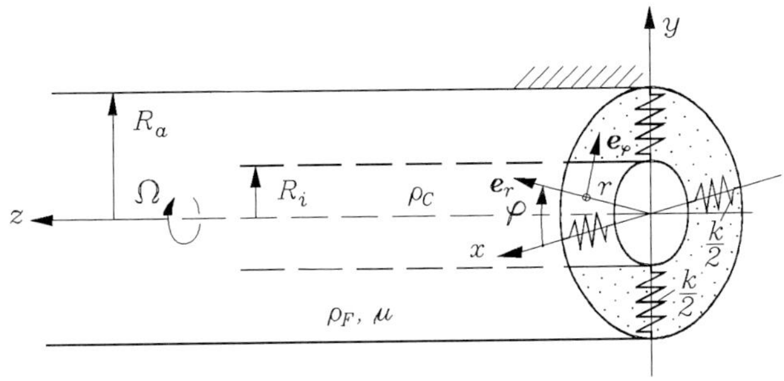

Fig. 1: Geometry of the both cylinders

The first one represents the classical Taylor vortex flow instability still possible also for flexibly supported cylinders; the second one is a real transverse interaction problem without axial perturbations Δw.

Technical background is the interaction of rotating shafts and the surrounding fluid in seals, for instance[7]. But then, different practically given facts, e. g., finite length and non–uniform geometry of the seals, and very often, such high angular speeds that usually a fully turbulent flow results, complicate the problem to be solved so extremely that purely numerical investigations are absolutely required.

2. Formulation

Consider an elastically supported (foundation stiffness per unit length k), rigid and perfectly balanced cylinder of radius R_i and mass density ρ_C. It rotates at a constant angular speed Ω in a centric position within a cylindrical rigid stator of radius R_o. Fig. 1 illustrates the basic geometry and shows an appropriate r, φ, z–coordinate system used in the following. The annular volume between the two infinitely long cylinders is filled with an incompressible viscous (dynamic viscosity μ) fluid of mass density ρ_F. Perturbations of the laminar, rotational symmetric, purely azimuthal flow field $(u_0, w_0 \equiv 0)$ with its velocity $v_0(r)$ in circumferential direction φ and pressure $p_0(r)$ for the centric position $x_0, y_0 \equiv 0$ appear in form of a dynamic state in which the rotating inner cylinder moves on a small time–dependent orbit composed by the transverse displacements $\Delta x(t), \Delta y(t)$ (torsional vibrations will be excluded) and the state variables of the fluid fluctuate $[\Delta u(r,\varphi,z,t), \Delta v(r,\varphi,z,t), \Delta w(r,\varphi,z,t), \Delta p(r,\varphi,z,t)]$ about their stationary values due to the eccentric motion of the inner cylinder. It is important to state that during the general motion in which a vortex flow may arise, also a non–vanishing axial velocity component $\Delta w(r,\varphi,z,t)$ has to be taken into consideration.

For mathematical description, the continuity equation, the momentum equations in radial, azimuthal and axial direction together with the dynamic transition conditions at the surface (located at $r = R_i + \Delta x \cos\varphi + \Delta y \sin\varphi$) between fluid and rotating inner cylinder (force balance for this cylinder) and the kinematical boundary and transition conditions (no–slip) at the surface between fluid and stator and fluid and rotor are the starting point.

In order to express all governing equations of motion in dimensionless form, non–dimensional variables

$$\begin{aligned} r &= \bar{r}(R_o - R_i), \quad z = \bar{z}(R_o - R_i), \quad t = \bar{t}\frac{R_o - R_i}{R_i\Omega} \\ x &= \bar{x}(R_o - R_i), \quad y = \bar{y}(R_o - R_i), \\ p &= \bar{p}R_i^2\Omega^2\rho_F, \quad u = \bar{u}R_i\Omega, \quad v = \bar{v}R_i\Omega, \quad w = \bar{w}R_i\Omega, \end{aligned} \tag{1}$$

and parameters

$$\alpha = \frac{R_o - R_i}{R_i}, \quad \mathrm{Re} = \frac{\alpha R_i^2 \Omega \rho_F}{\mu}, \quad \delta = \frac{\rho_F}{\rho_C}, \quad \gamma = \frac{k\rho_F R_i^2}{\pi\mu^2} \tag{2}$$

are introduced. Under this rescaling, the governing boundary value problem becomes, after dropping all overbars for convenience, its final form. In detail, it reads

$$\begin{aligned} \frac{\partial u}{\partial r} + \frac{u}{r} + \frac{1}{r}\frac{\partial v}{\partial \varphi} + \frac{\partial w}{\partial z} &= 0, \\ \left(\frac{\partial u}{\partial t} + u\frac{\partial u}{\partial r} + \frac{v}{r}\frac{\partial u}{\partial \varphi} + w\frac{\partial u}{\partial z} - \frac{v^2}{r}\right) &= -\frac{\partial p}{\partial r} + \frac{1}{\mathrm{Re}}\left(\Delta[u] - \frac{u}{r^2} - \frac{2}{r^2}\frac{\partial v}{\partial \varphi}\right), \\ \left(\frac{\partial v}{\partial t} + u\frac{\partial v}{\partial r} + \frac{v}{r}\frac{\partial v}{\partial \varphi} + w\frac{\partial v}{\partial z} + \frac{uv}{r}\right) &= -\frac{1}{r}\frac{\partial p}{\partial \varphi} + \frac{1}{\mathrm{Re}}\left(\Delta[v] - \frac{v}{r^2} + \frac{2}{r^2}\frac{\partial u}{\partial \varphi}\right), \\ \left(\frac{\partial w}{\partial t} + u\frac{\partial w}{\partial r} + \frac{v}{r}\frac{\partial w}{\partial \varphi} + w\frac{\partial w}{\partial z}\right) &= -\frac{\partial p}{\partial z} + \frac{1}{\mathrm{Re}}\Delta[w], \\ \frac{d^2x}{dt^2} + \frac{\delta\alpha^4\gamma}{\mathrm{Re}^2}x + \frac{\delta\alpha}{\pi}\int_0^{2\pi}\left(-\sigma_{rr}\cos\varphi + \sigma_{r\varphi}\sin\varphi\right)_{r=\frac{1}{\alpha}+\Delta x\cos\varphi+\Delta y\sin\varphi} d\varphi &= 0, \\ \frac{d^2y}{dt^2} + \frac{\delta\alpha^4\gamma}{\mathrm{Re}^2}y - \frac{\delta\alpha}{\pi}\int_0^{2\pi}\left(\sigma_{rr}\sin\varphi + \sigma_{r\varphi}\cos\varphi\right)_{r=\frac{1}{\alpha}+\Delta x\cos\varphi+\Delta y\sin\varphi} d\varphi &= 0, \\ u(\tfrac{1+\alpha}{\alpha}, \varphi, z, t) = v(\tfrac{1+\alpha}{\alpha}, \varphi, z, t) = w(\tfrac{1+\alpha}{\alpha}, \varphi, z, t) &= 0, \\ u(\tfrac{1}{\alpha} + \Delta x\cos\varphi + \Delta y\sin\varphi, \varphi, z, t) &= \frac{d\Delta x}{dt}\cos\varphi + \frac{d\Delta y}{dt}\sin\varphi, \\ v(\tfrac{1}{\alpha} + \Delta x\cos\varphi + \Delta y\sin\varphi, \varphi, z, t) &= 1 - \frac{d\Delta x}{dt}\sin\varphi + \frac{d\Delta y}{dt}\cos\varphi, \\ w(\tfrac{1}{\alpha} + \Delta x\cos\varphi + \Delta y\sin\varphi, \varphi, z, t) &= 0 \end{aligned} \tag{3}$$

where $\Delta[.]$ is the Laplacian operator

$$\Delta[.] = \frac{\partial^2[.]}{\partial r^2} + \frac{1}{r}\frac{\partial[.]}{\partial r} + \frac{1}{r^2}\frac{\partial^2[.]}{\partial \varphi^2} + \frac{\partial^2[.]}{\partial z^2} \tag{4}$$

and (for a viscous incompressible fluid)

$$\sigma_{rr} = -p + \frac{2}{\mathrm{Re}}\frac{\partial u}{\partial r}, \quad \sigma_{r\varphi} = \frac{1}{\mathrm{Re}}\left(\frac{\partial v}{\partial r} - \frac{v}{r} + \frac{1}{r}\frac{\partial u}{\partial \varphi}\right). \tag{5}$$

For a narrow gap, $\alpha \ll 1$, a shifted reference frame using the transformation

$$r = \frac{1}{\alpha} + \frac{1}{2} + \xi \tag{6}$$

is appropriate. Applying it, there are several simplifications. On the one hand, the location of the fluid surfaces in the centred position of the inner cylinder change,

$$r = \frac{1}{\alpha} \to \xi = -\frac{1}{2}, \quad r = \frac{1+\alpha}{\alpha} \to \xi = +\frac{1}{2}, \tag{7}$$

and on the other hand, the derivatives themselves modify:

$$\frac{\partial(.)}{\partial r} = \frac{\partial(.)}{\partial \xi}, \ \frac{\partial^2(.)}{\partial r^2} = \frac{\partial^2(.)}{\partial \xi^2}, \ \frac{1}{r}(.) = \alpha(.), \ \frac{1}{r^2}(.) = \alpha^2(.) \approx 0 \tag{8}$$

3. Analysis

Analogous to the solutions for the displacements of the cylinder

$$x(t) = x_0 + \Delta x(t), \quad y(t) = y_0 + \Delta y(t) \qquad x_0, y_0 \equiv 0, \tag{9}$$

also small disturbances $\Delta q(r,\varphi,z,t)$ $(q \in u, v, w, p)$ about the stationary flow state $q_0(r)$ $(q_0 \in v_0, p_0 \neq 0, u_0, w_0 \equiv 0)$ will be considered:

$$\begin{aligned} v(r,\varphi,z,t) &= v_0(r) + \Delta v(r,\varphi,z,t), \quad p(r,\varphi,z,t) = p_0(r) + \Delta p(r,\varphi,z,t), \\ u(r,\varphi,z,t) &= 0 + \Delta u(r,\varphi,z,t), \quad w(r,\varphi,z,t) = 0 + \Delta w(r,\varphi,z,t). \end{aligned} \tag{10}$$

The relations (9) and (10) are now substituted into the boundary value problem (3). Considering the stationary state first, all Δ–terms are ignored so that a time–independent boundary value problem

$$\begin{aligned} \frac{v_0^2}{r} = \frac{\partial p_0}{\partial r}, \quad \frac{\partial^2 v_0}{\partial r^2} + \frac{1}{r}\frac{\partial v_0}{\partial r} - \frac{v_0^2}{r^2} &= 0, \\ v_0\left(\tfrac{1}{\alpha}\right) = 1, \quad v_0\left(\tfrac{1+\alpha}{\alpha}\right) &= 0 \end{aligned} \tag{11}$$

for $v_0(r), p_0(r)$ remains. Obviously, (11) describes a circumferential Couette flow unchanged in comparison with the classical Taylor vortex flow problem and stable for sufficient low Reynolds numbers. Even for a finite gap, an analytical solution

$$\begin{aligned} v_0(r) &= \frac{r}{2+\alpha}\left[\frac{(1+\alpha)^2}{\alpha^2 r^2} - 1\right], \quad \frac{\partial p_0}{\partial r} = \frac{v_0^2}{r}, \\ p_0(r) &= p_{0i} + \frac{1}{2\alpha^2(\alpha+2)^2}\left[(\alpha^2 r^2 - 1)\left(1 + \frac{(\alpha+1)^4}{\alpha^2 r^2}\right) - 4(\alpha+1)^2 \ln(\alpha r)\right] \end{aligned} \tag{12}$$

can be given. It is remarkable that the stationary flow field remains the same also for a compressible fluid (but the pressure distribution modifies). For a narrow gap, the result simplifies to

$$v_0(\xi) = \frac{1}{2} - \xi, \quad \frac{\partial p_0}{\partial \xi} = \alpha v_0^2(\xi) \tag{13}$$

in form of a linear velocity profile.

To formulate the variational equations characterizing the superimposed small perturbations, a Taylor expansion of first order

$$q(\tfrac{1}{\alpha} + \Delta x \cos\varphi + \Delta y \sin\varphi, \varphi, z, t) = q(\tfrac{1}{\alpha}, \varphi, z, t) + \left.\frac{\partial q(r,\varphi,z,t)}{\partial r}\right|_{\frac{1}{\alpha}} (\Delta x \cos\varphi + \Delta y \sin\varphi) \tag{14}$$

is made in the transition conditions. Then the stationary quantities $q_0(r)$ are subtracted and the remaining equations are linearized in the Δ–quantities. This yields a boundary value problem

$$\frac{\partial \Delta u}{\partial r} + \frac{\Delta u}{r} + \frac{1}{r}\frac{\partial \Delta v}{\partial \varphi} + \frac{\partial \Delta w}{\partial z} = 0,$$

$$\frac{\partial \Delta u}{\partial t} + \frac{v_0}{r}\left(\frac{\partial \Delta u}{\partial \varphi} - 2\Delta v\right)$$
$$= -\frac{\partial \Delta p}{\partial r} + \frac{1}{\mathrm{Re}}\left(\frac{\partial^2 \Delta u}{\partial r^2} + \frac{1}{r}\frac{\partial \Delta u}{\partial r} - \frac{\Delta u}{r^2} + \frac{1}{r^2}\frac{\partial^2 \Delta u}{\partial \varphi^2} + \frac{\partial^2 \Delta u}{\partial z^2} - \frac{2}{r^2}\frac{\partial \Delta v}{\partial \varphi}\right),$$

$$\frac{\partial \Delta v}{\partial t} + \left(\frac{\partial v_0}{\partial r} + \frac{v_0}{r}\right)\Delta u + \frac{v_0}{r}\frac{\partial \Delta v}{\partial \varphi}$$
$$= -\frac{1}{r}\frac{\partial \Delta p}{\partial \varphi} + \frac{1}{\mathrm{Re}}\left(\frac{\partial^2 \Delta v}{\partial r^2} + \frac{1}{r}\frac{\partial \Delta v}{\partial r} - \frac{\Delta v}{r^2} + \frac{1}{r^2}\frac{\partial^2 \Delta v}{\partial \varphi^2} + \frac{\partial^2 \Delta v}{\partial z^2} + \frac{2}{r^2}\frac{\partial \Delta u}{\partial \varphi}\right),$$

$$\frac{\partial \Delta w}{\partial t} + \frac{v_0}{r}\frac{\partial \Delta w}{\partial \varphi} = -\frac{\partial \Delta p}{\partial z} + \frac{1}{\mathrm{Re}}\left(\frac{\partial^2 \Delta w}{\partial r^2} + \frac{1}{r}\frac{\partial \Delta w}{\partial r} + \frac{1}{r^2}\frac{\partial^2 \Delta w}{\partial \varphi^2} + \frac{\partial^2 \Delta w}{\partial z^2}\right),$$

$$\frac{d^2 \Delta x}{dt^2} + \frac{\delta \alpha^4 \gamma}{\mathrm{Re}^2}\Delta x + \frac{\delta \alpha}{\pi}\int_0^{2\pi}\left\{\left[\left(\Delta p - \frac{2}{\mathrm{Re}}\frac{\partial \Delta u}{\partial r}\right)\cos\varphi\right.\right.$$
$$\left. + \frac{1}{\mathrm{Re}}\left(\frac{\partial \Delta v}{\partial r} - \frac{\Delta v}{r} + \frac{1}{r}\frac{\partial \Delta u}{\partial \varphi}\right)\sin\varphi\right]_{\frac{1}{\alpha}} + \left[\frac{\partial p_0}{\partial r}\cos\varphi\right.$$
$$\left.\left. + \frac{1}{\mathrm{Re}}\left(\frac{\partial^2 v_0}{\partial r^2} - \frac{\partial}{\partial r}\left(\frac{v_0}{r}\right)\right)\sin\varphi\right]_{\frac{1}{\alpha}}(\Delta x \cos\varphi + \Delta y \sin\varphi)\right\} d\varphi = 0,$$

$$\frac{d^2 \Delta y}{dt^2} + \frac{\delta \alpha^4 \gamma}{\mathrm{Re}^2}\Delta y + \frac{\delta \alpha}{\pi}\int_0^{2\pi}\left\{\left[\left(\Delta p - \frac{2}{\mathrm{Re}}\frac{\partial \Delta u}{\partial r}\right)\sin\varphi\right.\right.$$
$$\left. - \frac{1}{\mathrm{Re}}\left(\frac{\partial \Delta v}{\partial r} - \frac{\Delta v}{r} + \frac{1}{r}\frac{\partial \Delta u}{\partial \varphi}\right)\cos\varphi\right]_{\frac{1}{\alpha}} + \left[\frac{\partial p_0}{\partial r}\sin\varphi\right.$$

$$- \frac{1}{\mathrm{Re}} \left(\frac{\partial^2 v_0}{\partial r^2} - \frac{\partial}{\partial r}(\frac{v_0}{r}) \right) \cos\varphi \Big]_{\frac{1}{\alpha}} (\Delta x \cos\varphi + \Delta y \sin\varphi) \Big\} d\varphi = 0,$$

$$\Delta u(\tfrac{1+\alpha}{\alpha}, \varphi, z, t) = \Delta v(\tfrac{1+\alpha}{\alpha}, \varphi, z, t) = \Delta w(\tfrac{1+\alpha}{\alpha}, \varphi, z, t) = 0,$$

$$\Delta u(\tfrac{1}{\alpha}, \varphi, z, t) = \frac{d\Delta x}{dt} \cos\varphi + \frac{d\Delta y}{dt} \sin\varphi, \quad \Delta w(\tfrac{1}{\alpha}, \varphi, z, t) = 0,$$

$$\Delta v(\tfrac{1}{\alpha}, \varphi, z, t) + \frac{\partial v_0}{\partial r}\Big|_{\frac{1}{\alpha}} (\Delta x \cos\varphi + \Delta y \sin\varphi) = -\frac{d\Delta x}{dt} \sin\varphi + \frac{d\Delta y}{dt} \cos\varphi,$$

$$\Delta u, \Delta v, \Delta w, \Delta p \ 2\pi\text{–periodic in } \varphi \tag{15}$$

underlying the following stability analysis.

To separate in (15) the t– and z–dependence, assume solutions

$$\begin{aligned} \Delta q(r,\varphi,z,t) &= Q(r,\varphi) e^{\alpha\lambda t + i\sigma z}, \quad \begin{matrix} q \\ Q \end{matrix} \in \left\{ \begin{matrix} u \\ U \end{matrix}, \begin{matrix} v \\ V \end{matrix}, \begin{matrix} p \\ P \end{matrix} \right\}, \\ \Delta w(r,\varphi,z,t) &= -iW(r,\varphi) e^{\alpha\lambda t + i\sigma z}, \quad \Delta x(t) = X e^{\alpha\lambda t}, \ \Delta y(t) = Y e^{\alpha\lambda t} \end{aligned} \tag{16}$$

The resulting boundary value problem is

$$\frac{\partial U}{\partial r} + \frac{U}{r} + \frac{1}{r}\frac{\partial \Delta V}{\partial \varphi} + \sigma W = 0,$$

$$\alpha\lambda U + \frac{v_0}{r}\left(\frac{\partial U}{\partial \varphi} - 2V \right)$$

$$= -\frac{\partial P}{\partial r} + \frac{1}{\mathrm{Re}} \left(\frac{\partial^2 U}{\partial r^2} + \frac{1}{r}\frac{\partial U}{\partial r} - \frac{U}{r^2} + \frac{1}{r^2}\frac{\partial^2 U}{\partial \varphi^2} - \sigma^2 U - \frac{2}{r^2}\frac{\partial \Delta V}{\partial \varphi} \right),$$

$$\alpha\lambda V + \left(\frac{\partial v_0}{\partial r} + \frac{v_0}{r} \right) U + \frac{v_0}{r}\frac{\partial V}{\partial \varphi}$$

$$= -\frac{1}{r}\frac{\partial P}{\partial \varphi} + \frac{1}{\mathrm{Re}} \left(\frac{\partial^2 V}{\partial r^2} + \frac{1}{r}\frac{\partial V}{\partial r} - \frac{V}{r^2} + \frac{1}{r^2}\frac{\partial^2 V}{\partial \varphi^2} - \sigma^2 V + \frac{2}{r^2}\frac{\partial U}{\partial \varphi} \right),$$

$$\alpha\lambda W + \frac{v_0}{r}\frac{\partial W}{\partial \varphi} = \sigma P + \frac{1}{\mathrm{Re}} \left(\frac{\partial^2 W}{\partial r^2} + \frac{1}{r}\frac{\partial W}{\partial r} + \frac{1}{r^2}\frac{\partial^2 W}{\partial \varphi^2} - \sigma^2 W \right),$$

$$\left[(\alpha\lambda)^2 + \frac{\delta\alpha^4\gamma}{\mathrm{Re}^2} \right] X + \frac{\delta\alpha}{\pi} e^{i\sigma z} \int_0^{2\pi} \Big\{ \Big[\left(P - \frac{2}{\mathrm{Re}}\frac{\partial U}{\partial r} \right) \cos\varphi$$

$$+ \frac{1}{\mathrm{Re}} \left(\frac{\partial V}{\partial r} - \frac{V}{r} + \frac{1}{r}\frac{\partial U}{\partial \varphi} \right) \sin\varphi \Big]_{\frac{1}{\alpha}} + \Big[\frac{\partial p_0}{\partial r} \cos\varphi$$

$$+ \frac{1}{\mathrm{Re}} \left(\frac{\partial^2 v_0}{\partial r^2} - \frac{\partial}{\partial r}(\frac{v_0}{r}) \right) \sin\varphi \Big]_{\frac{1}{\alpha}} (X\cos\varphi + Y\sin\varphi) \Big\} d\varphi = 0,$$

$$\left[(\alpha\lambda)^2 + \frac{\delta\alpha^4\gamma}{\mathrm{Re}^2} \right] Y + \frac{\delta\alpha}{\pi} e^{i\sigma z} \int_0^{2\pi} \Big\{ \Big[\left(P - \frac{2}{\mathrm{Re}}\frac{\partial U}{\partial r} \right) \sin\varphi$$

$$
\begin{aligned}
&- \frac{1}{\mathrm{Re}}\left(\frac{\partial V}{\partial r} - \frac{V}{r} + \frac{1}{r}\frac{\partial U}{\partial \varphi}\right)\cos\varphi\Big]_{\frac{1}{\alpha}} + \left[\frac{\partial p_0}{\partial r}\sin\varphi\right.\\
&\left.- \frac{1}{\mathrm{Re}}\left(\frac{\partial^2 v_0}{\partial r^2} - \frac{\partial}{\partial r}\left(\frac{v_0}{r}\right)\right)\cos\varphi\right]_{\frac{1}{\alpha}} (X\cos\varphi + Y\sin\varphi)\Big\}\, d\varphi = 0,\\
&U(\tfrac{1+\alpha}{\alpha},\varphi) = V(\tfrac{1+\alpha}{\alpha},\varphi) = W(\tfrac{1+\alpha}{\alpha},\varphi) = 0,\\
&U(\tfrac{1}{\alpha},\varphi)e^{i\sigma z} = \alpha\lambda(X\cos\varphi + Y\sin\varphi),\quad W(\tfrac{1}{\alpha},\varphi) = 0,\\
&V(\tfrac{1}{\alpha},\varphi)e^{i\sigma z} + \left.\frac{\partial v_0}{\partial r}\right|_{\frac{1}{\alpha}}(X\cos\varphi + Y\sin\varphi) = \alpha\lambda(-X\sin\varphi + Y\cos\varphi),\\
&U, V, W, P\ 2\pi\text{–periodic in } \varphi.
\end{aligned} \tag{17}
$$

Two cases have to be distinguished: 1. $\sigma \neq 0$, 2. $\sigma \equiv 0$.

If the wave number σ is finite, obviously the kinematical transition conditions between fluid and inner cylinder in (17) enforce that the displacements of the inner cylinder vanish: $X, Y \equiv 0$. As the consequence, the φ–dependence does not take place any longer and the dynamic transition conditions in (17) are identically fulfilled. The remaining boundary value problem coincides with the classical one governing the Taylor vortex flow instability which has been analyzed in detail also for a finite fluid gap (see a paper by Kirchgässner [8], for instance).

A vanishing wave number σ means that according to (16), all variables U, V, W and P will be independent of the axial space coordinate z. The velocity component W appears only in one equation which is independent from the other equations. Assuming solutions for the remaining variables Q ($Q \in U, V, P$) in product form

$$
\begin{aligned}
Q(r,\varphi) &= Q_0(r) + \sum_{n=1}^{\infty}[Q_n(r)\sin n\varphi + \bar{Q}_n(r)\cos n\varphi],\quad Q \in U, P,\\
V(r,\varphi) &= V_0(r) + \sum_{n=1}^{\infty}[\bar{V}_n(r)\sin n\varphi + V_n(r)\cos n\varphi]
\end{aligned} \tag{18}
$$

leads for every ordinal number n ($n = 0, 1, \ldots, \infty$) to a certain eigenvalue problem. In the dynamic transition condition within (17), the integral $\int_0^{2\pi} \ldots d\varphi$ vanishes identically for $n \neq 1$ because $\int_0^{2\pi} \sin\varphi \sin n\varphi d\varphi = 0$ for $n \neq 1$. This leads to $\Delta x, \Delta y \equiv 0$ in these cases. Only for $n = 1$, coupled fluid–cylinder vibrations are described so that in the following, exclusively this one is dealt with. Dropping the index $n = 1$ and taking notice that $\frac{d(.)}{dr} = (.)'$ yields

$$
\begin{aligned}
&U' + \frac{1}{r}U - \frac{1}{r}V = 0, &&\bar{U}' + \frac{1}{r}\bar{U} + \frac{1}{r}\bar{V} = 0,\\
&\alpha\lambda U - \frac{v_0}{r}(\bar{U} + 2\bar{V}) = -P' &&\alpha\lambda\bar{U} + \frac{v_0}{r}(U - 2V) = -\bar{P}'\\
&+ \frac{1}{\mathrm{Re}}(U'' + \frac{1}{r}U' - \frac{2}{r^2}U + \frac{2}{r^2}V), &&+ \frac{1}{\mathrm{Re}}(\bar{U}'' + \frac{1}{r}\bar{U}' - \frac{2}{r^2}\bar{U} - \frac{2}{r^2}\bar{V}),
\end{aligned}
$$

$$\alpha\lambda V + \frac{v_0}{r}(\bar U + \bar V) + v_0'\bar U = -\frac{1}{r}P \qquad \alpha\lambda\bar V + \frac{v_0}{r}(U - V) + v_0'U = \frac{1}{r}\bar P$$
$$+\frac{1}{\mathrm{Re}}(V'' + \frac{1}{r}V - \frac{2}{r^2}V + \frac{2}{r^2}U), \qquad + \frac{1}{\mathrm{Re}}(\bar V'' + \frac{1}{r}\bar V - \frac{2}{r^2}\bar V - \frac{2}{r^2}\bar U),$$

$$\left[\alpha\lambda^2 + \frac{\delta\alpha^3\gamma}{\mathrm{Re}^2}\right] X + \delta\left\{p_0'X + \frac{1}{\mathrm{Re}}\left[v_0'' - \left(\frac{v_0}{r}\right)'\right]Y\right\}_{\frac{1}{\alpha}}$$
$$+\delta\left[\bar P - \frac{1}{\mathrm{Re}}\left(2\bar U' - \bar V' + \frac{1}{r}\bar V + \frac{1}{r}\bar U\right)\right]_{\frac{1}{\alpha}} = 0,$$
$$\left[\alpha\lambda^2 + \frac{\delta\alpha^3\gamma}{\mathrm{Re}^2}\right] Y + \delta\left\{p_0'Y - \frac{1}{\mathrm{Re}}\left[v_0'' - \left(\frac{v_0}{r}\right)'\right]X\right\}_{\frac{1}{\alpha}}$$
$$+\delta\left[P - \frac{1}{\mathrm{Re}}\left(2U' + V' - \frac{1}{r}V + \frac{1}{r}U\right)\right]_{\frac{1}{\alpha}} = 0,$$
$$U(\tfrac{1+\alpha}{\alpha}) = \bar U(\tfrac{1+\alpha}{\alpha}) = V(\tfrac{1+\alpha}{\alpha}) = \bar V(\tfrac{1+\alpha}{\alpha}) = 0,$$
$$U(\tfrac{1}{\alpha}) = \alpha\lambda Y, \quad \bar U(\tfrac{1}{\alpha}) = \alpha\lambda X,$$
$$V(\tfrac{1}{\alpha}) = \left[\alpha\lambda Y - v_0'(\tfrac{1}{\alpha})X\right], \quad \bar V(\tfrac{1}{\alpha}) = -\left[\alpha\lambda X + v_0'(\tfrac{1}{\alpha})Y\right]. \tag{19}$$

Finally, eliminate X, Y and obtain in this way a complicated eigenvalue problem in $U, V, P, \bar U, \bar V, \bar P$ with space-dependent coefficients. Only an approximate solution seems to be possible. Promising is Galerkin's method using Ritz series expansions fulfilling all boundary conditions. But finding comparison functions for the present problem is a lengthy pre-calculation.

An alternative seems to be to consider the stability problem for a narrow gap. Here the coefficients are constant [using (6)–(8): $\frac{1}{r} = \alpha$, $\frac{1}{r^2} = \alpha \approx 0$, $v_0'(\xi) = -1$, $v_0'' = 0$] if the ξ dependence of v_0 is neglected because $|\xi| \le \frac{1}{2}$ [set $v_0(\xi) = v_0(0) = \frac{1}{2}$]. So, a further simplification of the governing eigenvalue problem (19) follows. In detail, use (6) — (8) and (13), eliminate X, Y and obtain

$$\begin{aligned}
U' + \alpha U - \alpha V = 0, \qquad \bar U' + \alpha\bar U + \alpha\bar V &= 0,\\
\alpha\lambda U - \frac{\alpha}{2}(\bar U + 2\bar V) + P' - \frac{1}{\mathrm{Re}}(U'' + \alpha U') &= 0,\\
\alpha\lambda\bar U + \frac{\alpha}{2}(U - 2V) + \bar P' - \frac{1}{\mathrm{Re}}(\bar U'' + \alpha\bar U') &= 0,\\
\alpha\lambda V + \left(\frac{\alpha}{2} - 1\right)\bar U + \frac{\alpha}{2}\bar V + \alpha P - \frac{1}{\mathrm{Re}}(V'' + \alpha V') &= 0,\\
\alpha\lambda\bar V + \left(\frac{\alpha}{2} - 1\right)U - \frac{\alpha}{2}V - \alpha\bar P - \frac{1}{\mathrm{Re}}(\bar V'' + \alpha\bar V') &= 0,\\
U(\tfrac{1}{2}) = V(\tfrac{1}{2}) = \bar U(\tfrac{1}{2}) = \bar V(\tfrac{1}{2}) &= 0,\\
\bar U(-\tfrac{1}{2}) + \alpha\lambda U(-\tfrac{1}{2}) - \alpha\lambda V(-\tfrac{1}{2}) = 0, \quad U(-\tfrac{1}{2}) - \alpha\lambda\bar U(-\tfrac{1}{2}) - \alpha\lambda\bar V(-\tfrac{1}{2}) &= 0,
\end{aligned}$$

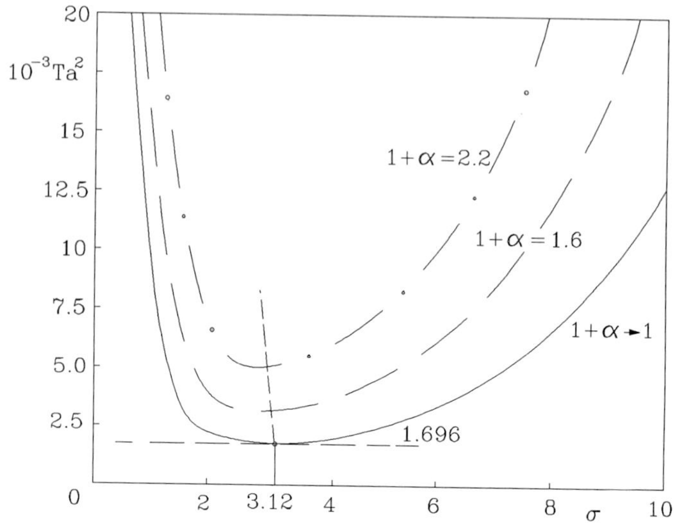

Fig. 2: Stability chart for the Taylor vortex flow problem

$$
\begin{aligned}
\left[\lambda + (2+\alpha)\frac{\delta\alpha}{\mathrm{Re}}\right] U(-\tfrac{1}{2}) - (2+\alpha)\frac{\delta\alpha}{\mathrm{Re}} V(-\tfrac{1}{2}) - \frac{\delta}{\mathrm{Re}} V'(-\tfrac{1}{2}) & \\
+ \alpha\delta\left(1 + \frac{\gamma\alpha^2}{\mathrm{Re}^2}\right)\left[\bar{U}(-\tfrac{1}{2}) + \bar{V}(-\tfrac{1}{2})\right] + \delta P(-\tfrac{1}{2}) &= 0, \\
\left[\lambda + (2+\alpha)\frac{\delta\alpha}{\mathrm{Re}}\right] \bar{U}(-\tfrac{1}{2}) + (2+\alpha)\frac{\delta\alpha}{\mathrm{Re}} \bar{V}(-\tfrac{1}{2}) + \frac{\delta}{\mathrm{Re}} \bar{V}'(-\tfrac{1}{2}) & \\
- \alpha\delta\left(1 + \frac{\gamma\alpha^2}{\mathrm{Re}^2}\right)\left[U(-\tfrac{1}{2}) - V(-\tfrac{1}{2})\right] + \delta \bar{P}(-\tfrac{1}{2}) &= 0,
\end{aligned}
\tag{20}
$$

where $\frac{d(.)}{d\xi)} = (.)'$. This eigenvalue problem in $U(\xi)$, $V(\xi)$, $P(\xi)$ and $\bar{U}(\xi)$, $\bar{V}(\xi)$, $\bar{P}(\xi)$ possesses constant coefficients and is much easier to solve. The classical solution in exponential form leads to a matrix eigenvalue problem solvable by standard methods.

4. Results

The Taylor vortex flow instability problem appearing also for rotating cylinders with a flexible support is completely discussed. Even for a finite gap, all essential results can be found in the literature. Fig. 2 shows the square of the Taylor number

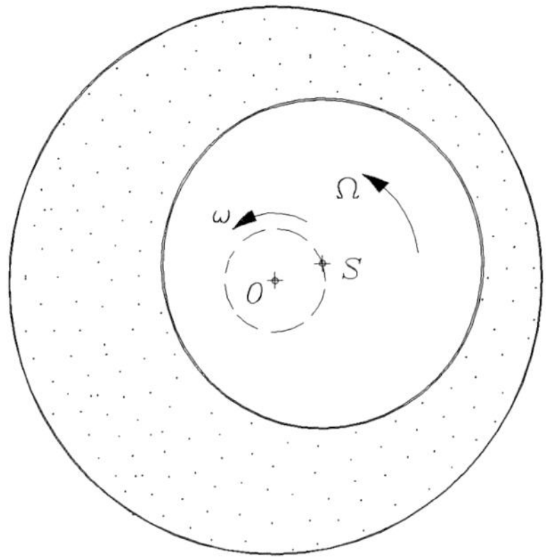

Fig. 3: Coupled fluid–structure vibrations ($\sigma \equiv 0$) at $\mathrm{Re} = \mathrm{Re}_c$

$\mathrm{Ta}^2 = \alpha \mathrm{Re}^2$ versus the wave number σ with the non–dimensional gap width α as parameter. The numerical results are based on a 4–term Galerkin approximation[9] and are in a good agreement with those calculated in an earlier paper[8] using another approach. As generally known, the local minimum of the depicted graphs characterizes the stability boundary. If this limit will not be exceeded, it is guaranteed that all perturbations die away and the stationary Couette flow remains stable. For a narrow gap, $\alpha \ll 1$, for instance, there results a critical Taylor number $\mathrm{Ta}^2 = 1696$ and a corresponding wave number $\sigma = 3.12$.

The eigenvalue problem (20) for a narrow gap (governing the stability behavior in the case of coupled fluid structure vibrations where $\sigma \equiv 0$) is solved numerically. Attention is focussed to the stability boundary where the real part of λ changes from its original negative value to a positive one. But it has to be noticed that the corresponding imaginary part of λ cannot be expected to be zero there. The threshold speed is characterized by the minimum of Re for which this condition is fulfilled as a function of the parameters α, δ and γ. A lengthy evaluation shows that this critical Reynolds number Re_c can be calculated as

$$\mathrm{Re}_c \approx 1.7\gamma^{\frac{1}{2}}\alpha^{\frac{3}{2}} \quad \text{for} \quad \delta > \alpha \quad (\delta < \alpha \text{ does not make sense}) \tag{21}$$

and

$$\Im(\lambda) \approx 0.5 \quad [\text{but note that always } \Im(\lambda) < 0.5\,!]. \tag{22}$$

This means (see Fig. 3) that the centre S of the inner cylinder rotates on a circle of constant radius $[\Re(\lambda) = 0]$ about the centre O of the stator. The angular speed is $\omega = \Im(\lambda)\Omega \approx \frac{1}{2}\Omega$.

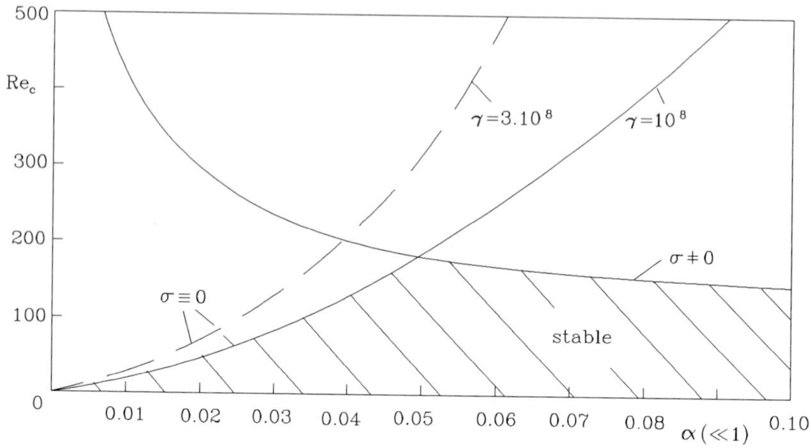

Fig. 4: Stability chart for a narrow gap

It is obvious that both the threshold speed and the type of instability ($\sigma \neq 0$ or $\sigma \equiv 0$) depend on the parameters α (gap width) $\ll 1$ and γ (stiffness), in general. In detail (see Fig. 4), there is

$$\begin{aligned} \mathrm{Re}_c &\approx \sqrt{1696}\alpha^{-\frac{1}{2}} \quad \text{for} \quad \sigma \neq 0, \\ \mathrm{Re}_c &\approx 1.7\gamma^{\frac{1}{2}}\alpha^{\frac{3}{2}} \quad \text{for} \quad \sigma \equiv 0. \end{aligned} \tag{23}$$

If the stiffness of the foundation tends to zero ($\gamma \to 0$), the whole operation range $\mathrm{Re} \geq 0$ is unstable. If it tends to infinity ($\gamma \to \infty$), only Taylor instability is possible.

5. Conclusions

The fluid flow dynamics between coaxial rotating cylinders with a flexible support have been analyzed. Without essential restrictions, the outer cylinder was constructed as a stator and the inner flexibly supported cylinder rotated at a constant speed. The fluid was assumed to be incompressible and viscous.

An accurate modeling in the sense of continuum mechanics was suggested. The governing boundary value problem demonstrates that a circumferential Couette flow between the coaxial rotating cylinders is the basic state also for a flexibly supported inner cylinder and the variational equations discussing the stability behavior lead to two independent eigenvalue problems. The first one describes the pure Taylor vortex flow instability about the centred, non–oscillating inner cylinder. The second one characterizes coupled transverse vibrations of the fluid–cylinder system.

In addition to the well-known results on Taylor vortex flow, the corresponding stability boundary characterized by a critical Reynolds number for the transverse fluid–structure oscillations was computed for the special case of a narrow gap. The discussion shows that both different types of instability can appear in practical applications; which is relevant and how big then the critical Reynolds number really is, depends on the parameters characterizing gap width (to be small here) and stiffness.

6. References

1. K. Bühler, J. E. R. Coney, M. Wimmer and J. Zierep. *Acta Mechanica* **62** (1986) 47.
2. S. Chandrasekhar. *Hydrodynamic and Hydromagnetic Stability* (Oxford University Press, Oxford. 1961).
3. R. C. Di Prima and J. T. Stuart. *J. Fluid Mech.* **67** (1975) 85.
4. R. C. Di Prima and A. Pridor. in *First Taylor Vortex Flow Working Party, Leeds, 1979.*
5. J. Wauer. *Zeitschr. Angew. Math. Mech.* **73** (1993) T592.
6. P. Hall. *J. Fluid Mech.* **146** (1984) 347.
7. H. S. Cheng and P. R. Trumpler. *J. J. Eng. Industry* **85** (1963) 274.
8. K. Kirchgässner. *Z. Angew. Math. Physik* **12** (1961) 14.
9. M. Breitling. *Study thesis* (Institut für Technische Mechanik, Universität Karlsruhe. 1992).

Author Index

Z

Subject Index

E

F

G

H

I

K

L

Q

R

S

T

V

W